CONSTRUCTIVISM IN MATHEMATICS

VOLUME I

STUDIES IN LOGIC

AND

THE FOUNDATIONS OF MATHEMATICS

VOLUME 121

ELSEVIER

AMSTERDAM • BOSTON • HEIDELBERG • LONDON • NEW YORK • OXFORD
PARIS • SAN DIEGO • SAN FRANCISCO • SINGAPORE • SYDNEY • TOKYO

CONSTRUCTIVISM IN MATHEMATICS

AN INTRODUCTION

VOLUME I

A.S. TROELSTRA

Universiteit van Amsterdam,
The Netherlands

D. VAN DALEN

Rijksuniversiteit Utrecht,
The Netherlands

ELSEVIER

AMSTERDAM • BOSTON • HEIDELBERG • LONDON • NEW YORK • OXFORD
PARIS • SAN DIEGO • SAN FRANCISCO • SINGAPORE • SYDNEY • TOKYO

ELSEVIER B.V.
Sara Burgerhartstraat 25
P.O. Box 211, 1000 AE
Amsterdam, The Netherlands

ELSEVIER Inc.
525 B Street, Suite 1900
San Diego, CA 92101-4495
USA

ELSEVIER Ltd
The Boulevard, Langford Lane
Kidlington, Oxford OX5 1GB
UK

ELSEVIER Ltd
84 Theobalds Road
London WC1X 8RR
UK

First edition 1988
Second impression 2004

Library of Congress Cataloging in Publication Data
Troelstra, A.S. (Anne Sjerp)
Constructivism in mathematics.
(Studies in logic and the foundations of mathematics; v. 121, 123)
Bibliography: p
Includes indexes.
1. Constructive mathematics. I. Dalen, D. van (Dirk), 1932- II. Title. III. Series. IV. Series: Studies in logic and the foundations of mathematics; v. 121, etc.
QA9.56.T74 1988 511.3 88-5240
ISBN 0-444-70266-0 (v. 1) HB ISBN 0-444-70506-6 PB
ISBN 0-444-70358-6 (v. 2)

British Library Cataloguing in Publication Data
A catalogue record is available from the British Library.

ISBN: 0-444-70266-0 (hardbound)
ISBN: 0-444-70506-6 (paperback)

Transferred to digital printing 2005.

This book is dedicated
to the memory of our teacher
Arend Heyting

This book is dedicated

to the memory of our teacher

Arend Heyting

PREFACE

The present volume is intended as an all-round introduction to constructivism. Here constructivism is to be understood in the wide sense, and covers in particular Brouwer's intuitionism, Bishop's constructivism and A.A. Markov's constructive recursive mathematics. The ending "-ism" has ideological overtones: "constructive mathematics is the (only) right mathematics"; we hasten, however, to declare that we do not subscribe to this ideology, and that we do not intend to present our material on such a basis.

The first successful introduction to constructive mathematics, more specifically to intuitionistic mathematics, was Heyting's "Intuitionism, an introduction", which first appeared in 1956 and went through three editions. At the moment of its appearance Heyting's book was unique; now there is a whole collection of introductory texts and monographs dealing with constructive mathematics in its various forms, as well as with the metamathematics of constructive mathematics. Let us therefore indicate the principal features of the present book.

(a) It treats constructive mathematics in its various forms, as well as the metamathematics of constructivism. Each can serve as useful background information for the other.

(b) The treatment of constructive mathematics does not attempt to be systematic, only to give some illustrative examples.

(c) We have tried to select topics we expected to retain their interest for some time to come; we have not tried to follow the latest fashions.

(d) The treatment of the metamathematics is intended to be essentially self-contained; therefore we have included certain details, such as the formalization of elementary recursion theory, which are not part of standard introductory texts on mathematical logic. We assume only some familiarity with classical first-order predicate logic, preferably in natural deduction style, such as may be obtained from many introductory texts. Some familiarity with elementary recursion theory, though not absolutely necessary, is also helpful.

The material suitable for a first introduction to the topic is properly contained in the first six chapters. More specifically, the introductory material consists of chapter 1, sections 2.1–6, 3.1–5, 4.1–7 and the chapters 5 and 6. In addition the chapters 1–4 contain more specialized sections, which have been placed there for systematic reasons. We have tried to avoid too much interdependence in the introductory material; hence there are several selections for an introductory course possible, for example:

Selection 1, with emphasis on examples of constructive mathematics, consists of the following sections and subsections: 1.1–3, 3.3.1–2, 3.3.6 (first proof), 3.4, 4.1–3, 4.5, 5.1–4, 6.1–2, 4.6.1–10, 4.6.13–14, 4.7, 6.3, 6.4.1–11, 6.4.13, 4.4 (sketchy), 6.4.12.

The logical aspects may be given somewhat more emphasis by also treating 2.1, 2.3.1–6, 2.5.1–11, and the whole section 3.3 except for the second proof in 3.3.6.

Another possible extension of this selection is 5.5–7.

Selection 2, with emphasis on logical and metamathematical aspects, consists in the following sections and subsections: 1.1–3, 2.1, 2.3.1–6, 2.4.1–4 (optional), 2.5.1–13, 2.6 (optional), 3.1–5, 4.1–7, possibly extended with 5.1–4.

Occasional reference to material outside these selections may be skipped. These selections can be varied in many different ways.

The treatment of the introductory material is more leisurely and detailed than in subsections and later chapters of a more specialized nature; there we often leave the routine proofs as an exercise for the reader.

Most of the material has been tried out in courses at Amsterdam and Utrecht.

All chapters, except the last one, have a final section entitled "Notes". These are reserved for credits, historical remarks, asides, suggestions for further reading etcetera. The historical remarks deal primarily with the constructivist tradition and its metamathematics. We did not attempt to trace systematically the history of all the topics treated, and there is no pretense of completeness.

The final chapter of volume 2, the "Epilogue", contains some observations of a general nature prompted by the technical developments in the earlier chapters, as well as a brief discussion of some controversial issues.

In order to make independent use of the first volume possible, we have provided a separate index, bibliography and list of symbols to volume 1.

Due to circumstances beyond our control the burden of authorship has not been equally divided. Specifically, the first author produced the bulk of the material and the second author wrote section 1.4 and chapter 8. Chapter

3 is the joint responsibility of both authors. Furthermore both authors accept the responsibility for the contents in the sense that the material was the subject of mutual discussions, criticisms and emendations.

H. Luckhardt read through an early draft of the manuscript and provided a long list with corrections.

We owe a special debt to I. Moerdijk, who over the past few years rapidly reversed the roles in the teacher–pupil relationship. In chapters 13–15 we made at several places liberal use of handwritten notes prepared by him. André Hensbergen tested the exercises of chapters 7 and 10, Rineke Verbrugge assisted us with research for some of the notes.

The job of preparing the final version of the manuscript was made considerably easier by Yvonne Voorn, who typed a draft of the manuscript on a wordprocessor.

Muiderberg/De Meern A.S. TROELSTRA
January 1987 D. VAN DALEN

PRELIMINARIES

1. *Internal references.* Within chapter n, "$k.m$" refers to the subsection numbered $k.m$ in chapter n; "section k" refers to section k of chapter n; "section $k.m$" refers to section m of chapter k ($k \neq n$), and "$k.l.m$" refers to subsection $l.m$ of chapter k ($k \neq n$). "Exercise $k.l.m$" or "E$k.l.m$" refers to the exercise numbered $k.l.m$ at the end of chapter k. An exercise numbered $k.l.m$ should be regarded as belonging to section $k.l$.

2. *Bibliographical references.* are given as author's name followed by year of publication, possibly followed by a letter in the case of more than one publication in the same year by the same author, e.g. "as shown by Brouwer (1923)...", or "...in Brouwer (1923A) it was proved that...", or "the bar theorem (Brouwer 1954)...". References to works by two authors appear as "Kleene and Vesley (1965)".

In the case of three or more authors we use the abbreviation "et al.", e.g. "Constable et al. (1986)". The bibliography at the end of the book contains only items which are actually referred to in the text; by the completion of the logic bibliography Müller (1987) has relieved us of the task of providing something approaching a complete bibliography of constructivism.

In the rest of this section some general notational conventions are brought together, to be consulted when needed. There may be local deviations from these conventions. At the end of the book there is an index of notations of more than purely local use.

3. *Definitions.* $\equiv$ indicates *literal identity*, modulo renaming bound variables. $:=$ is used as the *definition symbol*, the defining expression appears on the right hand side.

4. *Variables, substitution.* The concept of free and bound variable is defined as usual; for the sets of free and bound variables of the expression $\mathfrak{a}$ we use

FV(α) and BV(α) respectively. As a rule we regard expressions which differ only in the names of bound variables as isomorphic; that is to say, bound variables are used as position markers only.

For the result of *simultaneous substitution* of $t_1, \ldots, t_n$ for the variables $x_1, \ldots, x_n$ in the expression α we write $\alpha[x_1, \ldots, x_n/t_1, \ldots, t_n]$. We shall often use a looser notation: once $\alpha(x_1, \ldots, x_n)$ has appeared in a context, we write $\alpha(t_1, \ldots, t_n)$ for $\alpha[x_1, \ldots, x_n/t_1, \ldots, t_n]$. Using vector notation as an abbreviation we can also write $\alpha[\vec{x}/\vec{t}\,]$ for $\alpha[x_1, \ldots, x_n/t_1, \ldots, t_n]$.

In using the substitution notation we shall as a rule tacitly assume the terms to be free for the variables in the expression considered (or we assume that a suitable renaming of bound variables is carried out). In our description of logic in chapter 2 we are more explicit about these matters.

We shall frequently economize on parentheses by writing Ax or At (A a formula) instead of $A(x)$ or $A(t)$.

5. *Logical symbols.* As logical symbols we use

$$\bot, \neg, \vee, \wedge, \rightarrow, \leftrightarrow, \forall, \exists, \exists!$$

both formally and informally. In bracketing we adopt the usual convention that $\neg, \forall, \exists$ bind stronger than any of the binary operators, and that $\wedge, \vee$ bind stronger than $\rightarrow, \leftrightarrow$. Occasionally dots are used as separating symbols instead of parentheses. In discussing formal systems it will be usually clear from the context whether the symbol is used as part of the formalism or on the metalevel. Where it is necessary to avoid confusion we use $\Rightarrow$, $\Leftrightarrow$ on the metalevel; sometimes a comma serves as a conjunction, and "iff" abbreviates "if and only if". Unless stated otherwise, $\neg$, $\leftrightarrow$, $\exists!$ are regarded as abbreviations defined by

$$\neg A := A \rightarrow \bot\,;$$

$$A \leftrightarrow B := (A \rightarrow B) \wedge (B \rightarrow A);$$

$$\exists! x A := \exists x(A \wedge \forall y(x = y \leftrightarrow A[x/y])).$$

For repeated equivalences we sometimes write $A \leftrightarrow B \leftrightarrow C \leftrightarrow \cdots$ meaning $(A \leftrightarrow B) \wedge (B \leftrightarrow C) \wedge \cdots$. For iterated finite conjunctions and disjunctions we use

$$\bigwedge, \bigvee.$$

Notation for restricted quantifiers:

$$\forall x \in X, \exists x \in X.$$

For iterated quantifiers we use the abbreviations

$$\forall x_1 x_2 \ldots x_n := \forall x_1 \forall x_2 \ldots \forall x_n, \; \exists x_1 x_2 \ldots x_n := \exists x_1 \exists x_2 \ldots \exists x_n,$$

and for iterated restricted quantifiers

$$\forall x, y \in A := \forall x \in A \, \forall y \in A \text{ etcetera.}$$

For possibly undefined expressions $\mathfrak{a}$, $E\mathfrak{a}$ means "$\mathfrak{a}$ exists", or "$\mathfrak{a}$ is well-defined" (cf. the use of $t\downarrow$ in recursion theory).

6. *Set-theoretic notation.* Our standard set-theoretic symbols are

$$\emptyset, \in, \notin, \subset, \supset, \setminus, \cap, \cup, \bigcap, \bigcup.$$

Here $\subset$ and $\supset$ arc used for not necessarily proper *inclusion*, i.e. $X \subset Y := \forall x(x \in X \to x \in Y)$, etc. We also use the standard notations

$$\{x_1, x_2, \ldots, x_n\} \quad \text{(for finite sets)},$$

$$\{x : A(x)\}, \{f(x) : A(x)\}, \{x \in B : A(x)\} \quad (f \text{ a function}).$$

For *complements* of sets relative to some fixed set we often use ^{c}. The fixed superset involved will be clear from the context. Thus if X is a subset of $\mathbb{N}$, we write X^c for $\mathbb{N} \setminus X$.

For finite *cartesian products* we use $\times$, for arbitrary cartesian products $\prod$.

If X is a set and $\sim$ an equivalence relation on X we write $X/\sim$ for the set of *equivalence classes* of X modulo $\sim$. For the *equivalence class* of an $x \in X$ we write $x/\sim$, $x_\sim$, $(x)_\sim$ or $[x]_\sim$.

The *set of functions* from X to Y is written as $X \to Y$ or Y^X. *Restrictions* are indicated by $\upharpoonright$ or $|$. $\mathrm{P}(X)$ is the *powerset* of X.

"f is a function from X to Y" is written as $f \in X \to Y$ (and only occasionally as $f: X \to Y$); the use of this notation must not be regarded as a commitment to the set of all functions from X to Y as a well-defined totality. If t is a term we can also introduce "t regarded as a function of the parameter (variable) x" by one of the following notations:

$$\lambda x.t \quad \text{or} \quad f : x \mapsto t.$$

Notations (in diagrams) for *injections*, *surjections*, *bijections* and *embeddings* are $\rightarrowtail$, $\twoheadrightarrow$, $\twoheadrightarrow\!\!\!\!\!\rightarrowtail$, $\hookrightarrow$ respectively.

For the *characteristic function* of a relation R we use χ_R, where $\chi_R(t) = 0 \leftrightarrow R(t)$.

For the function f applied to the argument t we usually write $f(t)$, or even ft, if no confusion can arise; in general we drop parentheses whenever we can safely do so. In certain chapters we use $t(t')$ or tt' for term t applied to term t' (e.g. in chapter 9); in such cases we use square brackets instead of parentheses to refer to occurrences in a term, writing $t[x]$ instead of $t(x)$ etc.

Pairs, or *n-tuples* for fixed length, are usually also indicated by means of parentheses (,); in the case of finite sequences of variable length we use $\langle\ ,\ \rangle$: $\langle x_1, x_2, \ldots, x_n\rangle$ is a sequence of length n (notation $\mathrm{lth}\langle x_1, x_2, \ldots, x_n\rangle = n$). For concatenation we use $*$. A sequence of arguments may also be indicated by vector notation; thus we write $f(\vec{t})$ or $f\vec{t}$ for $f(t_1, \ldots, t_n)$, where $\vec{t} \equiv (t_1, \ldots, t_n)$; $\vec{t} = \vec{s}$ indicates equality of vectors.

We use the abbreviation $\lambda x_1 x_2 \ldots x_n.t$ for $\lambda x_1(\lambda x_2 \ldots (\lambda x_n.t)\ldots)$. If we wish to regard t as a function in n arguments $x_1, x_2, \ldots, x_n$ we use comma's: $\lambda x_1, x_2, \ldots, x_n.t$.

For the *graph* of the function f we write graph(f), dom(f) is its *domain*, range(f) its *range*.

7. *Mathematical constants.* The following constants

$$\mathbb{N}, \mathbb{Z}, \mathbb{Q}, \mathbb{R}, \mathbb{B}, \mathbf{C}$$

denote the natural numbers, the integers, the rationals, the reals, Baire space and the complex numbers respectively. Throughout the book, n, m, i, j, k, unless indicated otherwise, are supposed to range over $\mathbb{N}$, and $\alpha, \beta, \gamma, \delta$ over $\mathbb{B}$ or a subtree of $\mathbb{B}$. In metamathematical work we use $\bar{n}, \bar{m}$ for numerals.

For an infinite sequence $\mathfrak{a}_0, \mathfrak{a}_1, \mathfrak{a}_2, \ldots$ (i.e. a function with domain $\mathbb{N}$) we use the notation $\langle \mathfrak{a}_n\rangle_n$. A notation such as $\lim\langle x_n\rangle_n$ is self-explanatory.

The notation for arithmetical operations is standard; the multiplication dot $\cdot$ is often omitted.

8. *Formal systems and axioms.* Formal systems are designated by combinations of roman boldface capitals, e.g. **HA**, **IQC**, $\mathbf{EM}_0{\restriction}$, etc.

If **H** is a system based on intuitionistic logic, $\mathbf{H}^c$ is used for the corresponding system based on classical logic.

For the language of a formal system **H** we often write $\mathscr{L}(\mathbf{H})$. If $\mathbf{H}'$ extends **H**, and $\mathscr{L}(\mathbf{H}')$ extends $\mathscr{L}(\mathbf{H})$, while $\mathbf{H}' \cap \mathscr{L}(\mathbf{H}) = \mathbf{H}$, then $\mathbf{H}'$ is said to be a *conservative extension* of **H** ($\mathbf{H}'$ is *conservative over* **H**). A *definitional extension* is a special case of a conservative extension, where the extra symbols of $\mathscr{L}(\mathbf{H}')$ can be replaced by explicit definitions (cf. 2.7.1).

Axiom schemas and rules are usually designated by combinations of roman capitals: REFL, BI, BI_{D}, FAN, WC-N, CT_0 etc.

$\mathbf{H} \vdash A$ means "A is derivable in the system **H**", and $\mathbf{H} + \mathrm{XYZ} \vdash A$ "A is derivable in **H** with the axiom schema or rule XYZ added". Occasionally we use subscript notation: $\vdash_{\mathbf{H}} A$ instead of $\mathbf{H} \vdash A$.

9. *Validity.* For validity of a sentence A in a model $\mathscr{M}$ we use the notation $\mathscr{M} \models A$. It is to be noted that if the relations, functions and constants of the model $\mathscr{M}$ are defined constructively, then $\mathscr{M} \models A$ also makes sense constructively (cf. 2.5.1).

definitional extension is a special case of a conservative extension, where the extra symbols of $\mathcal{L}(\mathbf{H}')$ can be replaced by explicit definitions (cf. 2.7.1).

Axiom schemas and rules are usually designated by combinations of roman capitals, REFL, BI, BI_{D}, FAH, WC-N, CT_0 etc.

$\mathbf{H} \vdash A$ means "A is derivable in the system $\mathbf{H}$", and $\mathbf{H} + \mathrm{XYZ} \vdash A$ "A is derivable in $\mathbf{H}$ with the axiom schema or rule XYZ added". Occasionally we use subscript notation: $\vdash_{\mathbf{H}} A$ instead of $\mathbf{H} \vdash A$.

9. Validity. For validity of a sentence A in a model $\mathcal{M}$ we use the notation $\mathcal{M} \models A$. It is to be noted that if the relations, functions and constants of the model $\mathcal{M}$ are defined constructively, then $\mathcal{M} \models A$ also makes sense constructively (cf. 2.5.1).

CONSTRUCTIVISM IN MATHEMATICS: CONTENTS

CONTENTS OF VOLUME I

CHAPTER 1

INTRODUCTION

In this chapter we first present a brief characterization of the various forms of constructivism which shall play a role in the book later on. After the review of constructivist trends, we illustrate the idea of constructivity by means of some simple and quite elementary examples in section 2; most of the points discussed will receive a more thorough treatment in later chapters.

Section 3 introduces the Brouwer–Heyting–Kolmogorov interpretation of the logical operators, and the method of weak counterexamples.

Section 4 gives a brief historical survey.

1. Constructivism

1.1. Time and again, over the last hundred years certain mathematicians have defended an approach to mathematics which might be called "constructive" in the broad sense used in this book, in more or less explicit opposition to certain forms of mathematical reasoning used by the majority of their colleagues. Some of these critics of the mathematics of their (our) time not only criticized contemporary mathematical practice, but actually endeavoured to show how mathematics could be rebuilt on constructivist principles.

There are, however, considerable differences in outlook between the various representatives of constructivism; constructivism in the broad sense is by no means homogeneous, and even the views expressed by different representatives of one "school", or by a single mathematician at different times are not always homogeneous. Our descriptions below present a simplified picture of this complex reality. We shall attempt a brief characterization of the principal constructivist "trends" or "schools" playing a role in this book.

In our discussions we shall use the adjective "classical" for logical and mathematical reasoning based on the usual two-valued logic, in which every meaningful statement is assumed to be either true or false.

1.2. *Finitism.* The principal tenets of finitism are:

(a) only concretely (finitely) representable structures are objects of mathematics; operations on such structures are to be combinatorial in nature and hence, of course, effective.

(b) abstract notions such as (arbitrary) set, operation, construction etc. have no place in finitist mathematics.

Kronecker (1823–1891), who with some justification may be regarded as "the first constructivist", may perhaps more specifically be regarded as a precursor of the finitist approach. Finitism is characteristically represented by Skolem in his paper (1923) and in two books by Goodstein (1957, 1961).

Finitism also plays a role in Hilbert's programme: for methods regarded by all mathematicians as incontrovertible, to be used in a proof of freedom of contradiction of (the bulk of) existing mathematics, Hilbert wanted to draw on finitist mathematics; this of course does not mean that Hilbert is to be regarded as a finitist.

A formal system which is regarded as codifying a typical part of finitist mathematics is **PRA**, the system of primitive recursive arithmetic, to be discussed briefly in section 3.2.

1.3. *Predicativism.* This may be regarded as "constructivism with respect to definitions". In predicativism:

(a) Definitions of mathematical objects should be predicative, that is to say it is not permissible to define an object d by referring to a collection D of which the object d is to be an element; in particular, quantification over D in defining d is not permitted (avoidance of a "vicious circle" in definitions).

(b) The demand for predicative definitions is regarded as compatible with traditional logic, that is to say mathematical statements are assumed to be true or false – regardless of human knowledge (this in contrast to intuitionism, described in the next subsection).

As a rule the set of natural numbers $\mathbb{N}$ is regarded as unproblematic from a predicativist point of view. This means that one accepts quantification over $\mathbb{N}$ and thus arithmetically defined predicates as predicatively meaningful.

Having thus grasped the idea of an arithmetically defined set or predicate (sets of level 1), we may quantify over such sets in constructing predicatively defined sets of level 2, etcetera.

On the other hand, defining the least upper bound of a set X of reals (defined as Dedekind (left) cuts) as the intersection of all left cuts which are upper bounds for X is not permissible: the least upper bound itself occurs among the collection of upper bounds.

The first to show that parts of analysis could be developed predicatively was Weyl in his monograph "Das Kontinuum" (1918). More recently Lorenzen in his books (1955, 1965) has given a predicative development of analysis. Lorenzen (1955), however, does not accept classical logic outright, but justifies its use relative to constructive logic.

Poincaré and the French "empiricists" such as Borel may be regarded as precursors of predicativism, but they did not develop parts of mathematics predicatively in a systematic way (cf. also 4.2).

Predicativism as such does not play a role in this book, but the demands of predicativism may or may not be combined with the demands of intuitionism and Bishop constructivism and thus the issue of predicativity occasionally crops up in other constructivist trends as well.

1.4. *Bishop's Constructive Mathematics ("BCM").* This approach is typically represented by Bishop's book "Foundations of constructive analysis" (1967). In the work of this school there are few discussions of philosophical principles, the emphasis is entirely on the practice of constructive mathematics, instead of on its philosophy.

(a) The key phrase is that "mathematical statements should have numerical meaning". This means, in particular, that existential statements must, in principle, be capable of being made explicit (and also, in asserting $A \vee B$, the choice between A or B must be possible in principle): one can only show that an object exists by giving a finite routine for finding it.

(b) In BCM it is not assumed that all mathematical objects are to be given in the form of an algorithm (in the mathematically precise sense of recursive function theory). Choice sequences (cf. 1.6 below) are not admitted as legitimate objects. Bishop does not hesitate to accept abstract concepts such as "rule" or "operation".

In BCM no definite position is taken with respect to the admissibility of impredicative definitions; in practice, generalized inductive definitions are admitted (Bishop 1967, p. 68); for a discussion of this type of definition see section 4.8.

1.5. *Constructive Recursive Mathematics ("CRM").* This is a form of constructivism developed by Markov (1903–1979) from ca. 1950 onwards. In the sense we understand it here it is typically represented by the writings of Markov, Shanin and their students in the period 1950–1967.

A feature CRM has in common with finitism is that mathematical objects should be finitely representable. More specifically:

(a) The objects of mathematics are *algorithms*, understood in a *mathematically precise sense*: algorithms can be presented as "words" in some finite alphabet of symbols.

(b) Limitations due to finite memory capacity are disregarded, the length of symbol strings is unbounded (though always finite).

(c) Logically compound statements not involving $\exists$, $\vee$ are understood in a direct way, but existential statements and disjunctions always have to be made explicit.

(d) If it is impossible that an algorithmic computation does not terminate, we may assume that it *does* terminate ("Markov's principle").

1.6. *Intuitionism.* Intuitionism will be understood here as the constructive approach to mathematics in the spirit of Brouwer (1881–1966) and Heyting (1898–1980). The philosophical basis of this approach is already present in Brouwer's thesis (1907) but with respect to the mathematical consequences Brouwer (1918) is a more appropriate starting point. The basic tenets may be summarized as follows.

(a) Mathematics deals with mental constructions, which are immediately grasped by the mind; mathematics does not consist in the formal manipulation of symbols, and the use of mathematical language is a secondary phenomenon, induced by our limitations (when compared with an ideal mathematician with unlimited memory and perfect recall), and the wish to communicate our mathematical constructions to others.

(b) It does not make sense to think of truth or falsity of a mathematical statement independently of our knowledge concerning the statement. A statement is *true* if we have proof of it, and *false* if we can show that the assumption that there is a proof for the statement leads to a contradiction. For an arbitrary statement we can therefore not assert that it is either true or false.

(c) Mathematics is a *free* creation: it is not a matter of mentally reconstructing, or grasping the truth about mathematical objects existing independently of us (this is in contrast to e.g. the French empiricists; cf. 4.2).

It follows from (b) that it is necessary to adopt a different interpretation of statements of the form "there exists an x such that $A(x)$ holds" and "A or B holds". In particular, "A or not A" does not generally hold on the intuitionistic reading of "or" and "not"; we return to this in section 3.

In agreement with, but not necessarily following from (c), intuitionism permits consideration of unfinishable processes: the ideal mathematician may construct longer and longer initial segments $\alpha(0), \ldots, \alpha(n)$ of an infinite sequence of natural numbers α where α is not a priori determined by some fixed process of producing the values, so the construction of α is never finished: α is an example of a *choice sequence*. We return to this topic in section 4.6.

Unfinishable objects such as choice sequences clearly do not fit into the framework of CRM, the viewpoint described in the preceding subsection.

On the other hand, "Markov's principle", mentioned under (d) in 1.5, is not considered as intuitionistically valid. In fact, it is difficult to decide the validity of Markov's principle by informal analysis alone. We shall return to Markov's principle in section 4.5.

1.7. In this book we shall have frequent occasion to refer to intuitionism, CRM and BCM in particular. However, the present book is not written from the viewpoint of any particular school or trend mentioned above, inasmuch we shall not accept any of these views as *normative* for mathematics.

We do present constructive mathematics (in the wide sense) as a legitimate part of mathematics, containing material which is mathematically interesting, regardless of any philosophical bias.

Constructive mathematics, and especially its logic and its metamathematics are of considerable interest for the philosophy of mathematics. There are some indications that intuitionistic logic can be useful in other parts of mathematics too, such as topos theory (see e.g. Johnstone 1977, 1982), or theoretical computer science (see for example Constable et al. 1986).

2. Constructivity

2.1. In this section we shall attempt to illustrate the idea of constructivity by means of some quite elementary examples. Most of the points discussed will receive a more thorough treatment in later chapters.

Which objects can be said to exist as (mental) constructions? Natural numbers are usually regarded as unproblematic from a constructive point of view; they correspond to very simple mental constructions: start thinking of an abstract unit, think of another unit distinct from the first one and consider the combination ("think them together"). The indefinite repetition of this process generates the collection $\mathbb{N}$ of natural numbers.

It should be pointed out that already here an element of idealization enters. We regard 5, 1000 and $10^{10^{10}}$ as objects of "the same sort" though our mental picture in each of these cases is different: we can grasp "five" immediately as a collection of units, while on the other hand $10^{10^{10}}$ can only be handled *via* the notion of exponentiation; 1000 represents an intermediate case. Visualizing $10^{10^{10}}$ as a sequence of units is out of the question. Exponentiation as an always performable *operation* on the natural numbers involves a more abstract idea than is given with the generation of $\mathbb{N}$.

Certain mathematicians, such as A.S. Esenin-Vol'pin have tried to distinguish between "reasonably" small and "unreasonably" large numbers, and between different "natural number structures" according to the principles needed to generate them. Permitting exponentiation may result in a bigger number sequence with a more complex structure, than in the case where exponentiation is not considered, although addition and multiplication are accepted. This approach might be described as *ultra-finitism*. The name "ultra-intuitionism" has sometimes been used, but seems less appropriate. There are considerable obstacles to be overcome for a coherent and systematic development of ultra-finitism, and in our opinion no satisfactory development exists at present. In this volume we shall therefore refrain from any further discussion of ultra-finitism except for some remarks in 4.8.

Thus all the constructivist schools described in section 1 contain elements of idealization, that is to say their description of constructive mathematical principles contains "theoretical" elements. In particular, we regard Brouwer's intuitionism and Markov's constructivism as "theories about mathematics". It is also for this reason that one cannot identify intuitionism with a psychologistic approach to mathematics.

2.2. *A non-constructive definition.* Let A be any mathematical statement which at present has been neither proved nor refuted (e.g., $A \equiv$ "there are infinitely many twin primes"). Ordinarily one permits in mathematics descriptions (definitions) of natural numbers such as the following

$$p = \begin{cases} 1 & \text{if } A \text{ holds,} \\ 2 & \text{otherwise.} \end{cases}$$

Constructively this is unacceptable as the description of a natural number since, as long as the truth of A has not been decided, we cannot identify p

with either 1 or 2; in other words we do not know how to obtain p by "thinking some abstract units together", since we do not know whether to stop at one unit or to go on till we have two distinct units.

2.3. Once having accepted the natural numbers, there is also no objection to accepting pairs of natural numbers, pairs of pairs etc. as constructive objects; and this permits us to define $\mathbb{Z}$ (the integers) and $\mathbb{Q}$ (the rationals) in the usual way as pairs of natural numbers and pairs of integers respectively, each modulo suitable equivalence relations.

Infinite sequences of natural numbers, integers or rationals may be constructively given by some process enabling us to determine the nth term (value for n) for each natural number n; in particular, one may think of sequences given by a *law* (or recipe) for determining all its values (terms).

Thus a real number can be specified by a fundamental sequence of rationals together with a Cauchy modulus, that is to say we specify a sequence $\langle r_n \rangle_n \in \mathbb{N} \to \mathbb{Q}$ and a sequence $\alpha \in \mathbb{N} \to \mathbb{N}$ (the modulus) such that for all $k \in \mathbb{N}$ and all $m, m' \geq \alpha k$

$$|r_m - r_{m'}| \leq 2^{-k}.$$

Two such sequences $\langle r_n \rangle_n$, $\langle s_n \rangle_n$ are said to be *equivalent* (equal as real numbers) if for all $k \in \mathbb{N}$ there is an $n \in \mathbb{N}$ such that for all m

$$|r_{n+m} - s_{n+m}| \leq 2^{-k}.$$

Much of the elementary theory of real numbers can then be developed in exactly the same way as usual; we shall do this in chapter 5.

We now give an often quoted example of a non-constructive proof for the following

Proposition. *There exist two irrational real-numbers a, b such that a^b is rational.*

Proof. $(\sqrt{2})^{\sqrt{2}}$ is either rational, and then we can take $a = b = \sqrt{2}$, or $(\sqrt{2})^{\sqrt{2}}$ is irrational, and then we can take $a = (\sqrt{2})^{\sqrt{2}}$, $b = \sqrt{2}$. □

This proof does not enable us to construct a as a real number, that is to say we cannot compute a with any desired degree of accuracy, as required by our description of what it means to be *given* a real number. In other words, we do not know how to find an arbitrarily close rational approximation. Note however, that our objections to the proof depend on reading "there exist" as "one can construct".

A constructive proof of the proposition above is possible, e.g. by an appeal to Gelfond's theorem: if $a \notin \{0,1\}$, a algebraic, b irrational algebraic, then a^b is irrational (even transcendental). Cf. e.g. Gelfond (1960, p. 106, theorem 2).

As a second example of a non-constructive argument, we consider the classical proof of König's lemma.

2.4. PROPOSITION (*König's lemma*). *Let T be an infinite, finitely branching tree, then T has an infinite branch.*

PROOF. We "construct" the infinite branch $\alpha \in \mathbb{N} \to T$ as follows. For $\alpha 0$ we take the root of T. Suppose $\overline{\alpha}n$ to have been constructed such that $\overline{\alpha}n$ has infinitely many successors. Among the finite set of immediate successors $t_0, \ldots, t_k$ of $\overline{\alpha}n$ there is at least one t_i such that t_i has infinitely many successors; take $\overline{\alpha}(n+1) := t_i$. Clearly we can continue this process indefinitely and an infinite branch is generated. □

Given some ordering of the nodes, we can make the description of α in the proof more definite by taking for $\overline{\alpha}(n+1)$ the first t_i having infinitely many successors. Nevertheless, this "construction" of α is not constructive, since we do not know, in general, how to decide whether a finite sequence in T has infinitely many successors or not; so the proof above does not give us a recipe for producing $\overline{\alpha}n$ for each n.

The preceding examples of non-constructive proofs show that the natural constructive reading of "A or B" and "$\exists x A(x)$" as "we can decide between A and B" and "we can construct an x such that $A(x)$" respectively does not agree with all the laws of classical logic. In particular, "A or $\neg A$" does not hold if we read "or" as above: we have no reason to assume that we can always *decide*, for any assertion A, whether A is true or refutable.

In actual mathematical practice, one is often interested in constructivizing "locally", e.g. making existential statements of a *specific* theorem explicit by estimating bounds etc. In contrast, we shall mostly investigate *globally* constructive theories, where our logic can consistently be interpreted as permitting *explicit* realizations of all existential statements, and all theories should be read with a strong constructive meaning of $\exists$ and $\vee$.

3. Weak counterexamples

In 1908 Brouwer introduced for the first time "weak counterexamples", for the purpose of showing that certain classically acceptable statements are constructively unacceptable (Brouwer 1908). Too much emphasis on these examples has sometimes created the false impression that refuting claims of classical mathematics is the principal aim of intuitionism.

Nevertheless, the weak counterexamples are pedagogically useful and may serve to motivate e.g. Kripke semantics, which we shall introduce in the next chapter. Before we can explain these counterexamples we must explain in a more systematic way the constructive interpretation of the logical operators.

In discussing pure logic we shall treat "constructive" and "intuitionistic" as synonymous.

3.1. *Logical operations; the BHK-interpretation.* In the actual building of constructive mathematics we do not *need* logic; nevertheless we find it convenient to use logical symbolism. We shall explain the use of the logical operations in a constructive context by the following stipulations (going back to Heyting 1934), which tell us what forms proofs of logically compound statements take in terms of the proofs of the constituents.

(H1) A proof of $A \wedge B$ is given by presenting a proof of A and a proof of B.

(H2) A proof of $A \vee B$ is given by presenting either a proof of A or a proof of B (plus the stipulation that we want to regard the proof presented as evidence for $A \vee B$).

(H3) A proof of $A \to B$ is a construction which permits us to transform any proof of A into a proof of B.

(H4) Absurdity $\perp$ (contradiction) has no proof; a proof of $\neg A$ is a construction which transforms any hypothetical proof of A into a proof of a contradiction.

(H5) A proof of $\forall x A(x)$ is a construction which transforms a proof of $d \in D$ (D the intended range of the variable x) into a proof of $A(d)$.

(H6) A proof of $\exists x A(x)$ is given by providing $d \in D$, and a proof of $A(d)$.

This explanation is quite informal and rests itself on our understanding of the notion of construction and, implicitly, the notion of mapping; it is not hard to show that, on a very "classical" interpretation of construction and mapping, H1–6 justify the principles of two-valued (classical) logic (cf. E1.3.4). In clause H4 the notion of a contradiction is to be regarded as a primitive (unexplained) notion.

If, in the clauses H5, H6 the domain D is "simple" enough, it may happen that any d in D so to speak represents its own proof of belonging to D (d is presented to us as an object of D). $\mathbb{N}$ is an example: a natural number is given to us as such, we do not need a separate proof of this fact.

For such simple domains, the reference to "a proof of $d \in D$" in H5 and H6 may be dropped. The clauses H5 and H6 then read

(H5′) A proof of $\forall x A(x)$ is a construction which transforms any $d \in D$ into a proof of $A(d)$.

(H6′) A proof of $\exists x A(x)$ is given by presenting a $d \in D$ and a proof of $A(d)$.

In the sequel we shall refer to the clauses H1–6 above as the "Brouwer–Heyting–Kolmogorov interpretation" (BHK-interpretation, see 5.3).

3.2. Even if the explanations H1–6 leave a lot of questions open, they suffice to show that certain logical principles should be generally acceptable from a constructive point of view, while some other principles from classical logic are not acceptable.

As a positive example, consider the statement $A \to (B \to A)$, for arbitrary mathematical statements A, B. There is a very general proof, obtained as follows. Suppose a proves A. Then $\lambda b.a$, the constant function which assigns a to each argument, transforms a hypothetical proof of B always into a proof of A; so $\lambda b.a$ proves $B \to A$. Therefore if a proves A, then $\lambda b.a$ proves $B \to A$; and so the mapping $\lambda a.[\lambda b.a]$, which assigns to a the constant mapping $\lambda b.a$, transforms a proof of A into a proof of $B \to A$; hence $\lambda a.[\lambda b.a]$ is a proof of $A \to (B \to A)$.

Another consequence is that $\perp \to A$ is generally provable: since there is no proof of $\perp$, $\lambda a.a$ (or any other mapping) may count as a proof of $\perp \to A$, since it has to be applied to an empty domain. The principle $\perp \to A$ ("ex falso sequitur quodlibet") has sometimes been rejected as non-constructive (Johannson 1937); Heyting (1956, 7.1.3) regards it as an extra stipulation fixing the meaning and use of $\perp$, $\to$.

The rejection of $\perp \to A$ means that one regards $\perp$ as a proposition not *known* to be provable (but which may turn out to be provable, if our mathematical world is inconsistent). This position leads to *minimal* logic; the difference with intuitionistic logic based on H1–6 is slight (cf. 10.5.2).

On the negative side, consider the Principle of the Excluded Middle (PEM, or "tertium non datur")

PEM $\quad A \vee \neg A$,

which is generally valid in classical logic. Constructively, accepting PEM as a general principle means that we have a universal method for obtaining, for any proposition A, either a proof of A or a proof of $\neg A$, i.e. a method for obtaining a contradiction from a hypothetical proof of A.

But if such a universal method were available, we could also decide a statement A the truth of which has not been decided yet (e.g. the example of the infinitely many twin primes in subsection 2.2), which is not the case. Thus we cannot accept PEM as a universally valid principle on the BHK-interpretation.

The preceding argument is typically a "weak counterexample": PEM is not refuted, that is to say, we have not shown how to derive an actual contradiction from the assumption that PEM is valid. Instead, we have only shown PEM to be unacceptable as a BHK-valid principle of reasoning, since accepting it means that we ought to have certain knowledge (i.e. a decision as to the truth of A) which in fact we do not possess. One might also say that the "weak counterexample" makes the validity of PEM for the BHK-interpretation highly implausible.

3.3. In fact, we cannot hope to refute any individual instance of PEM, that is to say we cannot find a mathematical statement A such that $\neg(A \vee \neg A)$. This is impossible, since $\neg\neg(A \vee \neg A)$ holds universally: it is an intuitionistic logical law. This may be seen on the basis of the BHK-interpretation as follows. Suppose

$$c \text{ establishes } \neg(A \vee \neg A). \tag{1}$$

Hence,

$$\text{if } d \text{ establishes } A \vee \neg A,\ c(d) \text{ proves } \bot .$$

Clearly, there are operations $e \mapsto e_0$, $f \mapsto f_1$ such that

$$e \text{ establishes } A \Rightarrow e_0 \text{ establishes } A \vee \neg A,$$

$$f \text{ establishes } \neg A \Rightarrow f_1 \text{ establishes } A \vee \neg A.$$

Therefore if (1) holds, $g: e \mapsto c(e_0)$ (i.e. $g := \lambda e.c(e_0)$) is a construction for $A \to \bot \equiv \neg A$ and $h: f \mapsto c(f_1)$ (or $h := \lambda f.c(f_1)$) for $\neg A \to \bot$. Therefore $h(g)$ is a construction for $\bot$, so the mapping $c \mapsto h(g)$ establishes $\neg\neg(A \vee \neg A)$.

3.4. In the chapters dealing with examples from constructive mathematics we shall often appeal to a number of simple logical laws valid on the

BHK-interpretation; in particular the following are frequently encountered:

$$A \to \neg\neg A;$$

$$\neg A \leftrightarrow \neg\neg\neg A;$$

$$\neg(A \vee B) \leftrightarrow \neg A \wedge \neg B;$$

$$\neg(A \wedge B) \leftrightarrow (A \to \neg B) \leftrightarrow (B \to \neg A);$$

$$\neg\neg(A \to B) \leftrightarrow (\neg\neg A \to \neg\neg B) \leftrightarrow (A \to \neg\neg B) \leftrightarrow \neg\neg(\neg A \vee B);$$

$$\neg(A \to B) \leftrightarrow \neg(\neg A \vee B);$$

$$\text{if } A \to B, \quad \text{then} \quad \neg B \to \neg A,$$

and with quantifiers

$$\neg\exists x A(x) \leftrightarrow \forall x \neg A(x);$$

$$\neg\neg\forall x A(x) \to \forall x \neg\neg A(x);$$

$$\neg\neg\exists x A(x) \leftrightarrow \neg\forall x \neg A(x).$$

We leave it as an exercise to verify that these laws are valid on the BHK-interpretation.

On the other hand, besides PEM, the following are examples of principles which are valid in classical logic but not on the BHK-interpretation:

$$\neg\neg A \to A;$$

$$\neg A \vee \neg\neg A;$$

$$(A \to B) \vee (B \to A);$$

$$\neg(\neg A \wedge \neg B) \to A \vee B;$$

$$\neg(\neg A \vee \neg B) \to A \wedge B;$$

$$\forall x \neg\neg A(x) \to \neg\neg\forall x A(x);$$

$$\neg\neg\exists x A(x) \to \exists x \neg\neg A(x);$$

$$(A \to \exists x B(x)) \to \exists x (A \to B(x)) \; (x \notin \mathrm{FV}(A)).$$

We shall encounter methods to demonstrate their non-validity in later chapters (sections 2.5, 4.3, 4.6).

In the standard formalism, **IQC**, for intuitionistic predicate logic, discussed in the following chapter, terms are supposed to be always defined, that is to say terms always denote something, just as in the usual formalisms for classical predicate logic. However, in intuitionistic mathematics possibly undefined terms or expressions which are well-defined only for certain values of the parameters, crop up naturally (e.g. x^{-1} for reals x); while the classical way out, namely to guarantee that such expressions always denote something by means of arbitrary stipulations, is not available here (more about this in 5.3.11). Thus some care is needed in handling such expressions; if we use $\boldsymbol{E}t$ for "t is well-defined" (corresponding to $t\downarrow$ in recursion theory), we have in particular

$$\begin{aligned} &\forall x A(x) \wedge \boldsymbol{E}t \rightarrow A(t), \\ &A(t) \wedge \boldsymbol{E}t \rightarrow \exists x A(x). \end{aligned} \tag{1}$$

Replacing the usual principles $\forall x A(x) \rightarrow A(t)$ and $A(t) \rightarrow \exists x A(x)$ by the more cautious (1) corresponds on the formal level to the E^+-logic of 2.2.3. In handling possibly undefined expressions in our treatment of parts of constructive mathematics in chapters 5–8 we shall assume (1).

3.5. *Undecidability of equality for the reals.* A more sophisticated weak counterexample is the following. Let $A(n)$ be a predicate of natural numbers, such that $A(n)$ is decidable, but the truth of $\forall n A(n)$ is unknown, i.e.

$$\begin{aligned} &\forall n(A(n) \vee \neg A(n)), \\ &?\forall n A(n) \vee \neg \forall n A(n). \end{aligned} \tag{1}$$

At the time of writing this section we might take "Fermat's last theorem" as an example of such A, putting $A(n) :\equiv \forall m_1 m_2 m_3 k(m_1 + m_2 + m_3 + k \leq n \rightarrow (m_1 + 1)^{k+3} + (m_2 + 1)^{k+3} \neq (m_3 + 1)^{k+3})$, where m_1, m_2, m_3, k, n range over $\mathbb{N}$.

Suppose now that we define a real number x^A via a fundamental sequence of rationals $\langle r_n^A \rangle_n$ defined as follows:

$$r_n^A := \begin{cases} 2^{-n} & \text{if } \forall k \leq n A(k), \\ 2^{-k} & \text{if } \neg A(k) \wedge k \leq n \wedge \forall k' < k A(k'). \end{cases}$$

It is easy to see that this is indeed a fundamental sequence, since $\forall n \geq m(|r_n^A - r_m^A| < 2^{-m})$; also

$$x^A = 0 \leftrightarrow \forall n A(n).$$

The direction from right to left is obvious; conversely, if $x^A = 0$, then

$r^A_{n+1} < 2^{-n}$ necessarily follows for all n, hence $A(n)$ for all n. Therefore

$$x^A = 0 \lor x^A \neq 0$$

is equivalent to $\forall n A(n) \lor \neg\forall n A(n)$. Thus we have a weak counterexample to the decidability of equality between reals; this example, in its turn, can serve for the construction of weak counterexamples to many other statements, as we shall see below.

3.6. Let us call a set *finite* iff it is in 1–1 correspondence with an initial segment of $\mathbb{N}$, and consider for example the following subset of the reals:

$$X := \{0, x^A\}.$$

X is the image of the finite set $\{0,1\}$ (under the mapping φ given by $\varphi 0 = 0$, $\varphi 1 = x^A$), or in other words, X is *finitely indexed* (namely by $\{0,1\}$). However, we cannot show X to be finite, since this would require us to decide whether it has one or two distinct elements, and this in turn would require us to decide $\forall n A(n) \lor \neg\forall n A(n)$.

Another consequence of the weak counterexample to $x = 0 \lor x \neq 0$ is, that the usual examples of discontinuous functions from $\mathbb{R}$ to $\mathbb{R}$ fail to yield total functions. For example, the function f specified by

$$f(x) = 1 \quad \text{if } x \neq 0$$

$$f(x) = 0 \quad \text{if } x = 0$$

is classically everywhere defined, but not so constructively. For if we suppose f to be constructively defined everywhere on $\mathbb{R}$, we ought to be able to construct arbitrarily close approximations to $f(x)$ for any argument x; but this means in particular that we should be able to decide

$$f(x^A) < 1 \lor f(x^A) > 0,$$

hence $\neg x^A \neq 0 \lor x^A \neq 0$. This is equivalent to $x^A = 0 \lor x^A \neq 0$: for decidable $A(n)$ one readily sees $\neg\neg A(n) \leftrightarrow A(n)$, and since $x^A = 0 \leftrightarrow \forall n A(n)$, $\neg\neg\forall n A(n) \to \forall n \neg\neg A(n)$, we see that $\neg x^A \neq 0 \leftrightarrow x^A = 0$. Since $x^A = 0 \lor x^A \neq 0$ is undecided, we also cannot assert that f is everywhere defined on $\mathbb{R}$.

At first sight, it seems as if the failure of decidability of equality between the reals must have disastrous consequences for the development of real analysis. That this is not the case, is due to the fact that in most cases it suffices to make decisions which can be made from sufficiently good approximations. Thus, instead of using $x = 0 \lor x \neq 0$, we can often use the fact that for reals x, y, z, if $x < y$, then $(z < y) \vee (x < z)$ can be decided for all z: just choose sufficiently close rational approximations to x, y, and z.

In the weak counterexample of 3.5 $A(n)$ may be any predicate for which our state of knowledge is represented by (1) of 3.5.

We may reformulate 3.5 in terms of (characteristic) functions: let $\alpha \in \mathbb{N} \to \mathbb{N}$ and let $\langle r_n^\alpha \rangle_n$ be a fundamental sequence determining $x^\alpha \in \mathbb{R}$, defined by

$$r_n^\alpha := \begin{cases} 2^{-n} & \text{if } \forall k \leq n(\alpha k = 0), \\ 2^{-k} & \text{if } \alpha k \neq 0 \wedge k \leq n \wedge \forall k' < k(\alpha k' = 0); \end{cases}$$

thus x^A as defined in 3.5 corresponds to x^α if α is the characteristic function of A. We have $x^\alpha = 0 \leftrightarrow \forall n(\alpha n = 0)$, and for a suitable choice of α $x^\alpha = 0 \vee x^\alpha \neq 0$ is unknown; but as we saw in 3.3 we cannot show $\neg(\forall n(\alpha n = 0) \vee \neg\forall n(\alpha n = 0))$. On the other hand, it is possible to obtain proper refutations of the statement "PEM is generally true". For the universal validity of PEM implies in particular

$$\forall\text{-PEM} \quad \forall\alpha[\forall n(\alpha n = 0) \vee \neg\forall n(\alpha n = 0)],$$

or equivalently

$$\forall\alpha[\forall n(\alpha n \neq 0) \vee \neg\forall n(\alpha n \neq 0)],$$

and as we shall see in chapter 4, both in constructive recursive mathematics and in traditional intuitionism certain principles hold which exclude a classical interpretation of the logical operators, and in particular imply $\neg\forall$-PEM.

In CRM $\neg\forall$-PEM is a consequence of the assumption that all operations have to be algorithmic; in intuitionism $\neg\forall$-PEM is the result of the admission of choice sequences as legitimate mathematical objects.

Thus the reasons for accepting $\neg\forall$-PEM are quite different in these two cases: in the first case $\neg\forall$-PEM is the result of a severe restriction on the range of possible operations, in the second case it is the result of enlarging the domain of functions in $\mathbb{N} \to \mathbb{N}$ from predeterminate functions given by a law to the much wider class of functions determined by successive (more or less free) choices.

3.7. *Splitting of notions.* Since the BHK-interpretation causes us to reject certain classically valid logical principles such as PEM, a splitting of notions results: classically equivalent definitions are not always equivalent on the basis of constructive logic, as shown by the example of "finite" and "finitely indexed" in the preceding subsection. Constructively, one has also to distinguish between "finitely indexed set" and "subset of a finite set" (E1.3.2).

Yet another example is the distinction between "non-empty set" and "inhabited set": X is *non-empty* iff $\neg\neg\exists x(x \in X)$, and *inhabited* iff $\exists x(x \in X)$.

A more delicate example is the distinction between inequality and *apartness* between reals: $x \neq y$ simply means that x and y are not equal, but $x \# y$ ("x is apart from y") means that there is a positive distance between x and y:

$$x \# y := \exists r \in \mathbb{Q}(|x - y| > r > 0).$$

Intuitionistically these notions are distinct, but not in CRM (cf. 4.5.4).

One might expect that the splitting of notions leads to an enormous proliferation of results in the various parts of constructive mathematics when compared with their classical counterparts. In practice, usually only very few constructive versions of a classical notion are worth developing, since other variants do not lead to a mathematically satisfactory theory.

4. A brief history of constructivism

The conscious practice of constructive mathematics is a relatively young phenomenon. Before the introduction of highly abstract notions, going hand in hand with abstract proofs, i.e. proofs that went beyond well-trusted tangible procedures, there was no need for a separate domain of the constructive. Of course, with hindsight, we can locate non-constructive proofs in the older mathematics but it would be a historical injustice to project our mathematical culture on our forefathers.

To mention one example that from our point of view personifies non-constructive reasoning, proof by contradiction was universally accepted and employed. There is one particular domain where proof by contradiction is almost a trademark: geometry. From Euclid's *Elements* to Hilbert's *Grundlagen* proof by contradiction is required to set the deductive machinery working. It seems plausible that partly on the authority of Aristotle's logic, partly on idealized and imperfect knowledge of the mathematical universe, the principle of the excluded middle in particular for identity, but more generally for arbitrary notions was taken for granted.

In this chapter we will sketch some of the main trends in constructive mathematics. We will outline the basic ideas and the underlying philosophies, historic details of specific mathematical or logical notions or developments are to be found in the final section ("notes") of each chapter.

In our exposition we have as a rule adopted the terminology that prevails today, the reader should be aware however that in the older literature terms like neo-intuitionist (for intuitionist), empiricist (for semi-intuitionist), etc. occur.

4.1. *Kronecker and the arithmetization of mathematics.* A first clearcut manifest of constructive mathematics was issued by Kronecker (1823–1891), best known for his work in number theory and algebra.

Kronecker's constructivism is most easily recognized in his views on the status of certain mathematical objects. In his essay "Über den Zahlbegriff" (1887) he outlined the project of "*arithmetizing*" Algebra and Analysis; that is, to found these disciplines on the most fundamental notion of number. He explicitly, with a reference to Gauss, restricted this arithmetization to analysis and algebra; geometry and mechanics were left an independent existence, as not being purely determined by our mind, but referring to a certain reality outside our mind.

In a number of lectures Kronecker developed a unified theory of numbers, posthumously published by Hensel (Kronecker 1901), in which numbers, congruences, polynomials and rational functions are treated on an equal footing. In the spirit of this tradition he obtains the integers from the natural numbers by considering $\mathbb{N}[x]$ mod $(x + 1)$.

Similarly rational numbers are introduced by considering congruences (Kronecker 1887, p. 261).

The introduction of algebraic numbers is also based on modulus arithmetic, a procedure that is now universally adopted. One consequence of Kronecker's arithmetization project was that he considered a definition (in number theory and algebra) acceptable only if it could be checked in a finite number of steps whether a given number falls under it or not. This viewpoint naturally led to his criticism of "pure" existence proofs. He stated that an existence proof for a number could only then be considered to be totally exact if it also contained a method to find the number whose existence was proven, (Kronecker 1901, p. vi). In Kronecker's published works there is a minimum of philosophizing; apart from a few remarks on the nature of his programme he consistently sticks to actual mathematical practice. Some of his remarks belong to mathematical folklore, and they capture fairly well his basic outlook. Best known is his widely reported statement that, "the Lord made the natural numbers (ganze Zahlen), everything else is the work of men", and the following statement reiterates the same idea: "I consider mathematics only as an abstraction of the arithmetical reality".

Kronecker's programme was taken up by Jules Molk. In a long paper (1885) he makes a few remarks along the line of Kronecker. On the topic of definitions he says that "definitions should be algebraic and not just logical. It is not sufficient to say: something is or it is not". He points out that a logical definition, like "an irreducible function is by definition not reducible" is not much help. In algebra a definition of irreducible should allow one to test in a finite number of steps that a function does not factorize in nontrivial factors. Only such a test will give meaning to the words *reducible* and *irreducible*.

It must be pointed out that Molk does not contest the *truth* of statements like "a bounded set of numbers has a least upper bound", he views this statement as a logical truth that does not belong to algebra. In the light of Kronecker's remarks on the nature of mathematical objects and proofs it seems likely that Kronecker would reject the above statement as meaningless.

4.2. *The French semi-intuitionists.* The next stage in the evolution of constructive mathematics was the discussion about Zermelo's proof of the well-ordering theorem. Zermelo's use of the axiom of choice in its full strength raised a mixed variety of objections. In particular a group of French mathematicians expressed criticism based on more or less clearly articulated constructive views. We will not go into the details of the axiom-of-choice controversy, but instead concentrate on some of the constructivistic issues. The main representatives of constructive tendencies in this controversy were Baire, Borel, Lebesgue, Lusin and Poincaré. They did not offer a definite philosophical doctrine, but rather a mixture of views on mathematics with constructive tendencies.

Poincaré (1854–1913). Poincaré's views are well-known, he forcefully attacked set theory (Cantorism) and "logistics" (i.e. mathematical logic). Strictly speaking one cannot count him as a constructivist; his direct influence on later developments in constructive mathematics is mostly restricted to his views on intuition closely tied up with his criticism of logic. He was sceptical of both Aristotelian logic – "a syllogism cannot teach us anything essentially new" (1902, chapter I) – and the newer variants of Russell, Frege and Peano.

Poincaré argues that in mathematics one needs more than logic, one needs intuition. He mentions three kinds of *intuition* (1905, chapter I): the appeal to the senses and the imagination, generalization by induction, and finally the intuition of pure number. The latter one guarantees the exactness of mathematics. Mathematics has its specific methods of proof, and the method *par excellence* is the principle of induction, which is according to Poincaré a synthetic judgement a priori ("Les derniers efforts des logisticiens", in 1908).

In a number of ways Poincaré is a forerunner of the later constructivists: his recognition of the principle of mathematical induction as an unprovable fact, his criticism of the "creative role" of logic, his rejection of the Cantorian *actual infinite* and his embracement of intuition in mathematics.

One should not, however, expect too much consistency in Poincare's philosophical writings, (cf. Mooij 1966); in his essay "Les mathématiques et la logique" (in 1908) he defended, for example, the rather platonistic view that "Existence can mean only one thing: freedom from contradiction".

The paradoxes around the turn of the century were, in his view, caused by impredicative reasoning, vicious circles. In a number of publications he defended a predicative version of mathematics. Discussing Richard's paradox, he wrote "*E* is the set of *all* numbers one can define with finitely many words, *without introducing the set E itself*. Without this (restriction) the definition of E will contain a vicious circle; one cannot define E through the set E itself" ("Les derniers efforts des logisticiens", in 1908).

This is the familiar syntactic formulation of the predicative restriction. In "La logique de l'infini" (in 1913) he provided a mathematical criterion: the definition of a set should be such that adding more elements to the universe under consideration does change the set.

Emile Borel (1871–1956). Emile Borel was the most pronounced spokesman of the semi-intuitionists, he was closer to Kronecker's position on a number of issues than Poincaré. He maintained, for example, that only effectively defined (i.e. by finitely many words) objects exist in the domain of science (1914, p. 171, 175), consistency does not suffice for existence. However, compared to Kronecker's "verifiability in finitely many steps" this "finite definability" is clearly a more liberal notion. It allows one, for example, to consider the functions f with $f(x) = 0$ for rational x and $f(x) = 1$ for irrational x, as well-defined and existing.

Lebesgue shared Borel's views on this point. From Borel's point of view individual real numbers had to be given by a finite definition, so as a set they could never get beyond the countable. Realizing how awkward a countable continuum would be, Borel considered the continuum as independently given by intuition: "we accept that the collection c of real numbers between 0 and 1 is given, without investigating how they could effectively be given" (Borel 1914, p. 16); he spoke of the *geometric continuum*. The same view had already been defended by Hölder in 1892 (cf. 1924, p. 194) and Brouwer independently elaborated it in his Ph.D. Thesis (1907).

This concept of geometric continuum yields a "whole" that precedes the individual elements, and cannot be thought of as "the set of its reals" (Lusin 1930). For the practice of mathematics Borel nevertheless wished to have available a sufficiently large part of the continuum that could be considered as "made up of individual real numbers". For this purpose he introduced the *practical continuum* (1914, p. 166), consisting of the finitely definable numbers.

The discussion about Zermelo's proof of the well-ordering theorem had two aspects that are relevant from the constructive viewpoint: the well-ordering of the continuum and the axiom of choice. The first aspect gave rise to reflections on the nature and structure of the continuum, the second one gave rise to reflections on the notion of "infinite succession of choices". The topic of legitimacy of sequences of choices was taken up by Borel (cf. 1914, p. 168). We will return to it in the context of intuitionism.

Borel's philosophical position may be inferred from the following quotation (1914, p. 173):

"I do not understand what the abstract possibility of an act, which is impossible for the human mind can mean. For me this is a purely metaphysical abstraction without any scientific reality".

The foundational practice of the semi-intuitionists was mostly carried out in the margin of their main mathematical researches. One cannot say that a systematic philosophical (or foundational) basis for their constructive convictions had been put forward. In general the constructivistic activities of the semi-intuitionists remained "a gentleman's pastime"; their views have, however, played a role in the subsequent development of constructive mathematics.

4.3. *Brouwer and intuitionism.* In the early years of the century Brouwer (1881–1966) had formed his views on mathematics, both pure and applied. In 1907 he published his dissertation "On the foundations of mathematics". Since the dissertation was written in Dutch (an English translation appeared only in his "Collected Works" (1975)) the impact was marginal. Nonetheless it is worthwhile to discuss it since it contains much that is basic for understanding his later and riper views. Leaving aside the purely mathematical topics we find three main themes: the continuum, mathematics and experience (applications), mathematics and logic.

Brouwer recognized in his thesis – as far as we can see, independently from Borel – the continuum as a primitive notion given directly by intuition.

Brouwer rejected the reduction of the continuum by means of any of the approximation methods of irrationals by rationals, since he considers these approximations as a process that unfolds in stages and hence at each stage remains denumerable (observe the analogy with Borel's continuum, but note that Brouwer was probably not aware of Borel's views until the international congress of mathematicians in 1908 in Rome).

Brouwer's view of mathematics is basic for the understanding of the development of intuitionism, his views on *existence*, *mathematical objects*, *logic*, *language* are all consequences of this characterization of mathematics:

Mathematics is a free creation of the mind, it allows the construction of mathematical systems on the basis of intuition. Mathematical objects, hence, are mental constructs.

Some consequences drawn by Brouwer are:

(1) From the Kantian a priori only the "intuition of time" remains.

(2) Mathematics is independent of logic; logic is an application of mathematics.

(3) Mathematics cannot be founded upon the axiomatic method.

In particular Brouwer rejected in his thesis Hilbert's (early) formalism (1905) and Cantorian set theory. Brouwer's criticism of certain parts of traditional mathematical practice has led to the unfortunate mistaken view that Brouwer's "revolution" was mainly a negative one (cf. Hilbert 1922). Actually Brouwer introduced a number of innovations that lifted his intuitionistic mathematics well above the level of securing some safe corners in the building of mathematics.

The concept that is most closely tied to Brouwer's name is that of "choice sequence". Sequences of numbers determined by choice had already occurred in the literature before Brouwer's first exposition in 1918. Borel had discussed them in relation with Zermelo's paper on the well-ordering theorem (1914, pp. 168–169), and he had pointed out that in order to recover the "complete arithmetical concept of the continuum" one needs countable sequences of choices (1914, p. 161). Before him Du Bois Reymond had already considered real numbers with decimals determined by throwing a die (1882). However, with Borel and Du Bois Reymond these sequences remained curiosities. It was Brouwer who discovered how to exploit the choice feature in practical mathematics. After rejecting choice sequences as non-intuitionistic (1912), he came to accept them (1914), and in 1918 he published his first basic intuitionistic paper in which the notion of choice sequence was a corner stone. In that paper his *continuity principle* was also enunciated and immediately used to show the non-denumerability of $\mathbb{N}^{\mathbb{N}}$, for a historical discussion cf. Troelstra (1982A).

Further reflection on the nature of choice sequences led Brouwer to a proof of the "uniform continuity theorem" (i.e. all functions from $[0, 1]$ to $\mathbb{R}$ are uniformly continuous). Analysis of the proof shows that it is based on yet another intuitionist principle, the principle of *Bar Induction* (cf. section 4.8, and 4.10.5).

The negative aspects of Brouwer's foundational activities are to be found in his criticism of the mathematical practice of his day. From his point of view, which characterized mathematics as an inner, mental constructional activity, language and logic were secondary phenomena. Language was a highly fallible tool for communication, not to be trusted as a source of insight, and logic was a product derived from mathematics. For Brouwer's views on language cf. (1907, 1929, 1949).

As far as logic was concerned, Brouwer had rather conservative views. Sometimes he understood by logic the traditional Aristotelian theory of syllogisms, sometimes the logic of Peano and Russell.

Mathematical reasoning, according to Brouwer, consists of constructing mathematical structures, and the frequent appearance of chains of applications of logical rules (syllogisms) is a practice that is only justified by mathematical constructions accompanying each application of a rule. The above applies to intuitive logical arguments. Theoretical logic, according to Brouwer, is an application of mathematics, and an empirical science; it will never teach us anything concerning the organization of the human intellect.

We highlight a few of Brouwer's views.

On quantification (1907, cf. 1975, p. 76):

"the mistake which so many people, thinking that they could reason logically about other objects than mathematical structures built by themselves, and overlooked, that wheresoever logic uses the word all or every, this word, in order to make sense, tacitly involves the restriction: *insofar as belonging to a mathematical structure which is supposed to be constructed beforehand*".

On consistency (in particular Hilbert's early formalism, 1905). Brouwer's comments are roughly as follows (cf. 1975, p. 78): Not recognizing any intuitive mathematics, the consistency of (e.g.) number theory has to be proved in a linguistic framework by means of axioms (among which *certainly* the axiom of complete induction), next one has to prove the consistency of this system. However,

(1) this does not constitute any progress over the previous stage,

(2) consistency does not imply the existence of an accompanying mathematical structure,

(3) even if the mathematical system of reasoning exists, this does not entail that it is *alive*, i.e. that it accompanies a sequence of thoughts, and even if the latter is the case this sequence of thoughts need not be a mathematical development, so it need not be convincing.

On set theory: Cantor's justification of the existence of sets given by comprehension (or even separation) is incorrect. For a number of important notions in set theory no existence grounds exist, e.g. the second number class and $\mathbb{N}^{\mathbb{R}}$ do not exist. (1975, p. 81, 87).

On the levels of meta-theory: In the logico-linguistic treatment of mathematics one can discern certain stages (1975, p. 174):

(1) The construction of pure mathematical (mental) systems.
(2) The linguistic description of mathematics.
(3) The mathematical study of (2), including the logical structure.
(4) Abstraction from the meaning of the expressions of (3), and of the logical operations, thus creating a second-order mathematical system (this is the system of the logicists).
(5) The language of (4).
(6) The mathematical study of (5) (a step made by Hilbert, but not by Peano and Russell).
(7) Abstraction from the meaning of the expressions of (6), thus creating a mathematical system of the third order.
(8) The language of (7) (this is the last stage considered by Hilbert). Etc.

The well-known, and somewhat notorious issue of the *principle of the excluded middle* (PEM) has sidetracked many of Brouwer's contemporaries into believing that intuitionism was primarily seeking a revision of the logical apparatus of mathematics. The title of Brouwer's first full scale exposition (1918), may very well have fostered this misunderstanding.

In the paper "The unreliability of the logical principles" (1908), Brouwer drew the inevitable consequence from his views on mathematics and refuted PEM using an undecided property of the decimals of π.

Brouwer himself never elaborated in writing the meaning of the logical constants, that step was eventually made by Heyting and Kolmogorov, but in his writings a "proofs as constructions"-concept is clearly implicit (cf. 1907, chapter III; 1908).

In the late twenties Brouwer started to look for a way to exploit the peculiar properties inherent in the creative subject, the results appeared in print only after the second world war, (1948, 1949). The part of intuitionism based on the creative subject has remained rather marginal.

Brouwer's programme of reconstructing mathematics along intuitionistic lines found few adherents; a considerable part of the programme was carried out by his principal student A. Heyting, whose best-known contribution to intuitionism was the formalization of intuitionistic logic (and arithmetic) and the formulation of a natural semantics for it. This semantics is the BHK-interpretation of section 3; it is sometimes called the *proof interpretation* (Heyting 1934 and 1.5.3 below).

4.4. *Hermann Weyl (1885–1955) and predicativity.* The tradition of Poincaré and Russell, which laid the blame of inconsistencies and paradoxes on impredicative definitions, was taken up in 1918 by Weyl, who published a monograph, "Das Kontinuum" (1918). Like his predecessors, Weyl was out to avoid vicious circles; he rebuilt analysis using only a very restricted kind of sets and functions. Indeed, he only allowed arithmetically definable sets, i.e. sets defined by comprehension over formulas without bound set variables. Weyl's reconstruction was quite successful, a large portion of traditional mathematics was recovered.

Weyl's philosophical position in this monograph (and in foundational matters in general) was that of Husserl's Phenomenology.

However, after learning of Brouwer's intuitionistic foundation of mathematics he gave up his own programme and embraced intuitionism. In "Über die neue Grundlagen Krise der Mathematik" (1921) he unequivocally sided with Brouwer in the latter's revolution, but later in life he dissociated himself from intuitionism and foundations in general. Lorenzen 1955 can be viewed as a continuation of Weyl's original programme.

Both Weyl and Lorenzen were concerned with the actual revision of mathematical practice. Later work in predicativism was mainly of a metamathematical nature (cf. Feferman 1964, 1978, Kreisel 1960). So far predicative mathematics has remained a curiosity for logicians rather than a realistic alternative for the working mathematician. Recently, however, there has been a renewed interest in proof theoretically weak systems with strong expressive power, (Takeuti 1978, Friedman 1980, Friedman et al. 1983), e.g. as in Friedman's "reverse mathematics". In fact, a surprisingly large body of mathematics can be developed in quite weak systems, much weaker than so-called predicative analysis. In many cases even conservative extensions of first-order arithmetic allow the development of a good deal of traditional mathematics, in particular analysis.

4.5. *Finitism.* Brouwer's intuitionism contained unmistakable abstract notions; in particular sets (species), choice sequences, spreads are abstract objects in mathematics proper. The use of abstract notions has been recognized and criticized by a number of mathematicians, e.g. Kronecker, Skolem and Goodstein. They proposed a very narrow version of constructive mathematics, one in which only concrete combinatorial operations on (strictly) finite mathematical objects are allowed. The prototype of "finite mathematical object" is "natural number" and a good example of a "combinatorial operation" is provided by a "table of multiplication".

Skolem, already in 1919, systematically considered what kind of combinatorial operation on the natural numbers can be introduced (1923).

He developed a sizable part of primitive recursive arithmetic, i.e. he studied primitive recursive functions *avant la lettre* in a quantifier-free system. In fact, he worked in an informal mathematics, but he initiated a long tradition in the theory of primitive recursive arithmetic, cf. Goodstein (1957).

Skolem explicitly adopted Kronecker's foundational viewpoint: "a mathematical definition is a genuine definition if and only if it leads to the goal by means of a *finite* number of trials" (1923). As far as the foundations of mathematics are concerned, Kronecker is a forerunner of finitism.

The restricted combinatorial view also turned up in Hilbert's programme. Hilbert, in his improved formalist philosophy considered the finitistic, combinatorial mathematics as the truly meaningful part of mathematics. Mathematics dealing with the actual infinite, e.g. full arithmetic, was to be reduced to the finitary mathematics (arithmetic). This is the basic idea of Hilbert's programme: ideal statements are to be granted a meaning in terms of finitary statements. Finitistic mathematics, of course, is completely constructive and hence a part of intuitionistic mathematics. Some early writers on the foundations of mathematics have confused finitistic mathematics and intuitionistic mathematics, e.g. Herbrand (1931).

After Gödel's demonstration that classical arithmetic could be reduced to intuitionistic arithmetic via his "negative translation", (1933), it was clear that even intuitionistic arithmetic went beyond finitism.

4.6. *Constructive recursive mathematics and Markov's School.* Another concept of constructivity, based on the theory of *algorithms* or recursive functions, has presented itself in the thirties. This development, however, had a totally different philosophical motivation. At first primitive recursive functions and recursive functions were introduced as tools for metamathematical research. Gödel used (what we now call) primitive recursive functions in his proof of the incompleteness theorem (1931), and as pointed out above, Skolem had already in 1919 developed the basis of primitive recursive arithmetic (without introducing the class of primitive recursive functions explicitly) (Skolem 1923). Elaborating a suggestion of Herbrand, Gödel introduced the wider class of recursive functions in (1934).

Along completely different lines Church introduced the λ-calculus (1932), a theory that only used λ-abstraction and application. In this theory the natural numbers could be represented and Kleene showed that the recursive functions were λ-definable.

Curry, in the meantime, had developed combinatory logic (1930), following the early work of Schönfinckel.

The equivalence of λ-calculus and combinatory logic was established by Rosser, and the equivalence of the Herbrand–Gödel recursive functions with the λ-definable ones was proved by Church (1936) and by Kleene (1936).

An interesting conceptual characterization of algorithms was undertaken by Turing, who used abstract machines, now known as *Turing machines* (Turing 1937). Turing analyzed the intuitive concept of "mechanically computable by a human operator" and put forward a convincing argument that "computable" coincides with "computable by a Turing machine". Church had also stated in 1936 a characterization of "computable", but in terms of recursiveness, he equated "computable" and "recursive" (1936). This identification of "effectively computable" and "recursive" has become known as *Church's Thesis*.

The theory of (partial) recursive functions has become a cornerstone of metamathematics. The reader will find more information on the historical details in the relevant chapters.

Mathematics on the basis of Church's Thesis was propagated by Markov, who founded a school of constructive mathematics in Moscow. Markov laid down the following principles for constructive mathematics:

(1) the objects of constructive mathematics are words in various alphabets.

(2) abstraction through potential realizability is allowed, but not abstraction through the actual infinite.

(1) must be understood as including e.g. recursive functions, as given (or coded) by a set of equations or an index. An example of a potentially

realizable process is the addition of natural numbers, it does not involve the actually infinite set of natural numbers, but an algorithm that for any given n and m produces (a representation of) $n + m$.

Markov formulated his constructive mathematics in terms of *normal algorithms*, also known as *Markov algorithms* (cf. Markov 1954A, Mendelson 1964). We shall not use Markov's formalism, but use the more familiar recursive functions.

A consequence of Markov's view is that number theoretic functions are recursive; that is to say the potentially realizable version of this "infinite" object must allow a representation by an index. In this sense Markov's constructive mathematics can be viewed as a theory based on Church's Thesis.

On the one hand Markov is more strict than the intuitionists, e.g. he rejects choice objects, on the other hand he is more liberal. This is illustrated by his argument for justification of his *principle of constructive choices* (Markov 1962), nowadays called *Markov's Principle*. (Markov formulated his principle in lectures in Leningrad in 1952/1953, it appeared in print in Markov 1954).

In familiar terms, the principle states that if it is not the case that a Turing machine does not halt on a given input, then it does halt. Markov's argument, in terms of Turing machines, is simply as follows; if it is impossible that the Turing machine will compute forever, then there is a clear algorithm for obtaining the output: just continue the process until it halts. Markov added:

"I know that intuitionists reject this method of argument, since they do not consider it 'intuitively clear'. In connection with this I consider it necessary to make the following remarks. Firstly, my intuition finds this sufficiently clear. On the other hand, I can in no way agree to taking 'intuitively clear' as a criterion for truth in mathematics, for this criterion would mean the complete triumph of subjectivism and would lead to a break with the understanding of science as a form of social activity. If I defend this means of argument here, it is not because I find it without error according to my intuition, but rather, firstly because I see no reasonable basis for rejecting it, and secondly, because arguments of this type make it possible to construct a constructive mathematics that is well able to serve contemporary natural science" (Markov 1962).

As Markov indicates, his principle is not accepted by the constructivists at large; the principle has, however, found its place in the metamathematics of constructive mathematics (see e.g. sections 3.5, 4.5).

4.7. *Bishop and constructive mathematics for the working mathematicians.* The landscape of constructive mathematics suddenly changed in 1967 when Bishop (1928–1983) published his "Foundations of constructive analysis". "This book, he said, had a threefold purpose: first, to present the constructive point of view; second, to show that the constructive program can succeed; third, to lay a foundation for further work. These immediate ends tend to an ultimate goal – to hasten the inevitable day when constructive mathematics will be the accepted norm". Bishop's message was that the real business of constructive mathematics is *mathematics*; it came at a time when constructivism was mostly tending its foundations, and it was presented by a first-rate mathematician.

In contrast to Brouwerian intuitionism (even after Heyting's presentation), Bishop's mathematics used the language of the working mathematician. He attacked mathematics much in the spirit of Kronecker, be it that he introduced abstract objects in a more direct way.

A few examples may help to illustrate Bishop's position.

In his view the only way to show that an object exists is to give a finite routine for finding it.

Sets are introduced by him as "the totality of all mathematical objects constructed in accord with certain requirements" and they come with a given equivalence relation called "equality".

The basic higher type objects are *operators*, i.e. rules which assign elements of a set A to elements of a set B.

Such a rule f must afford an explicit, finite mechanical reduction of the procedure for constructing $f(a)$ to the procedure for constructing a. A *function* then, is an operator that respects equality.

Bishop does not specify the notion of "rule", but throughout rules could be taken to be recursive; that is, Bishop's mathematics is compatible with Church's thesis.

The fundamental thesis of Bishop's mathematics is that "all mathematics should have numerical meaning", which is closer to Kronecker (or even Hilbert) than to Brouwer. Bishop foregoes the abstract objects of classical mathematics as well as those of the intuitionists. His mathematics rests on neutral ground, without ontological or idealistic objects and principles. Thus he cannot, like Brouwer, show that all real functions are continuous, but as a pragmatist he restricts his attention to continuous functions.

The reader can see for himself how actual mathematics is done constructively in Bishop's instructive book. Bishop had a number of followers, and at present a number of mathematicians are constructivizing mathematics along the same lines as Bishop.

The work of Bishop also raised a number of metamathematical questions, the main question being: which principles suffice to found Bishop's mathematics on. There are to date two main approaches to the problem: a set theoretical one, proposed by Myhill and Friedman, and an "operations and classes" one, proposed by Feferman. The reader may consult chapter 9 and Beeson (1985). Bishop's constructive mathematics was intended as the neutral part of constructive mathematics and as compatible with traditional classical mathematics. Metamathematical analysis has born this out, one can (in principle) extend his mathematics both in the direction of Brouwer's intuitionism and in the direction of recursive mathematics *à la* Markov.

4.8. *Esenin-Vol'pin's ultra-finitism.* In the opinion of a number of mathematicians Brouwer's criticism of traditional mathematics stopped short of the potential infinity of mathematics – in particular, of the natural number sequence. Borel, for instance, declared that "the very large finite offers the same difficulties as the infinite", Borel (1947), and van Dantzig (1956) posed the question "Is $10^{10^{10}}$ a finite number?" in the tradition of Mannoury (1909). Wittgenstein also shared the view that mathematics has no business with an idealized infinite set of natural numbers. The first to develop the strict finitist view beyond a collection of philosophical reflections was A.S. Esenin-Vol'pin. From 1959 onward he undertook the founding of mathematics on a strictly finite basis. He challenged the view that there is, up to isomorphism, a unique sequence of natural numbers. Taking a realistic view of natural numbers, there does not seem to be a good reason to assume that the successor operation can indefinitely be iterated. So there may be universes of natural numbers that do not have the same length. As a consequence of the limitations as to the natural numbers that can actually be reached, the universe(s) of natural numbers may not be closed under the common operations, such as multiplication or exponentiation. The strictly finitist view also has its consequences for logic; the derivations of A and $A \to B$ may still be within reach, but in order to apply modus ponens one might have to exceed the available natural numbers necessary for the length of the derivation of B. In a sense Esenin-Vol'pin's program is even more restrictive than Hilbert's finitism, therefore "ultra-finitism" is a more fitting name than the original "ultra-intuitionism".

The program indicated in Esenin-Vol'pin (1961, 1970, 1981) so far has not progressed beyond the initial stage. There are, however, certain researches into the possibility of introducing *feasible numbers*, e.g. by Parikh (1971), which tie up with considerations concerning complexity. Geiser

(1974) has applied the technique of non-standard models to ultra-finitist arithmetic.

5. Notes

5.1. *Books on constructive mathematics and its metamathematics.* Heyting (1956) is still a very readable introduction to intuitionistic mathematics. Brouwer's own approach is illustrated in his Cambridge lectures (1981).

Dummett (1977) is an introduction to intuitionism with only a little mathematics, but it contains a good deal of logic and metamathematics as well as philosophical discussion.

For a development of constructive mathematics in the style of E. Bishop see Bishop and Bridges (1985) and Bridges (1979).

Aberth (1980) gives an introduction to recursive analysis; Kushner (1973) (recently translated) is an excellent introduction to constructive recursive mathematics in the spirit of A.A. Markov.

Of the monographs on the metamathematics of constructive mathematics we mention Beeson (1985), Gabbay (1981) (intuitionistic predicate logic), Kleene and Vesley (1965) (intuitionistic analysis), Lambek and Scott (1986) (for connections between intuitionistic logic and category theory), and Troelstra (1973). Further information can be obtained from the notes at the end of each chapter, and the bibliography by Müller (1987).

5.2. *Hypothetical reasoning.* is also essential for the development of constructive mathematics.

Certain passages in Brouwer's thesis might erroneously create the impression that he did not want to consider hypothetical constructions, in particular constructions which afterwards turn out to be impossible (1907, reprint, p. 22). In any case there is no doubt that in his later papers Brouwer accepted hypothetical reasoning, e.g. in Brouwer (1923) where he shows that the triple absurdity of a statement is equivalent to its absurdity (the logical law $\neg A \leftrightarrow \neg\neg\neg A$ in our notation).

The obvious intuitive objection is: how can we have a clear mental picture of a construction which afterwards turns out to be impossible? This objection has been raised at least twice in the history of intuitionism (cf. Troelstra 1983A).

Freudenthal (1937) observes "We first ask ourselves how we should intuitionistically understand a proposition ["Satz" in the German original]. The most obvious explanation is to regard a proposition as the assertion of a fact, verified by a proof [...]. This explanation must be treated with caution, since, if a proposition is only the assertion of a fact, then the separation between proof and proposition is not justified any longer". Also, "If we recall that the correct formulation of a proposition is its proof, we immediately see that the construction of a proof of b from a proof of a, or the reduction of the solution of problem b to the solution of problem a is possible only if a has been shown to be correct – and thus implication is useless".

Heyting (1937) in his reply remarks "The following simple example shows that the problem $a \supset b$ in certain cases can be solved *without* a solution for problem a being known. For a I take the problem 'find in the sequence of decimals of π a sequence 0123456789', for b the problem 'find in the sequence of decimals of π a sequence 012345678'. Clearly b can be reduced to a by a very simple construction".

A late echo of this discussion is found in the attempt by Griss (see e.g. 1946, 1955) to develop intuitionistic mathematics without negation.

5.3. *The BHK-interpretation.* may be regarded as implicit in Brouwer's writings (for example in his treatment of negation in Brouwer 1923). Heyting (1930C, 1931, 1934) made the interpretation explicit. In Heyting (1930C, 1931) a proposition is explained as "the expectation to find a certain condition fulfilled"; asserting a proposition is then the same as stating that a construction has been found which shows the condition to be fulfilled.

Kolmogorov (1932) gave an interpretation of intuitionistic propositional logic as a calculus of problems. Thus for example, $A \to B$ represents the problem to solve problem B when a solution to problem A is known, or equivalently, to reduce the solution of B to the solution of A. Heyting (1934) contains the clause for implication as presented here; he also extended Kolmogorov's interpretation to predicate logic. Later Heyting came to regard Kolmogorov's interpretation as essentially the same as his own (1958).

Kreisel (1962, 1965; for an exposition see also Troelstra 1969) proposed a version of the BHK-interpretation where the construction establishing an implication $A \to B$ consists of a pair (p_0, p_1) where p_0 is the construction transforming any proof of A into a proof of B, and p_1 a construction verifying this fact. In addition $\pi_A :=$ "p proves A", and $\pi(p_0, p_1, A, B) :=$ "p_1 shows that, for all q, q proves $A \Rightarrow p_0(q)$ proves B" are assumed to be decidable for arbitrary p, p_0, p_1, A, B. The clause for universal quantification is similarly modified.

Kreisel proposed this version in the hope of obtaining interesting new models for intuitionistic systems, but this hope was not fulfilled. N. Goodman developed a rather complicated theory of constructions based on Kreisel's modified interpretation of the clauses for $\rightarrow$ and $\forall$ (see e.g. 1970), where the constructions were divided into levels, but it turned out that for his principal technical result – the conservativeness of the axiom of countable choice over intuitionistic arithmetic – these features of Goodman's theory were not needed.

Sundholm (1983) argued at length that the insight that "for all q: q proves $A \Rightarrow p_1(q)$ proves B" should not itself be regarded as a mathematical object. Weinstein (1983) on the other hand regards the decidability of the proof predicate and the extra clauses in Kreisel's variant of the BHK-interpretation as essential for a foundational justification of intuitionistic logic.

As shown by Note A in Troelstra (1981), it is possible to give trivial models with decidable proof predicates π_A and decidable π. See also the discussion in chapter 16.

Exercises

1.3.1. Show that the first two groups of schemas listed in section 3.4 are valid on the BHK-interpretation.

1.3.2. A *subfinite* set is a subset of a finite set. Give weak counterexamples to the assertions: (a) each subfinite set is finitely indexed and (b) each finitely indexed set is subfinite (cf. 3.6).

1.3.3. Give a weak counterexample to König's lemma in 2.4. *Hint.* Cf. the use of the predicates A and B in 5.2.12.

1.3.4. Let D be an inhabited domain and let $R_0, R_1, R_2, \ldots$ be relations over D. Let A be any sentence constructed from primitive statements $\perp$, $R_i(\vec{d})$ and logical operations $\wedge, \vee, \rightarrow, \forall, \exists$.

Associate with each A a set of "proofs" $p[A]$ by induction on the number of logical operations in A such that:

(i) $p[R_i(\vec{d})]$ is inhabited iff the finite sequence $\vec{d}$ of elements of D satisfies the relation R_i, and otherwise empty; $p[\perp] = \emptyset$;

(ii) $p[B \wedge B'] := p[B] \times p[B']$;

(iii) $p[B \vee B'] := \{(0, x) : x \in p[B]\} \cup \{(1, x) : x \in p[B']\}$;

(iv) $p[B \rightarrow B'] := p[B] \rightarrow p[B']$, the set of mappings from $p[B]$ to $p[B']$;

(v) $p[\forall x B(x)] := \{f : \forall x \in D(f(x) \in p[B(d)])\}$;
(vi) $p[\exists x B(x)] := \{(d, x) : x \in p[B(d)]\}$.
Show, using classical reasoning, that $A \leftrightarrow \exists x(x \in p[A])$ for arbitrary assignments p of sets of proofs to $R_i(\vec{d})$ satisfying the clauses (i)–(vi). This exercise shows that the BHK-interpretation in itself has no "explanatory power": the possibility of recognizing a classically valid logical schema as being constructively unacceptable depends entirely on our interpretation of "construction", "function", "operation".

CHAPTER 2

LOGIC

For a reader who is primarily interested in seeing some examples of constructive mathematics, very little logic is needed to understand almost all the material in chapters 5–8 (the parts of chapter 4 essential to these chapters also require little or no knowledge of formal logic). What is presupposed however, is that the reader has made himself familiar with the BHK-interpretation discussed in chapter 1, and has acquired some facility in reading mathematical statements written with the help of logical symbols, since we shall use logical notation almost continually.

Description of the contents of the chapter. The first section is devoted to a system of natural deduction for intuitionistic predicate logic **IQC**. Familiarity with the basics of classical predicate logic, though not necessarily with systems of natural deduction, is presupposed.

The second section extends the treatment to cover logic with possibly undefined terms in two variants (E-logic and E^+-logic).

It is possible to gain more familiarity with constructive logic than is to be obtained from chapter 1, without getting too involved in the hardware of the formalism, by studying 1.2–9, 2.1–5.

Section 3 discusses the relations between classical and intuitionistic predicate logic, in particular the Gödel-Gentzen "negative translation" and its variants.

Section 4 discusses systems for intuitionistic predicate logic of the axiomatic type (sometimes called "Hilbert-type" systems), which are frequently encountered in the literature and are often convenient in metamathematical work.

Kripke's semantics for **IQC** is introduced in section 5; section 6 contains a Henkin-type completeness proof. Kripke models provide an intuitively very appealing semantics for **IQC** – not (obviously) related to the BHK-interpretation – which permits us to show quite easily that certain statements are not derivable in **IQC**.

Section 7 finally discusses definitional extensions of systems obtained by adding definable predicates, total or partial functions.

1. Natural deduction

For the development of constructive mathematics as such we do not need a formal system of logic. Nevertheless it is convenient to have a set of logical rules available: thus we do not have to go back to the Brouwer–Heyting–Kolmogorov interpretation (1.3.1) each time we want to justify the use of a logical principle in our arguments.

We have chosen natural deduction as our starting point, because it agrees well with informal reasoning. Moreover, its rules are motivated by the BHK-interpretation in a direct way.

1.1. *The language of predicate logic.* The language $\mathscr{L}$ of predicate logic is based on a (countably) infinite supply of variables $v_0, v_1, \ldots$ (as metavariables for variables we shall use x, y, z in this chapter), n-ary predicate (relation) symbols $R_0^n, R_1^n, \ldots$ (for all $n \in \mathbb{N}$; metavariables $R, R', \ldots$) and n-ary function symbols $f_0^n, f_1^n, \ldots$ (for all $n \in \mathbb{N}$; metavariables f, g). The zero-place function symbols are also called (individual) constants (metavariables: c, d).

This language can easily be generalized in two directions: (i) dropping the restriction to *countably* many symbols of each kind, and (ii) extension to a many-sorted language, with a collection of *sorts* I, and variables $v_0^i, v_1^i, v_2^i, \ldots$ for each sort $i \in I$, a supply of n-ary relation symbols $R_k^{i_0, \ldots, i_{n-1}}$, $k \in \mathbb{N}$ for each sequence $i_0, \ldots, i_{n-1}$ of sorts in I, a supply of function symbols $f_k^{i_0, \ldots, i_{n-1}, i}$, $k \in \mathbb{N}$ for each $i_0, \ldots, i_{n-1}, i$ from I.

The language of so-called higher-order logic will be treated as a special instance of a many-sorted language.

Terms, and formulas on the basis of $\wedge, \vee, \rightarrow, \perp, \forall, \exists$ as primitives, are defined as usual. t, s will be used as metavariables for terms, A, B, C, D as metavariables for formulas. P, Q usually denote atomic formulas.

The language of *propositional* logic $\mathscr{L}_0$ has only 0-place relation symbols; these are called *propositional* variables (metavariables P, Q).

1.2. *Examples of natural deduction rules.* Suppose we have established B, repeatedly appealing to assumption A. This means that we have shown how to construct a proof of B from a hypothetical proof of A; thus, on the BHK-interpretation this means that we have established the implication $A \to B$. In this conclusion A is not an assumption anymore (A has been *cancelled*, *discharged*, or *eliminated* as an assumption). Schematically we can render this as

$$\dfrac{\begin{array}{c}\not{A}\\ \vdots\\ B\end{array}}{A \to B}$$

The crossing-out of A indicates that A has been eliminated as an assumption in the final conclusion. The type of inference just described is called implication introduction ($\to$ I) since in the final conclusion an implication sign is introduced.

There is also an implication elimination rule ($\to$ E): if we have shown $A \to B$ and A, we can also prove B, since a proof of $A \to B$ must provide a construction to transform a proof of A into a proof of B. Schematically:

$$\dfrac{\begin{array}{cc}\vdots & \vdots\\ A \to B & A\end{array}}{B}$$

As an example of a deduction by means of $\to$ I, $\to$ E we show how to deduce $A \to C$ from assumptions $A \to B$, $B \to C$:

$$\dfrac{\dfrac{B \to C \qquad \dfrac{A \to B \quad A}{B}\,{\to}\,\mathrm{E}}{C}\,{\to}\,\mathrm{E}}{A \to C}\,{\to}\,\mathrm{I}\ (A \text{ cancelled})$$

At each horizontal line we have indicated the rule which has been applied.

For conjunction, we can similarly justify an introduction rule $\wedge I$ and two (left- and right-) elimination rules, $\wedge E_l$ and $\wedge E_r$:

$$\wedge \mathrm{I}\ \frac{\begin{matrix}\vdots \\ A\end{matrix} \qquad \begin{matrix}\vdots \\ B\end{matrix}}{A \wedge B} \qquad \wedge \mathrm{E}_l \frac{\begin{matrix}\vdots \\ A \wedge B\end{matrix}}{A} \qquad \wedge \mathrm{E}_r \frac{\begin{matrix}\vdots \\ A \wedge B\end{matrix}}{B}$$

An example of a deduction based on these rules is

$$\dfrac{\dfrac{\wedge \mathrm{E}_r \dfrac{A \wedge B}{B} \qquad \wedge \mathrm{E}_l \dfrac{A \wedge B}{A}}{B \wedge A}\ \wedge \mathrm{I}}{A \wedge B \rightarrow B \wedge A}\ \rightarrow \mathrm{I}\ (A \wedge B \text{ cancelled})$$

Finally we consider a quantifier rule, namely all-introduction $\forall$I. Schematically $\forall$I can be rendered as

$$\frac{\begin{matrix}\vdots \\ A(y)\end{matrix}}{\forall x A(x)}$$

That is, if we have derived $A(y)$ for a completely arbitrary y, we may in fact infer $\forall x A x$, since our derivation of $A(y)$ serves as a schema which can be applied to any particular object in the range of the variable y – so we have the construction required by the BHK-interpretation for a proof of $\forall x A(x)$. In an application of this rule, the proof of $A(y)$ should *not* depend on other assumptions containing y, because then we could not regard y as being completely arbitrary.

That without such a restriction we indeed arrive at false conclusions becomes clear by the following example of a "deduction" in which the condition is violated:

$$\dfrac{\dfrac{A(x)}{\forall y A(y)}\ \forall \mathrm{I}}{A(x) \rightarrow \forall y A(y)}\ \rightarrow \mathrm{I}\ (A(x) \text{ is cancelled})$$

The deductions in our examples above may be regarded as labelled trees, the label attached to a node consisting of a statement, together with the name of the rule which has been applied to obtain the statement. For reasons of readability we have put the name of the rule next to the horizontal lines instead of next to the formula. Thus the trees in our examples of deductions (of $A \to C$ from $A \to B$, $B \to C$ and of $A \wedge B \to B \wedge A$) are

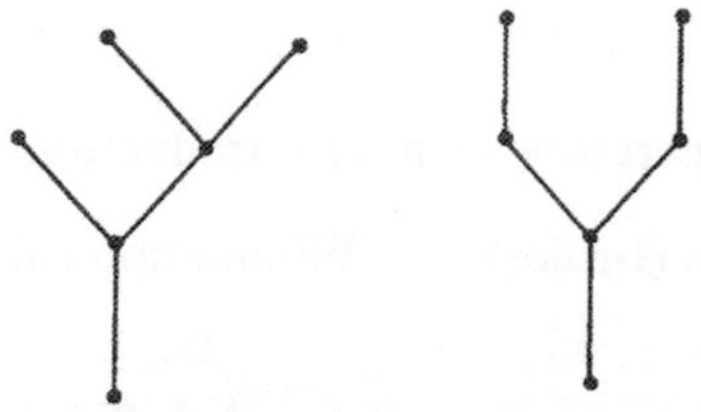

1.3. *Inductive definition of logical deductions.* We are now ready to give a more formal definition of deduction, open assumption, and cancelled (= discharged = eliminated) assumption. The definition will take the form of a simultaneous definition by recursion on the height of the deduction trees (or, equivalently, the form of a simultaneous inductive definition).

We shall use $\mathscr{D}$, possibly sub- or superscripted for arbitrary deductions. We write

$$\begin{array}{c}\mathscr{D}\\ B\end{array}$$

to indicate that B is the conclusion of $\mathscr{D}$ (so the occurrence B is part of $\mathscr{D}$ itself). We use $[A]$ for a (possibly empty) set of occurrences of a formula A in a deduction; thus

$$\begin{array}{c}[A]\\ \mathscr{D}\\ B\end{array}$$

is a deduction $\mathscr{D}$ with conclusion B, containing a set $[A]$ of occurrences of A, the elements of which are used as assumptions in $\mathscr{D}$. As a rule (cf. the "CDC" in 1.4(ii) below) we assume that $[A]$ contains *all* open assumptions of the form A in $\mathscr{D}$.

DEFINITION (of deduction, open and cancelled assumptions of a deduction).

Basis. The single-node tree with label A (i.e. a single occurrence of A) is a *deduction* from the open assumption A; there are no cancelled assumptions.

Inductive step. Now let $\mathscr{D}_1$, $\mathscr{D}_2$, $\mathscr{D}_3$ be deductions. A *deduction* $\mathscr{D}$ may be constructed according to one of the rules below. Some of these rules are subject to restrictions to be specified afterwards.

For $\bot$ we have the *intuitionistic absurdity rule*

$$\bot_i \frac{\begin{matrix}\mathcal{D}_1\\ \bot\end{matrix}}{A}$$

For the other logical operators we have introduction and elimination rules.

Introduction rules (I-rules) Elimination rules (E-rules)

$$\wedge \mathrm{I}\ \frac{\begin{matrix}\mathcal{D}_1 & & \mathcal{D}_2\\ A & & B\end{matrix}}{A \wedge B} \qquad \wedge \mathrm{E_r}\ \frac{\begin{matrix}\mathcal{D}_1\\ A \wedge B\end{matrix}}{A} \quad \wedge \mathrm{E_l}\ \frac{\begin{matrix}\mathcal{D}_1\\ A \wedge B\end{matrix}}{B}$$

$$\rightarrow \mathrm{I}\ \frac{\begin{matrix}[A]\\ \mathcal{D}_1\\ B\end{matrix}}{A \rightarrow B} \qquad \rightarrow \mathrm{E}\ \frac{\begin{matrix}\mathcal{D}_1 & \mathcal{D}_2\\ A \rightarrow B & A\end{matrix}}{B}$$

$$\vee \mathrm{I_r}\ \frac{\begin{matrix}\mathcal{D}_1\\ A\end{matrix}}{A \vee B} \quad \vee \mathrm{I_l}\ \frac{\begin{matrix}\mathcal{D}_2\\ B\end{matrix}}{A \vee B} \qquad \vee \mathrm{E}\ \frac{\begin{matrix} & [A] & [B]\\ \mathcal{D}_1 & \mathcal{D}_2 & \mathcal{D}_3\\ A \vee B & C & C\end{matrix}}{C}$$

$$\forall \mathrm{I}\ \frac{\begin{matrix}\mathcal{D}_1\\ A\end{matrix}}{\forall y A[x/y]} \qquad \forall \mathrm{E}\ \frac{\begin{matrix}\mathcal{D}_1\\ \forall x A\end{matrix}}{A[x/t]}$$

$$\exists \mathrm{I}\ \frac{\begin{matrix}\mathcal{D}_1\\ A[x/t]\end{matrix}}{\exists x A} \qquad \exists \mathrm{E}\ \frac{\begin{matrix} & [A]\\ \mathcal{D}_1 & \mathcal{D}_2\\ \exists y A[x/y] & C\end{matrix}}{C}$$

Open and *cancelled* assumptions are given by the following stipulations

(i) In $\rightarrow$ I-applications all open assumptions of the form A in $\mathcal{D}_1$, indicated by the set $[A]$ are cancelled; in an application of $\vee$E the sets $[A]$ in $\mathcal{D}_2$ and $[B]$ in $\mathcal{D}_3$ are cancelled; in $\exists$E the set $[A]$ in $\mathcal{D}_2$ is cancelled.

(ii) If $\mathcal{D}_i$ is a deduction of a premiss of the last rule application in $\mathcal{D}$, then the assumptions open in $\mathcal{D}_i$ remain open in $\mathcal{D}$ except as specified by (i). Thus assumptions which are not cancelled are open.

Instead of *cancelled* we also use *discharged* or *eliminated*.

The rules for the quantifiers are subject to the following restrictions.

(iii) In $\forall$E and $\exists$I t must be free for x in A.

(iv) In $\forall$I $\mathscr{D}_1$ must not contain open assumptions containing x free, and $y \equiv x$ or y is not free in A. In $\exists$E $\mathscr{D}_2$ must not contain open assumptions containing x free except the set $[A]$; x not free in C, $y \equiv x$ or y not free in A.

If A is among the open assumptions of a deduction $\mathscr{D}$ with conclusion B, the conclusion B in $\mathscr{D}$ is said to *depend on* A in $\mathscr{D}$. From now on we regard "assumption of $\mathscr{D}$" and "open assumption of $\mathscr{D}$" as synonymous. □

1.4. Remarks.

(i) We have assumed that the names of the rules are attached to the nodes of the tree. In most cases however, the form of the statements itself already determines the rule being applied, so that we can often omit the names of the rules when we are actually exhibiting deduction trees. Nevertheless, adding names of rules next to the horizontal lines marking the applications of the rules often facilitates following the proof.

(ii) Another device helping us to grasp the structure of a given deduction is to number the (occurrences of) assumptions at top nodes which are being discharged lower down in the tree, and to repeat the number near the node where the discharge takes place, either at the line marking the application of the rule causing the discharge, or placed next to the conclusion of this application. Assumptions which are discharged simultaneously may be given the same number.

Strictly speaking, the numbering of discharged assumptions is redundant information: by our definition as given above any assumption is discharged, if at all, at the earliest opportunity. From a given deduction tree we can therefore unambiguously read off, even without numbering, the node where an assumption is discharged (if at all). This discharge convention we call the *crude discharge convention* (CDC); we shall stick to it throughout except in section 10.8.

(iii) The rules $\forall$I and $\exists$E may be simplified to

$$\forall\text{I}\ \frac{\begin{matrix}\mathscr{D}_1\\ A\end{matrix}}{\forall x A} \qquad \exists\text{E}\ \frac{\begin{matrix}\mathscr{D}_1\\ \exists x A\end{matrix}\quad \begin{matrix}[A]\\ \mathscr{D}_2\\ C\end{matrix}}{C}$$

where in $\forall$I A does not depend on open assumptions in $\mathscr{D}_1$ containing x free, and in $\exists$E C does not contain x free and the deduction $\mathscr{D}_2$ of C does not contain open assumptions containing x except $[A]$. The more general form of the rules given before can be obtained from these special cases:

$$\dfrac{\dfrac{\dfrac{\begin{matrix}\mathscr{D}_1\\ A\end{matrix}}{\forall x A}}{A[x/y]}}{\forall y A[x/y]} \qquad \dfrac{\begin{matrix}\mathscr{D}_1\\ \exists y A[x/y]\end{matrix} \qquad \dfrac{\dfrac{(1)\ A[x/y]}{\exists x A} \qquad \begin{matrix}[A]\ (2)\\ \mathscr{D}_2\\ C\end{matrix}}{C\,(2)}}{(1)\ C}$$

We find it more convenient however to combine ∀I, ∃E with the possibility of renaming bound variables, since we regard formulas differing only in the names of their bound variables as isomorphic. □

The reader may now check for himself that all the rules given above are justified by the BHK-interpretation. The rules given are adequate in a practical sense: they cover all logical laws needed in a formalization of intuitionistic or constructive mathematical practice. It is natural to ask whether the rules are also *complete*, i.e. whether they permit us to derive all (intuitionistically or constructively) valid theorems. This delicate problem will be discussed in chapter 13, more particularly in sections 2 and 3 of that chapter.

1.5. EXAMPLES.

Example (a). We have argued before, in 1.3.3, that $\neg\neg(A \vee \neg A)$ should hold for our reading of $\neg$, $\vee$; here is a formal deduction, corresponding to the informal argument we gave before

$$\dfrac{\dfrac{(2)\ \neg(A \vee \neg A) \qquad \dfrac{(1)\ \dfrac{\dfrac{(2)\ \neg(A \vee \neg A) \qquad \dfrac{(1)\ A}{A \vee \neg A}\,\vee\mathrm{I}}{\bot}\,\to\mathrm{E}}{\neg A}\,\to\mathrm{I}}{A \vee \neg A}\,\vee\mathrm{I}}{\bot}\,\to\mathrm{E}}{(2)\ \neg\neg(A \vee \neg A)}\,\to\mathrm{I}$$

(recall that $\neg A \equiv A \to \bot$).

Example (b).

$$
\dfrac{\dfrac{(3)\ \neg\neg\forall x A x \qquad \dfrac{\dfrac{(2)\ \neg A y \qquad \dfrac{(1)\ \forall x A x}{A y}\ \forall\mathrm{E}}{\bot}\ \to\mathrm{E}}{(1)\ \neg\forall x A x}\ \to\mathrm{I}}{\dfrac{\dfrac{\bot}{(2)\ \neg\neg A y}\ \to\mathrm{I}}{\forall x \neg\neg A x}\ \forall\mathrm{I}}\ \to\mathrm{E}}{(3)\ \neg\neg\forall x A x \to \forall x \neg\neg A x}\ \to\mathrm{I}
$$

Confronted with the problem of constructing a deduction tree for a given statement, the best tactic is often to start from below, and to assume, if possible, that the final step was an instance of an I-rule. This is then repeated as long as possible. Thus, in attempting to construct a deduction for $\neg\neg\forall x A x \to \forall x \neg\neg A x$, we reduce the problem to the construction of a deduction for $\bot$ from assumptions $\neg\neg\forall x A x, \neg A y$. Now we attempt to apply E-rules to the assumptions. In order to apply an E-rule to $\neg\neg\forall x A x$, we need $\neg\forall x A x$, so we attempt to find a deduction

$$
\begin{array}{c}
[\neg\neg\forall x A x][\neg A y] \\
\mathscr{D} \\
\neg\forall x A x\,;
\end{array}
$$

again assuming the conclusion to be obtained from $\to$ I, we look for a deduction $\mathscr{D}'$ such that

$$
\begin{array}{c}
[\neg\neg\forall x A x][\neg A y][\forall x A x] \\
\mathscr{D}' \\
\dfrac{\bot}{\neg\forall x A x}
\end{array}
$$

and $\mathscr{D}'$ is easily constructed as

$$
\dfrac{\neg A y \qquad \dfrac{\forall x A x}{A y}}{\bot}
$$

Example (c).

$$\dfrac{\dfrac{\dfrac{\overset{(3)}{\neg\neg\neg A} \qquad \dfrac{\dfrac{\overset{(1)}{A} \quad \overset{(2)}{\neg A}}{\bot}}{\neg\neg A}\,(2)}{\bot}}{\neg A}\,(1)}{\neg\neg\neg A \to \neg A}\,(3) \qquad \text{and} \qquad \dfrac{\dfrac{\dfrac{\overset{(1)}{\neg B} \quad \overset{(2)}{B}}{\bot}}{\neg\neg B}\,(1)}{B \to \neg\neg B}\,(2)$$

If in the second derivation we take $\neg A$ for B, we see that $\neg A \to \neg\neg\neg A$; hence combining these results we find

$$\neg A \leftrightarrow \neg\neg\neg A.$$

Example (d). The tactics described under example (b) are not always so straightforward in their application; consider e.g. the following deduction

$$\dfrac{\dfrac{\dfrac{\overset{(5)}{A \vee \neg A} \qquad \dfrac{\dfrac{\overset{(4)}{A \to \exists x Bx} \quad \overset{(1)}{A}}{\exists x Bx} \qquad \dfrac{\dfrac{\overset{(3)}{By}}{A \to By}}{\exists x(A \to Bx)}}{\exists x(A \to Bx)}\,(3) \qquad \dfrac{\dfrac{\dfrac{\dfrac{\overset{(1)}{\neg A} \quad \overset{(2)}{A}}{\bot}}{By}}{A \to By}\,(2)}{\exists x(A \to Bx)}}{\exists x(A \to Bx)}\,(1)}{(A \to \exists x Bx) \to \exists x(A \to Bx)}\,(4)}{A \vee \neg A \to [(A \to \exists x Bx) \to \exists x(A \to Bx)]}\,(5)$$

A straightforward attempt to build from below suggests us to obtain "$\exists x(A \to Bx)$" from "$A \to By$", but then we would be stuck; so we must reckon with the possibility, while breaking down the conclusion, that $\vee$E- or $\exists$E-applications intervene when going upwards.

Example (e).

$$
\dfrac{\dfrac{\dfrac{\dfrac{{}^{(5)}\;\neg\neg A \to \neg\neg B \qquad \dfrac{\dfrac{{}^{(4)}\;\neg(A \to B) \qquad \dfrac{\dfrac{\dfrac{{}^{(2)}\;\neg A \quad A\;{}^{(1)}}{\perp}}{B}\,\perp_{\mathrm{i}}}{A \to B}\,(1)}{\perp}}{\neg\neg A}\,(2)}{\neg\neg B} \qquad \dfrac{\dfrac{{}^{(4)}\;\neg(A \to B) \qquad \dfrac{{}^{(3)}\;B}{A \to B}}{\perp}}{\neg B}\,(3)}{\perp}}{\neg\neg(A \to B)}\,(4)}{(\neg\neg A \to \neg\neg B) \to \neg\neg(A \to B)}\,(5)
$$

Some other logical laws which we shall need in the future are

(f) $\quad (A \to B) \to (\neg B \to \neg A)$

(g) $\quad \neg\neg(A \to B) \to (A \to \neg\neg B)$

(h) $\quad \neg\neg(A \wedge B) \leftrightarrow \neg\neg A \wedge \neg\neg B.$

We shall leave their proof as an exercise.

Note that all logical deductions are schematic in character: replacing the letters A, B etc. by compound statements throughout transforms correct proofs into correct proofs.

In practice one uses many short cuts in giving derivations; any logical law may be used as an assumption which is not discharged but does not count as a premiss for the conclusions; similarly we can use deductions from assumptions as additional (derived) rules, e.g. the derivation of $A \to C$ from $A \to B$, $B \to C$ (cf. 1.2) permits the introduction of a derived rule such as

$$
\mathrm{R}\;\dfrac{\begin{matrix}\mathcal{D}_1 & \mathcal{D}_2 \\ A \to B & B \to C\end{matrix}}{A \to C}
$$

As an example of a deduction involving derived rules, consider

$$\mathrm{R}'\;\cfrac{\cfrac{\begin{matrix}\mathscr{D}_1\\ (A\to\neg\neg B)\to(\neg\neg\neg B\to\neg A)\end{matrix}\quad \begin{matrix}\mathscr{D}_2\\ (\neg\neg\neg B\to\neg A)\to(\neg\neg A\to\neg\neg\neg\neg B)\end{matrix}}{(A\to\neg\neg B)\to(\neg\neg A\to\neg\neg\neg\neg B)}\,\mathrm{R}\qquad \begin{matrix}\mathscr{D}_3\\ \neg\neg\neg\neg B\leftrightarrow\neg\neg B\end{matrix}}{(A\to\neg\neg B)\to(\neg\neg A\to\neg\neg B)}$$

where $\mathscr{D}_1$, $\mathscr{D}_2$ exist by (f), $\mathscr{D}_3$ by (c), and where rule R′ is a schema of substitution for logically equivalent statements:

$$\mathrm{R}'\;\cfrac{\begin{matrix}\mathscr{D}_1\\ F(C)\end{matrix}\quad \begin{matrix}\mathscr{D}_2\\ C\leftrightarrow D\end{matrix}}{F(D)}$$

Combining this conclusion with (e), (g) we find

$$(A\to\neg\neg B)\leftrightarrow\neg\neg(A\to B)\leftrightarrow(\neg\neg A\to\neg\neg B).$$

1.6. *Classical logic* is obtained by strengthening the absurdity rule to the *classical absurdity rule*

$$\bot_c\;\cfrac{\begin{matrix}[\neg A]\\ \vdots\\ \bot\end{matrix}}{A}$$

The assumptions indicated by $[\neg A]$ are being discharged. Observe that by $\to$ I:

$$\to\mathrm{I}\;\cfrac{\begin{matrix}[\neg A]\\ \vdots\\ \bot\end{matrix}}{\neg\neg A}$$

and that therefore the adoption of $(\bot_c)$ is equivalent to having propositions of the form $\neg\neg A\to A$ as axioms:

$$\cfrac{\neg\neg A\to A\qquad \cfrac{\begin{matrix}[\neg A]\\ \mathscr{D}\\ \bot\end{matrix}}{\neg\neg A}\,\to\mathrm{I}}{A}\,\to\mathrm{E}$$

Classically, we can drop $\vee, \exists$ and their rules, by defining them as

$$A \vee B := \neg(\neg A \wedge \neg B),$$

$$\exists x A(x) := \neg\forall x \neg A(x).$$

We leave it as an exercise to show that the *rules* for $\vee, \exists$ can indeed be obtained as derived rules from the rules for $\wedge, \neg, \forall$.

1.7. *Deductions from axioms.* So far, we have only considered purely logical deductions. In mathematical theories there are also non-logical axioms. For example, in proving identities for abelian groups, we make use of the group axioms, such as commutativity:

$$\forall x \forall y (x + y = y + x). \tag{1}$$

Such axioms are often presented in open form as

$$x + y = y + x; \tag{2}$$

the universal quantifiers are tacitly omitted. Another way to look at this open form is as a schema: we may substitute t and s for x and y or "deduce" the form (1) from it.

There is a completely uniform way of accommodating such mathematical axioms in our natural deductions: we permit them to appear at the top nodes of our deductions. However, if we permit the use of open forms such as (2), we must strictly distinguish betweem axioms and open assumptions: an axiom appearing at a top node does not count as an open assumption. In particular, in the restrictions on $\forall$I and $\forall$E, we have to exclude only those variables which appear free in *assumptions*.

$\forall$I permits us to derive the universal closure of an axiom from the axiom: applying $\forall$I twice to (2) leads to (1); and from the universal closure by $\forall$E we obtain all special instances, such as $t + s = s + t$ in our example.

Another way of achieving the same effect is to treat axioms as "rules with an empty premiss"; in our proof trees this can be indicated for example as

$$\frac{\emptyset}{x + y = y + x} \quad \text{or even} \quad \frac{}{x + y = y + x}.$$

Now we do not have to distinguish between assumptions and axioms at top nodes; instead, we have a new type of rule.

If we agree to present axioms always in their universally closed form, there is no need to distinguish between axioms and assumptions any more: we simply permit axioms to appear as (undischarged) assumptions.

The choice of presentation is a matter of taste and/or convenience. Proofs in natural deduction style in mathematical theories are often presented in linear form, since tree presentations become unwieldy for complicated arguments; for an example of a proof in linear style see 3.2.4.

1.8. *Logic with equality.* In classical logic, equality is mostly regarded as part of the logic. In constructive mathematics, the equality of the language is often interpreted as a defined notion, not intuitively given as a primitive, and then it is more natural to think of equality as mathematical rather than logical. As axioms we can take (the universal closures of) *reflexivity*

REFL $\quad x = x$

and the *replacement schema*

REPL $\quad A(x) \wedge x = y \rightarrow A(y).$

If we take in this schema $A(x) := (x = t)$, $A(t)$ is true by REFL; so from $A(t) \wedge t = s \rightarrow A(s)$ we see that $t = s \rightarrow s = t$. Also, taking $A(x) := (x = r)$, we find

$$t = r \wedge t = s \rightarrow s = r.$$

Another way of accommodating equality in a natural deduction context is to have an axiom $x = x$, or a zero-premiss rule

REFL $\quad t = t$

and a *replacement rule*

REPL $$\frac{A[x/t] \quad t = s}{A[x/s]}$$

with as a special case the *transitivity rule*

TRANS $$\frac{t = r \quad t = s}{r = s}.$$

1.9. DEFINITION. We shall use **IPC** as the designation for the system of *intuitionistic propositional logic*, and **IQC** for *intuitionistic predicate logic*, with or without equality, with or without function symbols and constants. *Classical propositional logic* **CPC** and *classical predicate logic* **CQC** are obtained by adding the rule $\perp_c$ to the corresponding intuitionistic systems. □

We shall encounter in the sequel other formalizations (that is to say, different formalisms generating the same set of theorems) for **IPC**, **CPC**, **IQC** and **CQC**, but we shall not distinguish between these formalisms unless the structure of deductions is under discussion (as for example in section 4 below, and in chapter 10), or when the equivalence of these systems has to be proved. For example, we occasionally use N-**IPC**, N-**IQC** to distinguish the natural deduction versions of **IPC** and **IQC** described above from other versions under discussion.

1.10. *One-variable formulas in* **IPC**. As an illustration of the complexity of **IPC**, compared to classical propositional logic **CPC**, we consider the one-variable formulas in **IPC**, generated from a single proposition letter P by means of the logical operations $\to$, $\wedge$, $\vee$ and $\perp$. Modulo provable equivalence all such formulas occur in a sequence $\langle \mathrm{A}_n(P)\rangle_{n \in \mathbb{N} \cup \{\omega\}}$ given by

$$\mathrm{A}_0(P) := \perp, \quad \mathrm{A}_1(P) := P, \quad \mathrm{A}_2(P) := \neg P,$$

$$\mathrm{A}_{2n+1}(P) := \mathrm{A}_{2n-1}(P) \vee \mathrm{A}_{2n}(P),$$

$$\mathrm{A}_{2n+2}(P) := \mathrm{A}_{2n}(P) \to \mathrm{A}_{2n-1}(P),$$

$$\mathrm{A}_\omega(P) := P \to P.$$

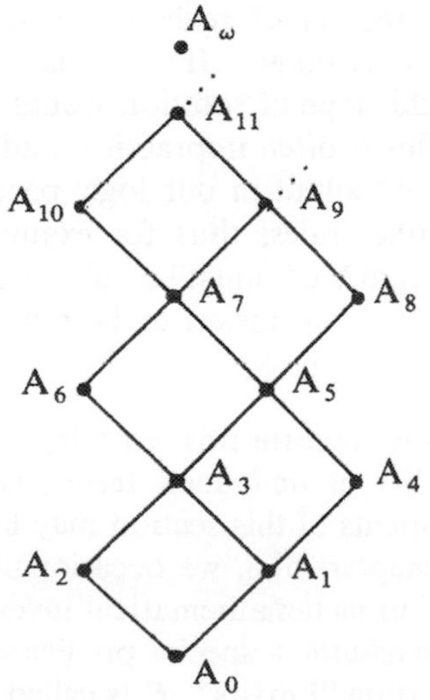

Implications hold between the A_n as indicated in the diagram above (A_i below A_j connected by edges means that $\mathrm{A}_i \to \mathrm{A}_j$ is derivable in **IPC**). The diagram is called the Rieger–Nishimura lattice, after Rieger (1949) who first investigated the one-variable formulas, and Nishimura (1960) who rediscovered the structure. See also E2.1.6 and subsection 13.5.5.

2. Logic with existence predicate

2.1. So far we have tacitly assumed that the expressions (terms) t which appear in rules such as ∀E, ∃I always *denote* some individual of the domain over which the variables are supposed to run. For example, in discussing arithmetic we can freely combine 0 and variables by $+, \cdot$ to obtain new expressions for natural numbers, since $+$ and $\cdot$ are total functions on $\mathbb{N} \times \mathbb{N}$.

However, in mathematics one frequently uses partial functions such as $x \mapsto x^{-1}$ on the real numbers, which is not defined for $x = 0$. On the basis of classical logic we can avoid the consideration of partial functions by an ad hoc stipulation, putting e.g. $0^{-1} = 0$ in the case of the inverse function. But, constructively, this does not give us a total function. Here a possible solution might be to introduce a new sort of variables ranging over $\mathbb{R}^+ := \{x \in \mathbb{R} : |x| > 0\}$, the set of reals positively distinct (apart) from zero, and to consider $^{-1}$ only on $\mathbb{R}^+$. If we consider many distinct partial functions from $\mathbb{R}$ to $\mathbb{R}$, this type of solution would require us to introduce many additional sorts. This is often impractical and inelegant.

If, on the other hand, we admit in our logic partially defined terms, we have to revise the quantifier rules; thus for example the instance of the ∀E-rule: infer $A[x/t^{-1}]$ from $\forall x A$, must be subject to a stipulation that t^{-1} exists, otherwise we take the conclusion to be either meaningless or false.

2.2. *E-logic.* We shall now indicate how to adapt our logic so as to take into account the possibility of undefined terms, i.e. terms which do not always "denote". The contents of this section may be skipped until needed. In the "mathematical" chapters 5–8, we occasionally need 2.2–4; the rest of this section is relevant to metamathematical investigations only.

As part of the logic we assume a special predicate constant E, where Et has the intuitive interpretation "t exists". E is called the *existence predicate*. Variables will be treated as purely schematic, that is to say, any term may be substituted for a free variable. Quantifiers on the other hand are supposed to range over *existing* objects only. This leads to the following modified quantifier rules:

$$\forall \mathrm{I}^{\mathrm{E}}\ \frac{\begin{matrix}[Ex]\\ \vdots \\ A\end{matrix}}{\forall y A[x/y]} \qquad \forall \mathrm{E}^{\mathrm{E}}\ \frac{\forall x A \quad Et}{A[x/t]},$$

$$\exists \mathrm{I}^{\mathrm{E}}\ \frac{A[x/t] \quad Et}{\exists x A} \qquad \exists \mathrm{E}^{\mathrm{E}}\ \frac{\exists y A[x/y] \quad \begin{matrix}[A][Ex]\\ \vdots \\ C\end{matrix}}{C}.$$

The variable conditions are straightforward modifications of the conditions in 1.3.

In addition, we have to express the schematic role of free variables. This can be done by adding a substitution rule

$$\text{SUB} \qquad \frac{A}{A[x/t]}$$

provided the deduction of A does not depend on assumptions containing x free.

It is easy to see that for purely logical deductions (i.e. without axioms), or deductions from sets of sentences, we do not need the substitution rule: it is derivable.

An open axiom cannot, in general, be replaced by its closure: $\forall x A(x)$ says something about "existing x" only, and $A(t)$ can be obtained from $\forall x A(x)$ only under the additional assumption Et; on the other hand, the rule SUB permits us to derive $A(t)$ from $A(x)$ without assuming Et. A typical example is the "theory of global existence", axiomatized by the single open axiom Ex. In particular, this axiom implies that all functions are total.

2.3. *E^+-logic.* In this variant we assume the *free variables* to range over existing objects, though *terms* need not denote anything.

Now we have axioms Ex for all variables x, and quantifier rules $\forall$I, $\exists$E, $\exists \mathrm{I}^{\mathrm{E}}$, $\forall \mathrm{E}^{\mathrm{E}}$. A substitution rule ought to have two premises

$$\frac{A \quad Et}{A[x/t]}$$

but this rule is in fact derivable from $\forall$I, followed by $\forall \mathrm{E}^{\mathrm{E}}$.

On the other hand, where formerly we had an axiom $A(x)$, which then by the substitution rule gave $A(t)$, we now have to treat all $A(t)$ as axioms.

As to the exact relationship between this and the preceding version, see exercise 2.2.4.

2.4. *E-logic and E^+-logic with equality.* We now add equality to E-logic and E^+-logic, with the intuitive interpretation: $t = s$ iff t and s both exist and are equal. Thus we have an axiom schema

$$t = t \leftrightarrow Et$$

(and so E could be defined in terms of $=$). Defining

$$t \simeq s := Et \vee Es \rightarrow t = s,$$

we have a replacement schema

$$A(t) \wedge t \simeq s \rightarrow A(s).$$

The intuitive meaning of $\simeq$ is: t is defined iff s is defined, and if they are both defined, they are equal.

Strictly speaking, this would suffice as a basis for logic with E and $=$. It will be convenient, however, to make an additional assumption of *strictness* for all primitive functions and relations, i.e.

$$\text{STR} \qquad \begin{cases} Ef(t_0, \ldots, t_{n-1}) \rightarrow Et_i \\ R(t_0, \ldots, t_{n-1}) \rightarrow Et_i \end{cases} \quad (i < n)$$

for each n-ary function symbol f of $\mathscr{L}$, and each n-ary relation symbol R of $\mathscr{L}$. For functions STR expresses the fact that the function value can exist only if all the arguments exist. As a result of STR, if a compound term refers to an existing object, then all its subterms must refer to existing objects. STR imposes a similar restriction on the basic relations of the language. But it should be noted that the property of strictness, by its very nature, does not extend to all logically compound predicates; that is to say, in general we do *not* have $A(t) \rightarrow Et$: one only has to take $\neg Ex$ for $A(x)$!

Once we have adopted STR, the meaning of Et and $t = s$ for compound terms can be explained in terms of equality between less complex terms, since the following are derivable (exercise)

$$\begin{aligned} &Et \leftrightarrow t = t \leftrightarrow \exists x(t = x), \\ &t = s \leftrightarrow \exists x(t = x \wedge s = x), \qquad (1) \\ &f\vec{t} = x \leftrightarrow \exists \vec{y}(\vec{y} = \vec{t} \wedge f\vec{y} = x). \end{aligned}$$

2.5. NOTATION. **IQCE** is intuitionistic logic with equality and E based on the propositional rules of 1.3, the quantifier rules of 2.2, equality axioms and STR as in 2.4. Finally, **IQCE**$^+$ is E^+-logic (cf. 2.3). □

2.6. In the presence of STR, we can simplify the equality axioms:

PROPOSITION. *The replacement schema*

$$A(t) \wedge t \simeq s \rightarrow A(s)$$

is derivable on the basis of the other axioms and rules of **IQCE** *or* **IQCE**$^+$ *from the special cases*:

$$R\vec{t} \wedge \vec{t} = \vec{s} \rightarrow R\vec{s}, \tag{1}$$

$$E(f\vec{s}) \wedge \vec{t} = \vec{s} \rightarrow f\vec{t} = f\vec{s}. \tag{2}$$

PROOF. By induction on A. We will only treat the atomic case, the other cases are left to the reader. For simplicity we shall only consider one-place predicates and functions. For a predicate P:

$t \simeq s \wedge Pt \Leftrightarrow$	(by definition)
$(Et \vee Es \rightarrow t = s) \wedge Pt \Rightarrow$	(by propositional logic)
$(Et \rightarrow t = s) \wedge Pt \Rightarrow$	(by STR)
$(Et \rightarrow t = s) \wedge Pt \wedge Et \Rightarrow$	(by propositional logic)
$t = s \wedge Pt \Rightarrow Ps$	(by assumption).

Similarly for function symbols:

$$t \simeq s \wedge Et \wedge E(ft) \rightarrow ft = fs \quad (\text{by } (1)),$$

hence since by STR $E(ft) \rightarrow Et$, also

$$E(ft) \wedge t \simeq s \rightarrow ft = fs,$$

and therefore $t \simeq s \rightarrow (E(ft) \rightarrow ft = fs)$; similarly $t \simeq s \rightarrow (E(fs) \rightarrow ft = fs)$, so $t \simeq s \rightarrow (E(ft) \vee E(fs) \rightarrow ft = fs)$, and thus $t \simeq s \rightarrow ft \simeq fs$. □

2.7. *Axiomatizing* **IQCE** *with* $\simeq$ *as a primitive.* We now consider E-logic with $\forall I^E, \exists I^E, \forall E^E, \exists E^E$ and a primitive binary predicate $\simeq$ satisfying

$$x \simeq y \wedge A(x) \rightarrow A(y),$$

$$\forall z(x \simeq z \leftrightarrow y \simeq z) \rightarrow x \simeq y.$$

We call the resulting system **IQCE***.

PROPOSITION. *Define in* **IQCE***

$$t = s := Et \wedge Es \wedge t \simeq s.$$

Then **IQCE*** *becomes a definitional extension of* **IQCE**; *conversely, defining in* **IQCE**

$$t \simeq s := Et \vee Es \to t = s,$$

IQCE *becomes a (definitional) extension of* **IQCE***. *The systems are equivalent in the sense that*

$$\mathbf{IQCE} \vdash t = s \leftrightarrow (Et \wedge Es \wedge (Et \vee Es \to t = s)),$$

$$\mathbf{IQCE}^* \vdash t \simeq s \leftrightarrow (Et \vee Es \to (Et \wedge Es \wedge t \simeq s)).$$

We can also formulate this by saying that **IQCE** *and* **IQCE*** *have a common definitional extension.*

PROOF. We consider **IQCE***. From $\forall z(x \simeq z \leftrightarrow y \simeq z) \to x \simeq y$ we obtain $x \simeq x$; hence by replacement $x \simeq y \wedge x \simeq x \to y \simeq x$, so $x \simeq y \to y \simeq x$, and also $x \simeq y \wedge y \simeq z \to x \simeq z$ (again by replacement).

Let $t \simeq s$; if Et, then also Es and vice versa, so $(Et \vee Es) \wedge t \simeq s \to Et \wedge Es \wedge t \simeq s$, hence

$$t \simeq s \to (Et \vee Es \to (Et \wedge Es \wedge t \simeq s)).$$

Conversely, if $Et \vee Es \to t = s$, and we assume Ez, $t \simeq z$ then Et. Therefore $t = s$, and hence $s \simeq z$ and vice versa, so $\forall z(t \simeq z \leftrightarrow s \simeq z)$, and therefore also $t \simeq s$.

We leave the argument for **IQCE** as an exercise. □

2.8. *The relation between* **IQCE** *and* **IQC**. The following proposition, proofs of which will be given in 10.5.6 and 13.6.16 settles the relation between **IQCE** on the one hand and **IQC** with special distinguished predicates E and $=$ on the other hand.

PROPOSITION. *Let* **IQCE** *be as above and let* **IQC*** *be* **IQC** *with the equality axioms for* $=$ *and a predicate E relative to which STR holds. We define a mapping* ^{E} *from the formulae of* **IQCE** *to* **IQC*** *by*

$$(Rt_0 \dots t_{n-1})^E := Rt_0 \dots t_{n-1};$$

$$(t_0 = t_1)^E := Et_0 \wedge Et_1 \wedge t_0 = t_1;$$

$$(A \circ B)^E := A^E \circ B^E \text{ for } \circ \in \{\wedge, \vee, \to\};$$

$$(\forall x A)^E := \forall x[Ex \to A^E];$$

$$(\exists x A)^E := \exists x[Ex \wedge A^E].$$

If $\simeq$ *appears as a primitive, we can add a clause*

$$(t_0 \simeq t_1)^E := Et_0 \vee Et_1 \to t_0 = t_1.$$

Then

$$\mathbf{IQCE} \vdash A \Leftrightarrow \mathbf{IQC}^* \vdash A^E. \quad \square$$

2.9. *Descriptions.* We can still further strengthen our logic (either E-logic or E^+-logic) by adding a *descriptor* or *description operator* $Ix.A(x)$, "the unique x such that $A(x)$". If there is no unique x such that $A(x)$, $Ix.A(x)$ is undefined; thus in particular $Ix.A(x)$ is undefined if there are several x such that $A(x)$.

The intended interpretation of $Ix.A(x)$ is adequately captured by the *description axiom*

$$\text{DESCR} \qquad \forall y(y = Ix.A(x) \leftrightarrow \forall x(A(x) \leftrightarrow x = y)).$$

For example, $E(Ix.x = x)$ iff the range of the variable x consists of a single element, and $\neg E(Ix.x \neq x)$.

In the case of E-logic the addition of a descriptor with axiom DESCR to *open* theories may result in inconsistency: let **T** be a theory (a fragment of first order arithmetic for example) in which we can show $A(x, y) \wedge A(x, y') \to y = y'$ and $\neg\forall x \exists y A(x, y)$ for a suitable A, and such that $\mathbf{T} \cup \{Ex\}$ is consistent (in other words, **T** is compatible with "global existence"); then $\mathbf{T}' \equiv \mathbf{T} \cup \{Ex, \text{DESCR}\}$ is *inconsistent*, at least if we interpret SUB in the new theory $\mathbf{T}'$ as applying to all terms of $\mathbf{T}'$ and not just to the terms of **T**.

Warning. In the case of **CQC** it is a wellknown fact that the addition of "Skolem-functions" (i.e. choice functions) is conservative, but this is not so in the case of **IQC**; see 5.13, proposition.

The essential properties of $Ix.A(x)$ in connection with E are given by the following proposition

2.10. Proposition. *Let* $y \notin \mathrm{FV}(A)$. *Then*
(i) $E(Ix.A(x)) \leftrightarrow \exists y \forall x[A(x) \leftrightarrow x = y]$;
(ii) $E(Ix.A(x)) \to A(Ix.A(x))$.

Proof. (i) Apply DESCR with $Ix.A(x)$ for y, then

$$\begin{aligned} E(Ix.A(x)) &\leftrightarrow \forall x[A(x) \leftrightarrow x = Ix.A(x)] \\ &\to \exists y \forall x[A(x) \leftrightarrow x = y]. \end{aligned} \tag{1}$$

Conversely if $\exists y \forall x[A(x) \leftrightarrow x = y]$, for some y $\forall x[A(x) \leftrightarrow x = y]$, Ey and with DESCR $y = Ix.A(x)$, therefore $E(Ix.A(x))$.

(ii) Apply (1): $E(Ix.A(x)) \to (A(Ix.A(x)) \leftrightarrow Ix.A(x) = Ix.A(x))$, and thus (ii) follows. □

As we shall see in section 7, under suitable conditions on the theories considered, the addition of a description operator is conservative, in fact a definitional extension.

2.11. Remark. It should be observed that the introduction of a description operator satisfying DESCR has the same effect as introducing function symbols for definable partial functions. For suppose

$$\vdash A(\vec{x}, y) \wedge A(\vec{x}, y') \to y = y',$$

then we may regard $Iy.A(\vec{x}, y)$ as a partial function in $\vec{x}$, satisfying

$$\vdash \exists y A(\vec{x}, y) \leftrightarrow A(\vec{x}, Iy.A(\vec{x}, y)) \wedge E(Iy.A(\vec{x}, y)).$$

Conversely, suppose that, whenever $\vdash A(x, y) \wedge A(x, y') \to y = y'$, we may introduce a constant φ_A for a partial function such that

$$\vdash \exists y A(\vec{x}, y) \leftrightarrow A(\vec{x}, \varphi_A \vec{x}) \wedge E\varphi_A \vec{x}.$$

Then we are also able to construct a term with the properties of $Iy.B(y)$, for any $B(y)$: let $B(\vec{x}, y)$ be given and consider

$$A(\vec{x}, y) := \forall z(B(\vec{x}, z) \leftrightarrow y = z).$$

Then clearly $\vdash A(\vec{x}, y) \wedge A(\vec{x}, y') \to y = y'$. Finally, take $\varphi_A(\vec{x})$ for $Iy.B(\vec{x}, y)$. We leave the verification of details to the reader. □

3. Relationships between classical and intuitionistic logic

In this section we shall show how **CQC** can be embedded into **IQC** by means of the various "negative translations". Furthermore, we shall obtain a number of conservation results of **CQC** over **IQC**, i.e. results of the form: for all A belonging to some syntactically defined class $\mathscr{F}$ of formulas

$$\mathbf{CQC} \vdash A \Rightarrow \mathbf{IQC} \vdash A.$$

The basic facts established in 3.2–3.8 will be frequently used later in the book. The remaining part deals with refinements, to which we shall refer only occasionally later on.

3.1. DEFINITION. A formula A in a first-order language is said to be in the *negative fragment* (or "A is *negative*", not to be confused with "A is a negation") if prime formulas P occur only negated (i.e. in a context $P \to \bot$) in A, and A does not contain $\vee$ or $\exists$. □

3.2. DEFINITION. Let **MQC**, the so-called *minimal predicate logic*, be the system **IQC** with the rule $\bot_i$ left out.

We use as abbreviations $\vdash_c$, $\vdash_m$, $\vdash_i$ for derivability in **CQC**, **MQC**, **IQC** respectively. □

N.B. In **MQC** $\bot$ is treated as an arbitrary proposition letter.

3.3. LEMMA. *For A negative* **MQC** $\vdash A \leftrightarrow \neg\neg A$.

PROOF. As seen by inspection of 1.5 and E2.1.1, the following are all provable in **MQC**

$$A \to \neg\neg A;$$
$$\neg\neg\neg A \leftrightarrow \neg A;$$
$$\neg\neg(A \wedge B) \to \neg\neg A \wedge \neg\neg B;$$
$$\neg\neg(A \to B) \to (\neg\neg A \to \neg\neg B) \leftrightarrow (A \to \neg\neg B);$$
$$\neg\neg\forall x A \to \forall x \neg\neg A.$$

The proof now proceeds by induction on the complexity of A: A has one of the forms $\neg P$ (P prime), $B \wedge C$, $B \to C$, $\forall x B$. Consider e.g. the case $A \equiv B \to C$. Then $\neg\neg A \equiv \neg\neg(B \to C) \to (B \to \neg\neg C) \to (B \to C) \to \neg\neg(B \to C)$; the second implication holds by the induction hypothesis. We leave the other cases to the reader. □

3.4. DEFINITION. For all formulas of predicate logic the (Gödel–Gentzen–) *negative translation* ^{g} is defined inductively by

(i) $P^g := \neg\neg P$ for P prime;
(ii) $(A \wedge B)^g := A^g \wedge B^g$;
(iii) $(A \to B)^g := A^g \to B^g$;
(iv) $(\forall x A)^g := \forall x A^g$;
(v) $(A \vee B)^g := \neg(\neg A^g \wedge \neg B^g)$;
(vi) $(\exists x A)^g := \neg\forall x \neg A^g$. □

REMARK. An inessential variant is obtained if (i) is replaced by
(i)′ $P^g := \neg\neg P$ for P prime, $P \not\equiv \bot$; $\bot^g \equiv \bot$.
Also we may systematically replace occurrences of $\neg\neg\neg$ by $\neg$. □

3.5. THEOREM. *For all A*
(i) $\vdash_c A \leftrightarrow A^g$;
(ii) $\Gamma \vdash_c A \Leftrightarrow \Gamma^g \vdash_m A^g$,
where $\Gamma^g := \{B^g : B \in \Gamma\}$.

PROOF. It is a matter of routine to establish the equivalence of A and A^g in **CQC**.

The proof of (ii) from right to left proceeds by induction on the height of deduction trees in **CQC**.

Each application of a rule for $\wedge, \rightarrow, \forall$ can be replaced, under the translation, by a corresponding instance of the same rule; for example, if we have already obtained deductions

$$\begin{matrix} \mathcal{D}_1^g & \mathcal{D}_1^g \\ A^g & B^g, \end{matrix}$$

corresponding to

$$\begin{matrix} \mathcal{D}_1 & \mathcal{D}_2 \\ A & B, \end{matrix}$$

then

$$\frac{\mathcal{D}_1 \quad \mathcal{D}_2}{A \wedge B} \quad \text{is translated into} \quad \frac{\mathcal{D}_1^g \quad \mathcal{D}_2^g}{A^g \wedge B^g}.$$

As for $\bot_c$, suppose that the deduction

$$\begin{matrix} [\neg A^g] \\ \mathcal{D}^g \\ \bot \end{matrix}$$

has already been constructed, and consider the following instance of $\bot_c$

$$\begin{matrix} [\neg A] \\ \mathcal{D} \\ \dfrac{\bot}{A} \end{matrix}$$

Then the proof tree below is a correct deduction of A^g in **MQC**:

$$\dfrac{(1)\dfrac{\begin{matrix} (1)\,[\neg A^g] \\ \mathcal{D}^g \\ \bot \end{matrix}}{\neg\neg A^g} \qquad \begin{matrix} \mathcal{D}' \\ \neg\neg A^g \rightarrow A^g \end{matrix}}{A^g}$$

where $\mathscr{D}'$ is the derivation in **MQC** of $\neg\neg A^{\mathrm{g}} \to A^{\mathrm{g}}$ which exists by lemma 3.3.

As for the rules for $\vee, \exists$, it is well-known that in **CQC** $\vee$ and $\exists$ are explicitly definable by $A \vee B := \neg(\neg A \wedge \neg B)$, $\exists x A := \neg\forall x \neg A$, in the sense that the rules for $\vee, \exists$ can be obtained as derived rules for these defined connectives. ^{g} replaces $\vee, \exists$ by their classical definition, and the rules for the other operations have already been shown to be valid under ^{g}.

The proof of (ii) from left to right follows from the first part of the theorem. □

3.6. COROLLARY. *For negative A*, **CQC** $\vdash A$ *iff* **IQC** $\vdash A$. □

This corollary shows that, in a sense, **CQC** is contained in **IQC**, since each formula of **CQC** is equivalent to a negative one.

3.7. *Other versions of the negative translation.* One of the best known variants is *Kolmogorov's negative translation* ^{k}. A^{k} is obtained by simultaneously inserting double negations in front of all subformulas of A (including A itself). Inductively we may define ^{k} by

$$P^{\mathrm{k}} := \neg\neg P \text{ for } P \text{ prime}, \quad \bot^{\mathrm{k}} := \bot;$$

$$(A \circ B)^{\mathrm{k}} := \neg\neg(A^{\mathrm{k}} \circ B^{\mathrm{k}}) \quad \text{for } \circ \in \{\wedge, \vee, \to\};$$

$$(QxA)^{\mathrm{k}} := \neg\neg Qx A^{\mathrm{k}} \quad \text{for } Q \in \{\forall, \exists\}.$$

Another variant A^{q} (*Kuroda's negative translation*), due to Kuroda (1951) ("q" from "quantifier") is obtained as follows: insert $\neg\neg$ *after* each occurrence of $\forall$, and in front of the whole formula.

3.8. PROPOSITION. **MQC** $\vdash A^{\mathrm{g}} \leftrightarrow A^{\mathrm{k}}$, **IQC** $\vdash A^{\mathrm{k}} \leftrightarrow A^{\mathrm{q}}$.

PROOF. By formula induction (exercise). □

The remainder of this section is devoted to refinements and generalizations.

3.9. *Subformula occurrences.* For the methods and results below we need the notion of a "(sub)formula occurrence" (sfo, for short), rather than the notion of a (sub)formula. An sfo of a formula A may be specified by a formula together with its position in the construction tree (parsing tree) of A, e.g. by the code of a node.

EXAMPLE. $A \equiv \forall x(P(x) \to Q) \to Q$, $P(x)$ and Q prime, has the following construction tree:

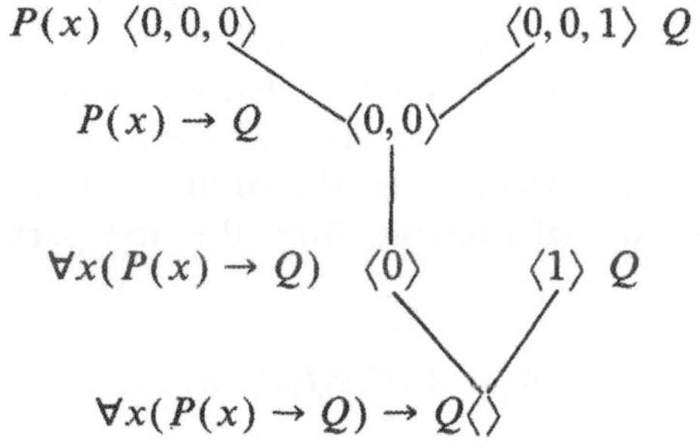

There are two distinct occurrences of Q, one at $\langle 0,0,1\rangle$, another at $\langle 1\rangle$. An elegant method of marking occurrences is the method of contexts or formula-occurrence schemas, in which a placeholder symbol $*$ marks the position of the sfo we are interested in. Syntactically $*$ is treated as a prime formula. Thus, for example, the two sfo's of Q in A correspond to the contexts $B[*] \equiv \forall x(P(x) \to *) \to Q$, and $C[*] \equiv \forall x(P(x) \to Q) \to *$ respectively. Note that $A \equiv B[Q] \equiv C[Q]$.

Formally we define the class $\mathscr{CON}$ of *formula contexts* as follows.

DEFINITION. Let A, F be metavariables for formulas and formula context respectively. Then $\mathscr{CON}$ is inductively generated by the clause

$$*,\ A \wedge F,\ F \wedge A,\ A \vee F,\ F \vee A,$$
$$A \to F,\ F \to A, \forall xF, \exists xF \in \mathscr{CON}.$$

In other words, $\mathscr{CON}$ is the least class of expressions $\mathscr{X}$ such that for all formulas A and all $F \in \mathscr{X}$ also $*, A \wedge F, F \wedge A, \ldots, \exists xF \in \mathscr{X}$. □

3.10. CONVENTION. For the remainder of this section we use the notation $F[*/C]$, $A[B/C]$ *permitting* that variables in C become bound in F, A (i.e. we permit $\mathrm{FV}(C) \cap \mathrm{BV}(F) \neq \emptyset$). Where the condition $\mathrm{FV}(C) \cap \mathrm{BV}(F) = \emptyset$ is needed, it will be stated explicitly. This represents a deviation from our customs elsewhere in this book.

For formula contexts $F \equiv F[*]$ we shall often write $F[B]$ instead of $F[*/B]$. □

Note that, for given A, each sfo in A corresponds to a uniquely determined $F \in \mathscr{CON}$ such that $A \equiv F[B]$ for a suitable B. An sfo in A is therefore specified by a pair (F, B) with $F \in \mathscr{CON}$ such that $F[*/B] \equiv A$ (or equivalently one might have taken the pair (F, A) such that for some B $F[B] \equiv A$).

3.11. Definition. We define simultaneously the *positive contexts* $\mathscr{POS}$ and the *negative contexts* $\mathscr{NEG}$. Let P, N be arbitrary elements of $\mathscr{POS}$, $\mathscr{NEG}$, and A an arbitrary formula. Then $\mathscr{POS}$, $\mathscr{NEG}$ are simultaneously inductively generated by the clauses

(i) $*$, $A \wedge P$, $P \wedge A$, $A \vee P$, $P \vee A$, $A \to P$, $N \to A$, $\forall xP$, $\exists xP \in \mathscr{POS}$;

(ii) $A \wedge N$, $N \wedge A$, $A \vee N$, $N \vee A$, $A \to N$, $P \to A$, $\forall xN$, $\exists xN \in \mathscr{NEG}$.

Note that $\mathscr{CON} := \mathscr{POS} \cup \mathscr{NEG}$ and that $\mathscr{POS} \cap \mathscr{NEG} = \emptyset$.

It is now obvious what is meant by a *positive* (*negative*) occurrence of B in A. □

Example. For the A in 3.9 we see that

$$C[*] \equiv \forall x(P(x) \to Q) \to * \in \mathscr{POS}, \quad C[Q] \equiv A;$$

$$B[*] \equiv \forall x(P(x) \to *) \to Q \in \mathscr{NEG}, \quad B[Q] \equiv A.$$

3.12. Lemma. *Let $\vec{z}$ be a list of variables free in B and C, and bound in $F[B]$ and $F[C]$, where $F[*] \in \mathscr{CON}$. Then*

(i) $F \in \mathscr{POS} \Rightarrow \forall \vec{z}(B \to C) \vdash_m F[B] \to F[C]$;

(ii) $F \in \mathscr{NEG} \Rightarrow \forall \vec{z}(B \to C) \vdash_m F[C] \to F[B]$.

Proof. We prove (i) and (ii) simultaneously by induction on the construction of F.

Case 1. $F \equiv *$: immediate.

Case 2a. $F \equiv \forall xD(x, *) \in \mathscr{POS}$. The following deduction is correct

$$\dfrac{\dfrac{\forall xD(x, B)}{[D(x, B)]} \quad \dfrac{\forall x\vec{z}(B \to C)}{[\forall \vec{z}(B \to C)]}}{\dfrac{\overset{\mathscr{D}}{D(x, C)}}{\forall xD(x, C)}}$$

$\mathscr{D}$ is given by the induction hypothesis.

Case 2b. $F \equiv \forall xD(x, *) \in \mathscr{NEG}$: completely similar.

Case 3a. $F \equiv N[*] \rightarrow A$, $N \in \mathcal{NEG}$. In the deduction below the subdeduction $\mathcal{D}$ exists by the induction hypothesis:

$$\dfrac{\dfrac{\dfrac{\begin{matrix}[\forall\vec{z}(B \rightarrow C)] \\ \mathcal{D} \\ N[C] \rightarrow N[B]\end{matrix} \quad (1)\ N[C]}{N[B]} \qquad (2)\ N[B] \rightarrow A \equiv F[B]}{\dfrac{A}{(1)\ N[C] \rightarrow A \quad \equiv F[C]}}}{(2)\ F[B] \rightarrow F[C]}$$

Case 3b. $F \equiv P[*] \rightarrow A$, $P \in \mathcal{POS}$,

Case 3c. $F \equiv A \rightarrow P[*]$, $P \in \mathcal{POS}$,

Case 3d. $F \equiv A \rightarrow N[*]$, $N \in \mathcal{NEG}$,
are treated similarly. We leave the remaining cases to the reader. □

3.13. We shall now describe a generalization of the conservative extension result of 3.6, which can be applied to many theories formulated in (many-sorted) first-order predicate logic.

We shall use *formula schema*, *axiom schema* for syntactic objects similar to formulas except for the possible presence of certain predicate symbols $\rho_1, \rho_2, \rho_3, \ldots$ (regarded as being outside our standard languages for predicate logic), called *place-holders*. For example, the "induction axiom" is characterized by a schema

$$\rho_1(0) \wedge \forall x(\rho_1(x) \rightarrow \rho_1(Sx)) \rightarrow \forall x \rho_1(x),$$

where ρ_1 is a unary (place-holder) predicate symbol.

A *formula* is a formula schema with zero (i.e. no) place-holders. If Γ is a collection of formula schemas, A a formula, we write $\Gamma \vdash A$ if A can be derived from instances of schemas in Γ; Γ^g is the collection of all A^g, A an instance of a schema in Γ.

3.14. Definition. Let $A[\rho_1, \ldots, \rho_n]$ be an axiom schema. $B_1, \ldots, B_n$ are supposed to be free for substitution for $\rho_1, \ldots, \rho_n$ respectively in A, i.e. without loss of generality we may assume $\mathrm{FV}(B_i) \cap \mathrm{BV}(A) = \emptyset$. The schema is *spreading* if

$$\mathbf{MQC} \vdash A[B_1^g, \ldots, B_n^g] \rightarrow A[B_1, \ldots, B_n]^g;$$

the schema is *wiping* if

$$\mathbf{MQC} \vdash A[B_1, \ldots, B_n]^g \to A[B_1^g, \ldots, B_n^g],$$

and *isolating* if

$$\mathbf{MQC} \vdash A[B_1, \ldots, B_n]^g \to \neg\neg A[B_1^g, \ldots, B_n^g].$$

A set of schemas is *M-closed under* ^{g} if $\Gamma \vdash_m A^g$ for each instance A of a schema in Γ. □

N.B. The notions of spreading, wiping and isolating are relative to **MQC**. We may distinguish the corresponding notions for **IQC** as *I-spreading*, *I-wiping* and *I-isolating*.

Since we may regard formulas simply as schemas without place-holders, the notions of spreading, wiping and isolating also apply to formulas.

A formula of the form $\forall x(A \to \forall y B)$, with A spreading, B isolating is called *essentially isolating*. □

We note the following easy lemma.

3.15. Lemma. *If* $\mathrm{FV}(A) \cap \mathrm{BV}(\Gamma \cup \{B\}) = \emptyset$, *then* $\Gamma \vdash_m B \Rightarrow \Gamma[\bot/A] \vdash_m B[\bot/A]$ *where* $\Gamma[\bot/A] := \{C[\bot/A] : C \in \Gamma\}$.

Proof. Obvious, since in **MQC** $\bot$ is treated as an arbitrary proposition. □

3.16. Theorem. *Suppose* Γ *to be a set of schemas* M*-closed under* ^{g}, *and let* $\Gamma \vdash_c A$. *Then*

(i) *If* A *is isolating and* $\bot$ *does not occur in* Γ, A *then* $\Gamma \vdash_m A$.

(ii) *If* $\bot$ *occurs in* Γ *only positively* (*i.e. for all* $C \in \Gamma$, $C \equiv F[*/\bot] \Rightarrow F \in \mathcal{POS}$) *and negatively in* A, *and* A *is isolating, then* $\Gamma \vdash_i A$.

(iii) *As* (*i*) *for essentially isolating* A.

(iv) *As* (*ii*) *for essentially isolating* A.

(v) *If* A *is* I*-wiping then* $\Gamma \vdash_i A$.

Proof. (i) Let A be isolating; assume $\Gamma \vdash_c A$, then $\Gamma^g \vdash_m A^g$, therefore $\Gamma \vdash_m A^g$ (Γ is M-closed), hence $\Gamma \vdash_m \neg\neg A$ (since A is isolating). Hence by the lemma 3.15 $\Gamma[\bot/A] \vdash_m (\neg\neg A)[\bot/A]$, and since $\bot$ does not occur in $\Gamma \cup \{A\}$, we have $\Gamma \vdash_m (A \to A) \to A$, i.e. $\Gamma \vdash_m A$.

(ii) Let $B \in \Gamma$, $B^* \equiv B[\bot/A]$. $\bot$ occurs only positively in Γ, hence by 3.12 $B[p_1, \ldots, p_n/A_1, \ldots, A_n] \to B^*[p_1, \ldots, p_n/A_1, \ldots, A_n]$. Since A contains $\perp$ only negatively, we have by 3.12 $\vdash_i A[\perp/A] \to A$. Thus from $\Gamma \vdash_m \neg\neg A$ we now obtain $\Gamma[\perp/A] \vdash_m (A[\perp/A] \to A) \to A$, so $\Gamma \vdash_i A$.

(iii) Let $A \equiv \forall x(B \to \forall yC)$, B spreading, C isolating. Then if $\Gamma \vdash_c A$, also $\Gamma, B \vdash_c C$ (y not free in Γ, B), and $\Gamma \cup \{B\}$ is M-closed under ^{g} since B is spreading. This reduces (iii) to (i).

(iv) is similarly reduced to (ii), (v) is trivial. □

3.17. Definition. We define simultaneously classes of axiom schemas $\mathscr{S}, \mathscr{W}, \mathscr{I}$ (from "spreading", "wiping" and "isolating") as follows. Let P range over prime formulas distinct from $\perp$, ρ over expressions $\rho_k t_1 \ldots t_n$ (ρ_k an n-ary place-holder symbol), S_1 and S_2 over $\mathscr{S}$, W_1 and W_2 over $\mathscr{W}$, J_1 and J_2 over $\mathscr{I}$. Then $\mathscr{S}, \mathscr{W}, \mathscr{I}$ are inductively generated by the clauses

(i) $\perp, P, \rho, S_1 \wedge S_2, S_1 \vee S_2, \forall x S_1, \exists x S_1, J_1 \to S_1 \in \mathscr{S}$;

(ii) $\perp, \rho, W_1 \wedge W_2, W_1 \vee W_2, \forall x W_1, S_1 \to W_1 \in \mathscr{W}$;

(iii) $P, W_1, J_1 \wedge J_2, J_1 \vee J_2, \exists x J_1 \in \mathscr{I}$.

If we extend (iii) with an extra clause

$$S_1 \to J_1 \in \mathscr{I},$$

we obtain wider classes denoted by $\mathscr{S}_i, \mathscr{W}_i, \mathscr{I}_i$. □

3.18. Proposition.

(i) *If* $A[\rho_1, \ldots, \rho_n]$ *is a schema in* $\mathscr{S}(\mathscr{W}, \mathscr{I})$ *then* A *is spreading* (*wiping, isolating*).

(ii) *Similarly with* $\mathscr{S}_i(\mathscr{W}_i, \mathscr{I}_i)$ *and* I-*spreading* (I-*wiping*, I-*isolating*).

Proof. By induction on $\mathscr{S}, \mathscr{W}, \mathscr{I}$, using the same identities as in the proof of 3.3. □

By means of this proposition we can easily show for many schemas and formulas that they are spreading, wiping or isolating.

3.19. Definition. A *d-formula* is either a prime formula distinct from $\perp$, a disjunction or an existential formula. □

The following proposition is a corollary to 3.18:

3.20. Theorem.

(i) *If there is no positive* (*negative*) *occurrence of a subformula of the form* $\forall xB$ *in* A, *then* A *is isolating* (*spreading*).

(ii) *If A is the negation of a prenex formula, then*

$$\mathbf{CQC} \vdash A \Rightarrow \mathbf{MQC} \vdash A.$$

(iii) *Suppose that all positive (negative) occurrences of* d-*formulas in A occur in fact negated. Then A is wiping (spreading).*

(iv) *If all positive occurrences of* d-*formulas in A occur negated, then*

$$\mathbf{CQC} \vdash A \Rightarrow \mathbf{MQC} \vdash A.$$

PROOF. (i) by simultaneous induction, using 3.18.

(ii) is an immediate consequence of (i).

(iii) The assumption states in fact that whenever $A \equiv P[B](A \equiv N[B])$, B a d-formula, then $P[*] \equiv N[*/* \to \bot]$ for suitable N ($N[*] \equiv P[*/* \to \bot]$ for suitable P), where $P \in \mathscr{POS}$, $N \in \mathscr{NEG}$. Now use simultaneous induction, with help of 3.18.

(iv) is an immediate consequence of (iii). □

We shall carry our refinements still further.

3.21. LEMMA. *Let Γ_A^+ (Γ_A^-) be the set of universal closures of all $\forall x \neg\neg B \to \neg\neg\forall xB$ such that $\forall xB$ occurs positively (negatively) in A, then*

$$\Gamma_A^+ \vdash_i A^g \to \neg\neg A, \quad \Gamma_A^- \vdash_i \neg\neg A \to A.$$

PROOF. Simultaneously by induction on the complexity of A. For example, let $A \equiv C \to D$. Then $\Gamma_A^+ = \Gamma_C^- \cup \Gamma_D^+$. By induction hypothesis

$$\Gamma_C^- \vdash_i \neg\neg C \to C^g, \quad \Gamma_D^+ \vdash_i D^g \to \neg\neg D,$$

and so

$$\Gamma_A^+ \vdash_i (C \to D)^g \to (\neg\neg C \to \neg\neg D) \to \neg\neg(C \to D).$$

If $A \equiv \forall xC$, $\Gamma_A^+ = \Gamma_C^+ \cup H$, where H is the universal closure of $\forall x \neg\neg C \to \neg\neg\forall xC$. By induction hypothesis $\Gamma_C^+ \vdash_i C^g \to \neg\neg C$, hence $\Gamma_A^+ \vdash_i (\forall xC)^g \to \neg\neg\neg\forall xC$, etcetera. □

3.22. LEMMA. *Suppose $A \equiv F[\neg\neg C]$, $F \in \mathscr{CON}$, $\mathrm{FV}(C) \cap \mathrm{BV}(\forall xA) = \emptyset$; then* $\mathbf{IQC} \vdash \forall xF[\neg\neg C] \to \neg\neg\forall xF[C]$.

PROOF. If $F \in \mathcal{NEG}$, $\forall\vec{z}(C \to \neg\neg C)$ holds and we can apply lemma 3.12(ii). If $F \in \mathcal{POS}$, we can find $\mathcal{D}_1$ (by lemma 3.12(i)) and $\mathcal{D}_2$ (by intuitionistic logic) such that the following is a correct deduction

$$\dfrac{\dfrac{\dfrac{\begin{array}{c}\mathcal{D}_2\\ \neg\neg(\neg\neg C \to C)\end{array} \qquad \dfrac{\dfrac{\begin{array}{c}(3)\ [\forall xF[\neg\neg C]]\ [\neg\neg C \to C]\ (1)\\ \mathcal{D}_1\\ \forall xF[C]\end{array} \qquad \neg\forall xF[C]\ (2)}{\bot}}{\neg(\neg\neg C \to C)}\,(1)}{\bot}}{\neg\neg\forall xF[C]}\,(2)}{\forall xF[\neg\neg C] \to \neg\neg\forall xF[C]}\,(3) \qquad \square$$

3.23. DEFINITION. The *strictly positive contexts* $\mathcal{SPOS}$ are inductively defined by

$$*,\ A \wedge S,\ S \wedge A,\ A \vee S,\ S \vee A,\ A \to S,\ \forall xS,\ \exists xS \in \mathcal{SPOS},$$

where S is a metavariable for elements of $\mathcal{SPOS}$.

A *strictly positive occurrence of A in B* refers to a pair (S, A) with $S \in \mathcal{SPOS}$, $S[*/A] \equiv B$.

A *strictly positive subformula* (or *strictly positive part, s.p.p.*) *of B* is a subformula which has a strictly positive occurrence in B. □

Note that $\mathcal{SPOS}$ is properly contained in $\mathcal{POS}$: $\forall x(* \to Q) \in \mathcal{POS} \setminus \mathcal{SPOS}$.

3.24. DEFINITION. We define a special class of multiple formula contexts $\mathcal{F}$ as follows. We have a countable set of symbols $*_1, *_2, *_3, \ldots$; F, F' are metavariables ranging over $\mathcal{F}$, A a metavariable for formulas. $\mathcal{F}$ is inductively generated by the clauses

(i) $*_n$, $\forall xF$, $A \to F \in \mathcal{F}$;

(ii) suppose for all n and m, that $n \neq m$ if $*_n \in F$ and $*_m \in F'$; then $F \wedge F' \in \mathcal{F}$. □

N.B. The condition in (ii) guarantees that each $*_n$ occurs in $F \in \mathscr{F}$ at most once. Without loss of generality we can always assume an $F \in \mathscr{F}$ to be of the form $F[*_1, \ldots, *_n]$.

The intuitive significance of $\mathscr{F}$ is easily explained. Suppose $F[*_1, \ldots, *_n] \in \mathscr{F}$ and $A \equiv F[B_1, \ldots, B_n]$, and let T be the tree of strictly positive occurrences in A; T is a subtree of the formula construction tree of A. Then any branch in T meets one of the occurrences marked by $*_i$ without encountering a d-formula between A itself and the occurrence marked by $*_i$.

3.25. Lemma. *Let* $F[*_1, \ldots, *_n] \in \mathscr{F}, \mathrm{BV}(F) \cap \mathrm{FV}(B_i) = \emptyset$ *for* $1 \leq i \leq n$. *Then*

$$\mathbf{IQC} \vdash \forall x \neg\neg F[B_1, \ldots, B_n] \rightarrow \neg\neg \forall x F[B_1, \ldots, B_n].$$

Proof. One first shows, by a straightforward induction over $\mathscr{F}$: $\mathbf{IQC} \vdash$

$$\neg\neg F[B_1, \ldots, B_n] \rightarrow F[\neg\neg B_1, \ldots, \neg\neg B_n].$$

Then we apply 3.22 n times. □

The following proposition is now an easy corollary:

3.26. Theorem.

(i) *Suppose that for each positive occurrence of a formula* $\forall x B$ *in* A *there is an* $F \in \mathscr{F}$ *with* $\forall x B \equiv F[B_1, \ldots, B_n]$, $\mathrm{FV}(B_i) \cap \mathrm{BV}(\forall x F) = \emptyset$, *then* A *is* I-*isolating, i.e.* $\mathbf{IQC} \vdash A^{\mathrm{g}} \rightarrow \neg\neg A$.

(ii) (*Mints, Orevkov*) *If, for any positive occurrence of a formula* $\forall x B$ *in* A, *there is no strictly positive occurrence of a* d-*formula in* B, *then* A *is* I-*isolating.*

(iii) (*Cellucci*). *If, for any positive occurrence of a formula* $\forall x B$ *in* A, $\forall x B$ *is in fact of the form* $\forall x(C(x) \rightarrow D)$, $x \notin \mathrm{FV}(D)$, *then* A *is* I-*isolating.*

(iv) *Suppose that* Γ *is* M-*closed under* ^{g}. *Then, if* $A \equiv \neg A'$ *satisfies the conditions under* (*ii*) *or* (*iii*), *we have* $\Gamma \vdash_{\mathrm{c}} A \Rightarrow \Gamma \vdash_{\mathrm{i}} A$.

Proof. (i) Immediate, combining 3.25 with 3.21.

(ii) Use (i) and note that the condition on A implies that each positive sfo of the form $\forall x B$ is such that $B \equiv B'[*_1, \ldots, *_n / \bot, \ldots, \bot]$ for some $B' \in \mathscr{F}$.

(iii) Use (i) and note that the condition on A implies that each positive sfo in A of the form $\forall xB$ is such that $B \equiv (C(x) \to *_1)[*_1/D]$, and that $C(x) \to *_1 \in \mathscr{F}$.

(iv) A is I-wiping since A is I-isolating and $\neg\neg\neg A' \leftrightarrow \neg A' \leftrightarrow A$ hence 3.16(v) applies. □

4. Hilbert-type systems

We shall now introduce codifications of intuitionistic predicate logic of a somewhat different type, to be denoted by the generic name of Hilbert-type systems. Deductions are performed from axioms by means of rules, but there is no cancelling of open assumptions. Hilbert-type systems are sometimes easier to handle than natural deduction in proofs of metamathematical results. The contents of this section are entirely technical, and the reading can be postponed until called for.

The designations H_i-**IQC** etc. for the various systems introduced below are entirely ad hoc and are used only locally.

4.1. *The system* H_1-**IQC**.

The first Hilbert-type system we shall consider is based on the following six groups of axiom schemata:

$$\wedge\text{-Ax} \quad \begin{cases} A \wedge B \to A, \quad A \wedge B \to B, \\ A \to (B \to (A \wedge B)); \end{cases}$$

$$\vee\text{-Ax} \quad \begin{cases} A \to A \vee B, \quad B \to A \vee B, \\ (A \to C) \to ((B \to C) \to (A \vee B \to C)); \end{cases}$$

$$\to\text{-Ax} \quad \begin{cases} A \to (B \to A), \\ (A \to (B \to C)) \to ((A \to B) \to (A \to C)); \end{cases}$$

$$\bot\text{-Ax} \quad \bot \to A;$$

$$\exists\text{-Ax} \quad \begin{cases} A[x/t] \to \exists xA \quad (t \text{ free for } x \text{ in } A), \\ \exists x(A \to B) \to (\exists yA[x/y] \to B) \\ \qquad (x \notin \mathrm{FV}(B); \quad y \equiv x \text{ or } y \notin \mathrm{FV}(A)); \end{cases}$$

$$\forall\text{-Ax} \quad \begin{cases} \forall xA \to A[x/t] \quad (t \text{ free for } x \text{ in } A), \\ \forall x(B \to A) \to (B \to \forall yA[x/y]) \\ \qquad (x \in \mathrm{FV}(B), \quad y \equiv x \text{ or } y \notin \mathrm{FV}(A)). \end{cases}$$

In addition there are two rules: $\to$ E (modus ponens) and $\forall$I (rule of generalization):

$\forall$I $\quad \vdash A \Rightarrow \vdash \forall x A.$

This corresponds to the rule $\forall$I of the natural deduction system in the case where no open assumptions are present.

If we also permit assumptions (which of course cannot be cancelled) we use $\forall$I in the same way as in the natural deduction system. That is to say, we have

$\forall$I $\quad \Gamma \vdash A \Rightarrow \Gamma \vdash \forall x A$

under the restriction that $x \notin \mathrm{FV}(\Gamma)$. This corresponds to Kleene's notion of a "deduction from assumptions with variables held constant" (1952, section 22). If we regard the set Γ as a collection of axioms, possibly in open form, and permit the $\forall$I-rule without the restriction $x \notin \mathrm{FV}(\Gamma)$, we have Kleene's "deduction from assumptions". This corresponds to the treatment of *non-closed axioms* in 1.7.

Below we shall only consider $\forall$I from hypotheses (assumptions) under the restriction $x \notin \mathrm{FV}(\Gamma)$. Therefore, if we consider deductions from axioms, we assume the axioms to be given in closed form.

Let us use $\Gamma \vdash_1 C$ to denote deducibility of C in $\mathrm{H_1}$-**IQC** from assumptions Γ, and let $\Gamma \vdash C$ be used for deducibility in the natural deduction formulation N-**IQC**. Then we have a

4.2. Theorem. *$\Gamma \vdash C$ iff $\Gamma \vdash_1 C$.*

Proof. $\Gamma \vdash_1 C \Rightarrow \Gamma \vdash C$ is easy: all the axioms of $\mathrm{H_1}$-**IQC** are derivable in N-**IQC**, and $\to$ E, $\forall$I are part of the rules of N-**IQC**.

For the converse implications we must show how to transform a deduction in N-**IQC** into a deduction in $\mathrm{H_1}$-**IQC**. We do this in two steps: we first show how to transform a deduction in N-**IQC** into a deduction in an intermediate system **H** with as only rules $\exists$I, $\to$ I, $\to$ E, but permitting axioms from $\wedge$-Ax, $\vee$-Ax, $\exists$-Ax, and axioms of the form $\forall x A(x) \to A(t)$.

The proof proceeds by induction on the height of a deduction tree in N-**IQC**. We have to show how we can successively remove all applications of $\wedge$I, $\wedge$E, $\vee$I, $\vee$E, $\exists$E, $\exists$I and $\forall$E. Suppose $\mathscr{D}_1, \mathscr{D}_2, \mathscr{D}_3$ to be deductions in N-**IQC** of height less than k, and let $\mathscr{D}_1', \mathscr{D}_2', \mathscr{D}_3'$ be the corresponding deductions in **H**.

To the left we show possible forms of a derivation $\mathcal{D}$ of height k: to the right we indicate its transformation into a deduction **H**.

$$\frac{\begin{matrix}\mathcal{D}_1 & \mathcal{D}_2\\ A & B\end{matrix}}{A\wedge B} \qquad\qquad \dfrac{\dfrac{A\to(B\to(A\wedge B)) \quad \begin{matrix}\mathcal{D}'_1\\ A\end{matrix}}{B\to(A\wedge B)} \quad \begin{matrix}\mathcal{D}'_2\\ B\end{matrix}}{A\wedge B}$$

$$\frac{\begin{matrix}\mathcal{D}_1\\ A\wedge B\end{matrix}}{A} \qquad\qquad \frac{A\wedge B\to A \quad \begin{matrix}\mathcal{D}'_1\\ A\wedge B\end{matrix}}{A}$$

$$\frac{\begin{matrix}\mathcal{D}_1\\ A\end{matrix}}{A\vee B} \qquad\qquad \frac{A\to A\vee B \quad \begin{matrix}\mathcal{D}'_1\\ A\end{matrix}}{A\vee B}$$

$$\frac{\begin{matrix} & [A] & [B]\\ \mathcal{D}_1 & \mathcal{D}_2 & \mathcal{D}_3\\ A\vee B & C & C\end{matrix}}{C} \qquad\qquad \dfrac{\dfrac{\dfrac{(A\to C)\to((B\to C)\to(A\vee B\to C)) \quad \dfrac{\begin{matrix}[A]\\ \mathcal{D}'_2\\ C\end{matrix}}{A\to C}}{(B\to C)\to(A\vee B\to C)} \quad \dfrac{\begin{matrix}[B]\\ \mathcal{D}'_3\\ C\end{matrix}}{B\to C}}{A\vee B\to C} \quad \begin{matrix}\mathcal{D}'_1\\ A\vee B\end{matrix}}{C}$$

$$\frac{\begin{matrix}\mathcal{D}_1\\ \forall xA(x)\end{matrix}}{A(t)} \qquad\qquad \frac{\forall xA(x)\to A(t) \quad \begin{matrix}\mathcal{D}'_1\\ \forall xA(x)\end{matrix}}{A(t)}$$

$$\frac{\begin{matrix}\mathcal{D}_1\\ A(t)\end{matrix}}{\exists xA(x)} \qquad\qquad \frac{A(t)\to\exists xA(x) \quad \begin{matrix}\mathcal{D}'_1\\ A(t)\end{matrix}}{\exists xA(x)}$$

$$\frac{\begin{matrix} & [A(y)]\\ \mathcal{D}_1 & \mathcal{D}_2\\ \exists xA(x) & C\end{matrix}}{C} \qquad\qquad \dfrac{\dfrac{\forall x(A\to C)\to(\exists xA\to C) \quad \dfrac{\dfrac{\begin{matrix}[A(y)]\\ \mathcal{D}'_2\\ C\end{matrix}}{A(y)\to C}}{\forall x(A\to C)}}{\exists xA\to C} \quad \begin{matrix}\mathcal{D}'_1\\ \exists xA\end{matrix}}{C}$$

The cases where $\mathscr{D}$ ends with an application of $\rightarrow$ E, $\rightarrow$ I, $\forall$I are of course trivial.

Now we shall show, by induction on the height of a derivation in **H**, how to eliminate all applications of $\rightarrow$ I.

First of all we observe that $A \rightarrow A$ can be established in **H** by means of the following standard deduction $\mathscr{D}_A$ using the schemas of $\rightarrow$ -Ax, where "AA" abbreviates "$A \rightarrow A$":

$$\dfrac{\dfrac{[A \rightarrow (AA \rightarrow A)] \rightarrow [(A \rightarrow AA) \rightarrow AA] \quad A \rightarrow (AA \rightarrow A)}{(A \rightarrow AA) \rightarrow AA} \qquad A \rightarrow AA}{A \rightarrow A}$$

Consider any derivation $\mathscr{D}$ in **H** ending with $\rightarrow$ I, say $\mathscr{D}$ is of the form

$$\dfrac{\begin{matrix}[A] \\ \mathscr{D}_1 \\ B\end{matrix}}{A \rightarrow B}$$

and let $\mathscr{D}_1'$ be the result of eliminating $\rightarrow$ I from $\mathscr{D}_1$ (induction hypothesis). We have to show how to eliminate the final $\rightarrow$ I from

$$\dfrac{\begin{matrix}[A] \\ \mathscr{D}_1' \\ B\end{matrix}}{A \rightarrow B}$$

We shall do this by showing how to transform each subdeduction $\mathscr{D}_0$ (with conclusion C say) of $\mathscr{D}_1'$ into a deduction

$$\begin{matrix}\mathscr{D}_0^* \\ A \rightarrow C,\end{matrix}$$

by induction on the height of $\mathscr{D}_0$.

Basis-step: a top formula (occurrence) C not in the class indicated by $[A]$ in $\mathscr{D}_1'$ is transformed into

$$\dfrac{C \rightarrow (A \rightarrow C) \quad C}{A \rightarrow C}$$

and the top formulas in $[A]$ are replaced by $\mathscr{D}_A$.

Induction step: $\mathscr{D}_0$ can end with $\to$E or $\forall$I. We again show to the left $\mathscr{D}_0$, to the right $\mathscr{D}_0^*$:

$$\dfrac{\begin{matrix}\mathscr{D}_2 & \mathscr{D}_3\\ C\to D & C\end{matrix}}{D}\qquad \dfrac{\dfrac{(A\to(C\to D))\to((A\to C)\to(A\to D)) \quad \begin{matrix}\mathscr{D}_2^*\\ A\to(C\to D)\end{matrix}}{(A\to C)\to(A\to D)} \quad \begin{matrix}\mathscr{D}_3^*\\ A\to C\end{matrix}}{A\to D}$$

$$\dfrac{\begin{matrix}\mathscr{D}_2\\ C(x)\end{matrix}}{\forall xC(x)}\qquad \dfrac{\forall x(A\to Cx)\to(A\to\forall xCx) \quad \dfrac{\begin{matrix}\mathscr{D}_2^*\\ A\to Cx\end{matrix}}{\forall x(A\to Cx)}}{A\to\forall xCx}$$

Note that we have used our restriction on $\forall$I in $\mathscr{D}_2^*$. Repeating this, we end up with

$$\begin{matrix}\mathscr{D}_1^*\\ A\to B\end{matrix}$$

and we have eliminated the final $\to$I in $\mathscr{D}$. By this method all $\to$I-applications in a derivation in **H** can be removed, starting with the uppermost ones. □

4.3. *The system* H_2-**IQC**. The following system is almost the same as the one used in Kleene (1952): based on the axioms $\wedge$-Ax, $\vee$-Ax, $\to$-Ax, $\bot$-Ax and

$$A[x/t]\to\exists xA,\quad \forall xA\to A[x/t]\ (t \text{ free for } x \text{ in } A),$$

and the rules $\to$E and

$$\forall\mathrm{I}_2\,\dfrac{A\to B}{A\to\forall yB[x/y]}\qquad\qquad \exists\mathrm{E}_2\,\dfrac{B\to A}{\exists yB[x/y]\to A}$$

where $x\notin \mathrm{FV}(A)$, and $y\equiv x$ or $y\notin\mathrm{FV}(B)$.

THEOREM. *Deducibility in* H_2-**IQC** *is equivalent to deducibility in* H_1-**IQC** (*and hence to deducibility in* N-**IQC**).

PROOF. Left as an exercise to the reader. Some care is needed in transforming applications of $\forall$I into applications of $\forall\mathrm{I}_2$. □

4.4. *The system* H_3-**IQC**. If we are interested in sentences only, we can also remove the rule of generalization. H_3-**IQC** has as axioms all universally closed instances of $\wedge$-Ax, $\vee$-Ax, $\to$-Ax, $\forall$-Ax, $\bot$-Ax, $\exists$-Ax; as its only rule, we have a form of modus ponens

$$\rightarrow \mathrm{E}_3 \frac{\forall \vec{x}(A \rightarrow B) \quad \forall \vec{x} A}{\forall \vec{x} B}.$$

For sentences we can formulate the following

THEOREM. *Deducibility in* H_3-**IQC** *is equivalent to deducibility in* H_1-**IQC**.

PROOF. Exercise. □

4.5. *Hilbert-type systems for* **IQCE** *and* **IQCE**$^+$. Let us first consider E^+-logic. H_2-**IQCE**$^+$ will be the system based on the axioms $\wedge$-Ax, $\vee$-Ax, $\rightarrow$-Ax, $\perp$-Ax and

$\exists$-$\mathrm{Ax}_2^{\mathrm{E}}$ $\quad A[x/t] \wedge Et \rightarrow \exists x A \quad$ (t free for x in A);

$\forall$-$\mathrm{Ax}_2^{\mathrm{E}}$ $\quad \forall x A \wedge Et \rightarrow A[x/t] \quad$ (t free for x in A),

$$\mathrm{EAX} \quad \begin{cases} Ex, \quad Et \leftrightarrow \exists x(x = t), \\ \quad \forall x(x = x), \quad t = s \wedge t' = s \rightarrow t = t', \end{cases}$$

strictness for predicate symbols R and function symbols f

$$\mathrm{STR} \quad \begin{cases} R(t_1, \ldots, t_n) \rightarrow Et_i, \\ Ef(t_1, \ldots, t_n) \rightarrow Et_i, \end{cases}$$

and finally replacement for atoms

$$\mathrm{REPL} \quad \begin{cases} R\vec{s} \wedge \vec{s} = \vec{t} \rightarrow R\vec{t}, \\ E(f\vec{s}) \wedge \vec{s} = \vec{t} \rightarrow f\vec{s} = f\vec{t}. \end{cases}$$

The only rules are $\rightarrow \mathrm{E}, \forall \mathrm{I}_2, \exists \mathrm{E}_2$.

The corresponding version H_2-**IQCE** of **IQCE** is obtained from H_2-**IQCE**$^+$ by dropping Ex from EAX and replacing $\forall \mathrm{I}_2, \exists \mathrm{I}_2$ by

$$\forall \mathrm{I}_2^{\mathrm{E}} \frac{A \wedge Ex \rightarrow B}{A \rightarrow \forall y B[x/y]},$$

$$\exists \mathrm{E}_2^{\mathrm{E}} \frac{B \wedge Ex \rightarrow A}{\exists y B[x/y] \rightarrow A},$$

where $x \notin \mathrm{FV}(A)$ and $y \equiv x$ or $y \notin \mathrm{FV}(B)$.

N.B. For theories based on **IQCE**$^+$ plus axioms we may have to add SUB, the substitution rule.

4.6. THEOREM. *Deducibility in* H_2-**IQCE**$^+$ *and* H_2-**IQCE** *is equivalent to deducibility in the natural deduction versions* N-**IQCE**$^+$, N-**IQCE** *respectively.*

Proof. Exercise. □

4.7. $\mathbf{H_4}$**-IQCE**$^+$ *and* $\mathbf{H_4}$**-IQCE**. If we drop the axioms Ex in $\mathbf{H_2}$**-IQCE**$^+$ and replace $\exists$-$\mathrm{Ax}_2^{\mathrm{E}}$, $\forall$-$\mathrm{Ax}_2^{\mathrm{E}}$ by the special cases

$$\exists\text{-Ax}_4 \qquad A[x/y] \to \exists xA,$$

$$\forall\text{-Ax}_4 \qquad \forall xA \to A[x/y],$$

we obtain the system $\mathbf{H_4}$**-IQCE**$^+$.

If in $\mathbf{H_2}$**-IQCE** we replace the axioms $\exists$-$\mathrm{Ax}_2^{\mathrm{E}}$, $\forall$-$\mathrm{Ax}_2^{\mathrm{E}}$ by

$$\exists\text{-Ax}_4^{\mathrm{E}} \qquad A[x/y] \wedge Ey \to \exists xA,$$

$$\forall\text{-Ax}_4^{\mathrm{E}} \qquad \forall xA \wedge Ey \to A[x/y],$$

we obtain $\mathbf{H_4}$**-IQCE**.

Theorem. *Deducibility in* $\mathbf{H_2}$**-IQCE**$^+$, $\mathbf{H_2}$**-IQCE** *is equivalent to deducibility in* $\mathbf{H_4}$**-IQCE**$^+$, $\mathbf{H_4}$**-IQCE** *respectively.*

Proof. We give an outline and leave the details as an exercise.

(i) Ex is derivable in $\mathbf{H_4}$**-IQCE**$^+$. For $x = x \to x = x$ holds by propositional logic, and thus with $\exists$-Ax_4 $x = x \to \exists y(x = y)$. On the other hand $\forall x(x = x) \to x = x$ from $\forall$-Ax_4, and thus with the axiom $\forall x(x = x)$ and modus ponens $\exists y(x = y)$, and therefore by $\exists y(x = y)$ also Ex.

(ii) For any prime formula A and any term t occurring in A, one gets $A \to Et$ by induction over term complexity, repeatedly applying strictness.

(iii) For any prime formula A by induction on the complexity of the terms occurring in A one shows with help of replacement and (ii)

$$A[x/s] \wedge s = t \to A[x/t]. \tag{1}$$

(iv) We establish (1) for arbitrary A by induction on the logical complexity of A.

(v) We deduce $\exists$-$\mathrm{Ax}_2^{\mathrm{E}}$, $\forall$-$\mathrm{Ax}_2^{\mathrm{E}}$ in $\mathbf{H_4}$**-IQCE**$^+$ as follows. By some logic in $\mathbf{H_4}$**-IQCE**$^+$

$$Et \to \exists y(t = y \wedge \forall xA \wedge Ey \to A[x/y]),$$

and therefore with (iv)

$$Et \to \exists y(t = y \wedge \forall xA \wedge Et \to A[x/t]),$$

hence $\forall xA \wedge Et \to A[x/t]$, etc. □

5. Kripke semantics

5.1. In the first section of this chapter we have presented a set of rules for deriving intuitionistically correct logical laws; these rules were also valid for the classical truth-value interpretation of the logical operations.

The notion "intuitionistically correct" or "intuitionistically valid" can be expressed more precisely as follows. Consider, for simplicity, a language $\mathscr{L}$ with relation symbols only, say $R_1, \ldots, R_n$, where R_i has $n(i)$ arguments, and let $A(R_1, \ldots, R_n)$ be a sentence in $\mathscr{L}$. An intuitionistic structure (model) for $\mathscr{L}$ is a $n+1$-tuple $\mathscr{M} \equiv (\mathrm{M}, \mathrm{R}_1, \ldots, \mathrm{R}_n)$, where M is an intuitionistically meaningful domain, and R_i is an $n(i)$-place relation over M, i.e. $\mathrm{R}_i \subset \mathrm{M}^{n(i)}$.

$A(R_1, \ldots, R_n)$ is *(intuitionistically) valid* in $\mathscr{M}$ (notation $\mathscr{M} \models A$) iff $A^{\mathrm{M}}(\mathrm{R}_1, \ldots, \mathrm{R}_n)$ holds intuitionistically; here $A^{\mathrm{M}}(\mathrm{R}_1, \ldots, \mathrm{R}_n)$ is obtained from A by replacing all occurrences of R_i by R_i and relativizing all quantifiers in A to M (i.e. $\forall x, \exists x$ are replaced by $\forall x \in \mathrm{M}$ and $\exists x \in \mathrm{M}$ respectively).

It is to be noted that formally $\mathscr{M} \models A$ is defined in exactly the same way as for classical structures; the only difference is that we interpret $A^{\mathrm{M}}(\mathrm{R}_1, \ldots, \mathrm{R}_n)$ intuitionistically.

A is intuitionistically valid if $\mathscr{M} \models A$ for all intuitionistic structures $\mathscr{M}$ for $\mathscr{L}$. We write $\models A$ for A is (intuitionistically) valid; if it is necessary to distinguish the intuitionistic reading of $\models$ from the classical one, we use a subscript: $\models_{\mathrm{i}}$, $\models_{\mathrm{c}}$.

From the weak counterexamples it became clear that not all classical logical laws are correct in the sense of $\models_{\mathrm{i}}$, and thus we do not expect such laws to be provable in **IQC**.

Showing directly, i.e. by means of combinatorial investigation of the possible deductions in **IQC**, that a certain schema is *not* derivable in **IQC** is usually far from easy. In classical logic underivability is mostly shown by giving a countermodel, but $\models_{\mathrm{i}}$ is technically not so easy to manage.

Here we shall introduce a different semantical interpretation for showing underivability (which, moreover has many other applications to boot). For this semantics all rules of **IQC** are valid, and thus a schema invalid for this semantics cannot be derived in **IQC**. In 6.6 we will show that this semantics exactly fits **IQC**.

This so-called Kripke semantics is not directly connected with the BHK-interpretation of the intuitionistic logical operators, but may be motivated via weak counterexamples.

Recall the weak counterexample to PEM of 1.3.2, based on an as yet undecided proposition A. The argument depended on the following information situation:

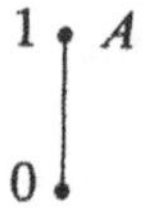

the bottom node 0 represented our *present* knowledge concerning A: A has neither been established nor refuted; 1 represents a possible future information situation where A has been established. Thus in 0 we cannot assert A, nor can we assert $\neg A$, since a future situation 1 where A holds is not excluded. Hence at 0 we cannot assert $A \vee \neg A$.

If the diagram above were a *complete* picture of the future information possibilities, seen from stage 0, we would be justified in asserting $\neg\neg A$ at 0; for $\neg\neg A$ means that whenever we discover $\neg A$ to be true, we must be able to obtain a proof of $\perp$ (absurdity); but since in the diagram above we will never be forced to accept $\neg A$, we can for trivial reasons assert $\neg\neg A$ at 0.

However, since A is an undecided proposition, the diagram is not a complete picture: there is also the possibility to discover that $\neg A$ holds, i.e. the situation is rather as in the picture below

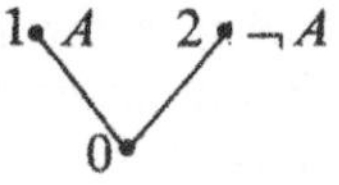

Now we certainly cannot assert $\neg\neg A$ at 0, since later we may arrive at node 2 which together with $\neg\neg A$ would force us to accept $\perp$; but it is tacitly assumed that we shall never obtain inconsistent information. If this picture is regarded as exhaustive w.r.t. A, we may in fact simplify it to the following

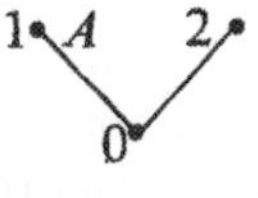

since at node 2 we will never be forced to accept A in a later stage. So, for trivial reasons, we may assert $\neg A$ at node 2 ("whenever in the future we discover A to be true, we can prove $\perp$ "). This picture also shows that we cannot assert $\neg A \vee \neg\neg A$ at 0.

A more complex situation arises if we combine information regarding A and another undecided, independent proposition B; the diagram of possibilities is

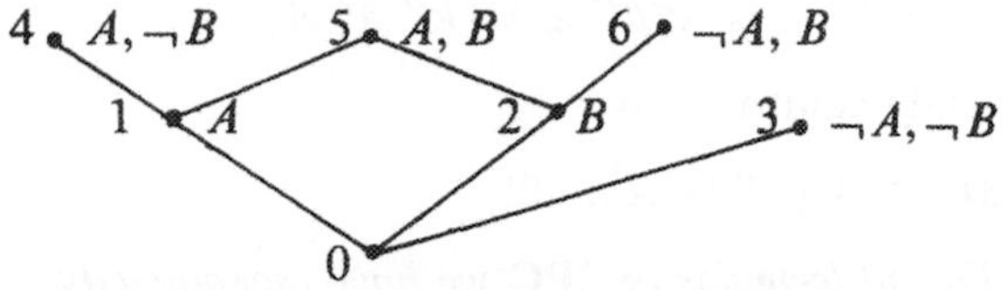

One may readily convince oneself that in this diagram we cannot accept $(A \to B) \vee (B \to A)$ at node 0; for suppose we could prove $A \to B$ at 0. Now node 4 represents a possible extension of information where we have A and $\neg B$, contradicting the assumption that already at 0 we had $A \to B$. Actually, $(A \to B) \vee (B \to A)$ is unacceptable at 0 already in the following simpler information diagram:

We leave it to the reader to convince himself of this fact.

Now we are ready for a formal definition of Kripke semantics for **IPC**.

5.2. Definition. A (*propositional*) *Kripke model* is a triple $\mathscr{K} \equiv (\mathrm{K}, \leq, \Vdash)$, where $(\mathrm{K}, \leq)$ is an inhabited, partially ordered set (poset), and $\Vdash$ a binary relation on $\mathrm{K} \times \mathscr{P}$ ($\mathscr{P}$ the set of proposition letters) such that

$$k \Vdash P \text{ and } k' \geq k \Rightarrow k' \Vdash P. \tag{1}$$

We shall use the expressions "*k forces P*" or "*P is true at k*" for $k \Vdash P$.

$\Vdash$ is then extended to logically compound formulas by the following clauses

Kr1 $k \Vdash A \wedge B := k \Vdash A$ and $k \Vdash B$,

Kr2 $k \Vdash A \vee B := k \Vdash A$ or $k \Vdash B$,

Kr3 $k \Vdash A \to B :=$ for all $k' \geq k$, if $k' \Vdash A$ then $k' \Vdash B$,

Kr4 not $k \Vdash \bot$ (i.e. $k \nVdash \bot$: no element of K forces $\bot$).

The elements of K are called *nodes* (of $\mathscr{K}$). □

Remark. As a consequence of this definition

$$k \Vdash \neg A \Leftrightarrow \forall k' \geq k (k' \nVdash A).$$

Also

$$k \Vdash \neg\neg A \Leftrightarrow \forall k' \geq k \neg \forall k'' \geq k'(k'' \nVdash A),$$

which is classically equivalent to

$$\forall k' \geq k \exists k'' \geq k'(k'' \Vdash A). \quad \square$$

5.3. LEMMA. *For all formulas of* **IPC** *we have monotonicity:*

$$\forall k, k' \in \mathscr{K}(k \Vdash A \quad and \quad k' \geq k \Rightarrow k' \Vdash A).$$

PROOF. By formula induction. Consider for example the case $A \equiv B \to C$. Assume $k \Vdash B \to C$, $k' \geq k$. If $k'' \geq k'$, $k'' \Vdash B$, then also $k'' \geq k$ and $k'' \Vdash B$, hence by $k \Vdash B \to C$ also $k'' \Vdash C$; thus $\forall k'' \geq k'(k'' \Vdash B \Rightarrow k'' \Vdash C)$, i.e. $k' \Vdash B \to C$. $\square$

5.4. DEFINITION. A formula A is *valid at* k in a Kripke model $\mathscr{K}$ iff $k \Vdash A$. A is *valid* in $\mathscr{K} \equiv (\mathrm{K}, \leq, \Vdash)$ iff for all $k \in \mathrm{K}$, $k \Vdash A$; notation $\mathscr{K} \Vdash A$. If Γ is a set of formulas, we say that $\Gamma \Vdash A$ ("A is a Kripke consequence of Γ") iff in each model $\mathscr{K}$ such that if for all $B \in \Gamma$ $\mathscr{K} \Vdash B$, then also $\mathscr{K} \Vdash A$. A is *Kripke valid* (*K-valid*) iff $\emptyset \Vdash A$ which will be written as $\Vdash A$. $\square$

REMARK. Suppose $\mathscr{K} \equiv (\mathrm{K}, \leq, \Vdash)$ is a Kripke model, $k \in \mathrm{K}$. The *truncated* model $\mathscr{K}_k$ is $(\mathrm{K}', \leq', \Vdash')$ where $\mathrm{K}' = \{k' : k' \geq k\}$, $\leq'$ is $\leq \restriction \mathrm{K}'$ (the restriction of $\leq$ to K'), and $\Vdash' := \Vdash \restriction (\mathrm{K}' \times \mathscr{P})$ (the restriction of $\Vdash$ to $\mathrm{K}' \times \mathscr{P}$).

It is easy to see that $k \Vdash A$ iff $\mathscr{K}_k \Vdash' A$ iff $k \Vdash' A$, since the definition of $k \Vdash A$ depends on $\Vdash$ for nodes $k' \geq k$ only.

Note also that validity in $\mathscr{K}$ is equivalent to validity at the bottom node of $\mathscr{K}$, if there is one.

5.5. REMARKS. Observe that $k \Vdash A$ is not assumed to be decidable, not even for A prime. Even if $k \Vdash P$ is decidable for $P \in \mathscr{P}$ the implication clause may prevent $k \Vdash A$ to be decidable for complex A.

The definition of Kripke model may be read either constructively or classically. As a rule we shall treat Kripke models constructively, unless stated otherwise. In this connection we often use the terms "externally" and "internally"; *external* reasoning is reasoning on the meta-level *about* the models; something holds *internally* if it is (Kripke-) valid in the (or all) models under consideration; internal reasoning is "arguing by means of internally valid principles".

If our external reasoning is intuitionistic, it might seem at first sight as if nothing is gained over the method of weak counterexamples: e.g. let $\mathscr{K}$ be a Kripke model with a single node, k_0, where $k_0 \Vdash P$ iff A holds, and A is an as yet undecided proposition. Then we cannot assert $k_0 \Vdash P \vee \neg P$. However, a different Kripke model (see examples below) shows (classically and constructively) that $\nVdash P \vee \neg P$; by the soundness theorem proved below in 5.10 this implies **IPC** $\nvdash P \vee \neg P$ (and not just: we cannot show that **IPC** $\vdash P \vee \neg P$).

If $\mathscr{K}$ consists of a single node and our external reasoning is classical, $\mathscr{K}$ is equivalent to a classical valuation with truth values in $\{0, 1\}$.

5.6. EXAMPLES. $\neg\neg P \vee \neg P$ is not K-valid; to see this, we specify a Kripke model by indicating the partially ordered structure as a diagram, writing next to each node the propositional variables forced at that node

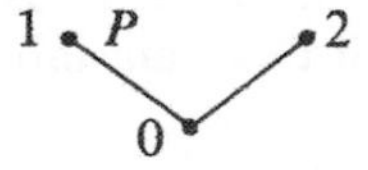

Clearly, $2 \Vdash \neg P$; hence $0 \nVdash \neg\neg P$; and since $1 \Vdash P$, also $0 \nVdash \neg P$ and therefore $0 \nVdash \neg\neg P \vee \neg P$.

Similarly, $(P \to Q) \vee (Q \to P)$ is not K-valid, by the following Kripke model.

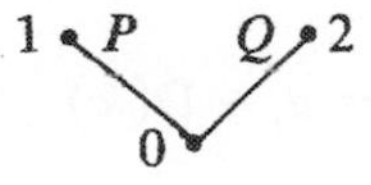

For a more complicated counterexample, consider

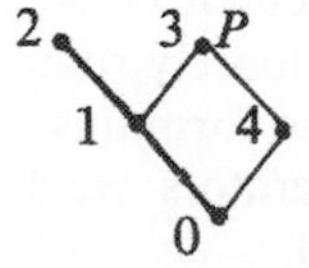

In this model $0 \nVdash (\neg\neg P \to P) \vee (\neg P \vee \neg\neg P)$. For assume the contrary; then either

$$0 \Vdash \neg P \quad \text{or} \quad 0 \Vdash \neg\neg P \quad \text{or} \quad 0 \Vdash \neg\neg P \to P.$$

$0 \Vdash \neg P$ is excluded since $3 \Vdash P$; $0 \Vdash \neg\neg P$ is excluded since $2 \Vdash \neg P$; and $0 \Vdash \neg\neg P \to P$ is excluded since $4 \Vdash \neg\neg P$, but $4 \nVdash P$; hence $4 \nVdash \neg\neg P \to P$ and thus also $0 \nVdash \neg\neg P \to P$.

As a result of the soundness theorem (5.10) this means that in **IPC** we cannot prove $\neg\neg P \vee \neg P$, $(P \to Q) \vee (Q \to P)$ or $(\neg\neg P \to P) \vee (\neg P \vee \neg\neg P)$.

5.7. *Kripke models for pure predicate logic.* The notion of a Kripke model can be extended to **IQC**; we describe below the extension for a language with predicate symbols only.

DEFINITION. A *Kripke model for* **IQC** is a quadruple $\mathscr{K} \equiv (\mathrm{K}, \leq, \mathrm{D}, \Vdash)$, K inhabited, such that

(i) $(\mathrm{K}, \leq)$ is a partially ordered set.

(ii) D is a function (the *domain function*) assigning inhabited sets to the elements of K such that all $\mathrm{D}(k)$ are inhabited and

$$\forall k \forall k'(k \leq k' \to \mathrm{D}(k) \subset \mathrm{D}(k')),$$

i.e. D is monotone.

(iii) Let the language be extended with constant symbols for each element of $\mathbf{D} = \bigcup\{\mathrm{D}(k) : k \in \mathrm{K}\}$; $\Vdash$ is a relation from K to the set of prime formulas in the extended language, such that

$$k \Vdash R^n(d_1, \ldots, d_n) \Rightarrow d_i \in \mathrm{D}(k) \quad \text{for } 1 \leq i \leq n,$$

$$k \Vdash R^n(d_1, \ldots, d_n) \wedge k \leq k' \Rightarrow k' \Vdash R^n(d_1, \ldots, d_n).$$

We now define $k \Vdash A$ for all sentences A in the extended language with constants in $\mathrm{D}(k)$. We will often loosely refer to the constants for elements of $\mathrm{D}(k)$ as *parameters*. For prime formulas, the definition is given by (iii) above; for the propositional operators we have the clauses Kr1–Kr4 as before; for the quantifiers we add,

Kr5 $\quad k \Vdash \forall x A(x) := \forall k' \geq k \forall d \in \mathrm{D}(k')(k' \Vdash A(d))$,

Kr6 $\quad k \Vdash \exists x A(x) := \exists d \in \mathrm{D}(k)(k \Vdash A(d))$. $\square$

As to the intuitive motivation, we may think of $D(k)$ as representing the elements of the domain which are known to exist at stage of information k; at later stages we may have discovered more elements belonging to the domain, hence $D(k) \subset D(k')$ for $k' \geq k$.

DEFINITION. For sentences in the language with constants for all $d \in \mathbf{D} = \bigcup\{D(k) : k \in K\}$ we can define "A *is valid at* k" (= k *forces* A = A *is true* at k), A *is valid* in K, A is *Kripke-valid*, and $\Gamma \Vdash A$ as in 5.4. □

In dealing with Kripke models in the sequel, we shall tacitly assume that we are dealing with *sentences* (possibly containing parameters), unless explicitly stated otherwise.

REMARK. We note also the fact that lemma 5.3 generalizes to **IQC**:

$$k \Vdash A \quad \text{and} \quad k' \geq k \Rightarrow k' \Vdash A. \quad \square$$

5.8. REMARK. There are other ways of presenting what is essentially the same notion of forcing in a Kripke model.

(A) Instead of extending the language with constants for the elements of $\mathbf{D} = \bigcup\{D(k) : k \in K\}$ one assigns elements of $D(k)$ to the variables of the language by a mapping φ, then defines "$k \Vdash A$ under assignment φ"; this method is well-known from classical model theory.

(B) $k \Vdash R(x_1, \ldots, x_n)$ for a predicate letter R is given by a subset of $D(k)^n$, corresponding to

$$\{(d_1, \ldots, d_n) : d_1 \in D(k) \wedge \ldots \wedge d_n \in D(k) \wedge k \Vdash R(d_1, \ldots, d_n)\}.$$

This variant suggests the definition of a Kripke model $(K, \leq, \Vdash)$ for pure predicate logic as a partially ordered collection of (classical) structures $(K, \leq, \{\mathcal{M}_k : k \in K\})$, where $\mathcal{M}_k$ is a (classical) structure for pure predicate logic for each $k \in K$. That is to say, if $\{R_i : i \in I\}$ is the collection of predicate symbols of the language, R_i with $n(i)$ arguments, then $\mathcal{M}_k \equiv (D(k), \{\mathrm{R}_{i,k} : i \in I\})$, $\mathrm{R}_{i,k} \subset D(k)^{n(i)}$, and $k \Vdash R_i(d_1, \ldots, d_n) := \mathrm{R}_{i,k}(d_1, \ldots, d_n)$. The model must satisfy the monotonicity conditions: if $k \leq k'$, then

$$D(k) \subset D(k') \quad \text{and} \quad \mathrm{R}_{i,k} \subset \mathrm{R}_{i,k'}.$$

Warning: the monotonicity condition is strictly weaker than "$\mathcal{M}_k$ is a submodel of $\mathcal{M}_{k'}$".

5.9. EXAMPLES. Next to each node k we indicate $\mathrm{D}(k)$ and the atomic sentences true at the node.

First example.

$$\begin{array}{l} 1 \bullet \ \mathrm{D}(1) = \{a, b\}\ R, R'(a) \\ \quad | \\ 0 \bullet \ \mathrm{D}(0) = \{a\} \quad R'(a) \end{array}$$

In this model

$$0 \nVdash \forall x(R \vee R'(x)) \to R \vee \forall x R'(x).$$

For we have $0 \Vdash \forall x(R \vee R'(x))$, since $0 \Vdash R'(a)$, and hence $0 \Vdash R \vee R'(a)$; and also $1 \Vdash R$, so $1 \Vdash R \vee R'(\mathrm{a})$, $1 \Vdash R \vee R'(b)$. But we do not have $0 \Vdash R \vee \forall x R'(x)$, since $0 \nVdash R$, and $0 \nVdash \forall x R'(x)$, since $1 \nVdash R'(b)$.

Second example.

$$\begin{array}{llll} & & \vdots & \\ k_4 & \{0,1,2,3,4\} & & R0, R1, R2, R3 \\ & & | & \\ k_3 & \{0,1,2,3\} & & R0, R1, R2 \\ & & | & \\ k_2 & \{0,1,2\} & & R0, R1 \\ & & | & \\ k_1 & \{0,1\} & & R0 \\ & & | & \\ k_0 & \{0\}. & & \end{array}$$

This model is a counterexample to $\neg\neg\forall x(R(x) \vee \neg R(x))$. Assume $k_0 \Vdash \neg\neg\forall x(R(x) \vee \neg R(x))$; then $\forall k(k \nVdash \neg\forall x(R(x) \vee \neg R(x)))$, so

$$\forall k \neg\neg\exists k' \geq k(k' \Vdash \forall x(R(x) \vee \neg R(x))). \tag{1}$$

Assume $k_p \Vdash \forall x(R(x) \vee \neg R(x))$, then in particular $k_p \Vdash R(p) \vee \neg R(p)$; $R(p)$ does not hold at k_p, but also $k_p \nVdash \neg R(p)$, since $k_{p+1} \Vdash R(p)$. This shows $\neg\exists k' \geq k(k' \Vdash \forall x(R(x) \vee \neg R(x)))$, therefore (1) is false, hence $k_0 \nVdash \neg\neg\forall x(R(x) \vee \neg R(x))$. Observe that even $k_0 \Vdash \neg\forall x(R(x) \vee \neg R(x))$.

Again by soundness (5.10), it follows that the schemas $\forall x(A \vee B(x)) \to A \vee \forall x B(x)$ and $\neg\neg\forall x(A(x) \vee \neg A(x))$ are not derivable in **IQC**.

5.10. THEOREM (*soundness for pure* **IQC**). $\Gamma \vdash A \Rightarrow \Gamma \Vdash A$.

PROOF. The proof is by induction on derivations in the system of natural deduction for **IQC**.

Let a derivation $\mathscr{D}$ terminate with a rule application

$$\frac{\Gamma_1 \vdash A_1 \quad \Gamma_2 \vdash A_2 \ \ldots \ \Gamma_n \vdash A_n}{\Gamma \vdash A}$$

where $\Delta \vdash B$ expresses that B has been derived from assumptions Δ. Let $\mathscr{K}$ be any Kripke model, $k \in \mathrm{K}$.

The induction hypothesis is as follows. For any $k \in \mathrm{K}$ let Γ_i', A_i' be obtained from Γ_i, A_i by some substitution of elements of $D(k)$ for the free variables in Γ_i, A_i. Assume that for all $k \in \mathrm{K}$ and for all substitutions, if $k \Vdash \Gamma_i'$ for $1 \le i \le n$, then $k \Vdash A_i'$; $k \Vdash \Gamma_i'$ abbreviates $k \Vdash B'$ for all $B \in \Gamma_i$.

We shall consider two cases.

Case 1. Suppose the last rule applied is $\vee$E:

$$\frac{\Gamma_1 \vdash A \vee B \quad \Gamma_2, A \vdash C \quad \Gamma_3, B \vdash C}{\Gamma_1 \cup \Gamma_2 \cup \Gamma_3 \vdash C}$$

where $A \notin \Gamma_2$, $B \notin \Gamma_3$. We have to show for any substitution, and any $k \in \mathrm{K}$ that if $k \Vdash \Gamma_1' \cup \Gamma_2' \cup \Gamma_3'$, then $k \Vdash C'$. Suppose $k \Vdash \Gamma_1' \cup \Gamma_2' \cup \Gamma_3'$; by the induction hypothesis $k \Vdash A' \vee B'$, $k \Vdash A' \Rightarrow k \Vdash C'$, $k \Vdash B' \Rightarrow k \Vdash C'$. Since $k \Vdash A' \vee B' \Rightarrow k \Vdash A'$ or $k \Vdash B'$, we have $k \Vdash C'$.

Case 2. Suppose the last rule applied is $\forall$I:

$$\frac{\Gamma \vdash A(y)}{\Gamma \vdash \forall x A(x)} \quad (y \text{ not free in } \Gamma).$$

For simplicity assume Γ and $\forall x A(x)$ to be closed. We have to show for each $k \in \mathrm{K}$: if $k \Vdash \Gamma$ then $k \Vdash \forall x Ax$. Assume $k \Vdash \Gamma$; by induction hypothesis, for all $k' \ge k$ and all $d \in \mathrm{D}(k')$ we have $k' \Vdash A(d)$. Hence $k \Vdash \forall x A(x)$.

The other cases we leave as an exercise to the reader. □

5.11. *Kripke models for the full language of* **IQC**. Let us now consider a language with equality, predicate symbols $\{R_i : i \in I\}$, function symbols and symbols for constants $\{f_j : j \in J\}$, where R_i has $n(i)$ arguments and f_j has $m(j)$ arguments. A classical structure $\mathscr{M}$ for this language is

$$(\mathrm{D}, \sim, \{\mathrm{R}_i : i \in I\}, \{\mathrm{f}_j : j \in J\}),$$

where D is an inhabited set, $\sim$ an equivalence relation on D interpreting equality, $R_i \subset D^{n(i)}$ such that

$$\vec{d} \sim \vec{d'} \quad \text{and} \quad \vec{d} \in R_i \Rightarrow \vec{d'} \in R_i,$$

$f_j \in D^{m(j)} \to D$ such that

$$\vec{d} \sim \vec{d'} \to f_j\vec{d} \sim f_j\vec{d'}.$$

This slightly deviates from the usual definition of the classical model where $\sim$ is replaced by $=$; for technical reasons we prefer retaining $\sim$.

DEFINITION. A *Kripke model* for the above language is a partially ordered collection of such classical structures:

$$\mathscr{K} \equiv \left(\mathrm{K}, \leq, \left\{\left(\mathrm{D}(k), \sim_k, \{\mathrm{R}_{i,k} : i \in I\}, \{\mathrm{f}_{j,k} : j \in J\}\right) : k \in \mathrm{K}\right\}\right)$$

satisfying monotonicity conditions: if $k \leq k'$ then

$$\mathrm{D}(k) \subset \mathrm{D}(k');$$

$$d \sim_k d' \Rightarrow d \sim_{k'} d';$$

$$\mathrm{R}_{i,k} \subset \mathrm{R}_{i,k'};$$

$$\mathrm{graph}(\mathrm{f}_{j,k}) \subset \mathrm{graph}(\mathrm{f}_{j,k'}). \quad \square$$

We can also describe Kripke models for the full language as quadruples $(\mathrm{K}, \leq, \mathrm{D}, \Vdash)$ where the forcing relation $\Vdash$ for prime formulas has to satisfy a number of conditions at each node. First of all

$$d \sim_k d' := k \Vdash d = d'$$

has to be an equivalence relation on $\mathrm{D}(k)$. Secondly, writing $k \Vdash \vec{d} = \vec{d'}$ for $\forall i < n(k \Vdash d_i = d_i')$ where $n = \mathrm{lth}\ d = \mathrm{lth}\ d'$, we must require

$$k \Vdash \vec{d} = \vec{d'} \Rightarrow k \Vdash f\vec{d} = f\vec{d'},$$

$$k \Vdash \vec{d} = \vec{d'} \quad \text{and} \quad k \Vdash R_i\vec{d} \Rightarrow k \Vdash R_i\vec{d'},$$

plus the obvious monotonicity conditions.

In the sequel, if we present a Kripke model of a language with equality and function symbols, we shall tacitly assume these conditions to be met. Note that the conditions just listed permit us to construct the partially ordered collection of classical structures of the definition above. The soundness theorem extends to the full language, as the reader can easily check:

THEOREM (*soundness for* **IQC**). $\Gamma \vdash A \Rightarrow \Gamma \Vdash A$. $\square$

5.12. *Kripke models with restrictions.* In dealing with languages with equality, and especially when function symbols are present, it is sometimes advantageous to use a different (but essentially equivalent) definition of Kripke model: a Kripke model is now defined as a quintuple $(\mathrm{K}, \leq, \mathrm{D}, \Vdash, \varphi)$, where $(\mathrm{K}, \leq)$ is as before; D assigns inhabited sets to elements of K, and φ is a collection $\{\varphi_{kk'} : k \leq k'\}$, $\varphi_{k,k'} \in \mathrm{D}(k) \to \mathrm{D}(k')$, of *transition functions* (*restrictions*) such that

$$\varphi_{kk} \text{ is the identity on } \mathrm{D}(k),$$

$$\varphi_{k'k''} \circ \varphi_{kk'} = \varphi_{kk''} \quad \text{for } k \leq k' \leq k'',$$

and $\varphi_{kk'}$ preserves forcing, i.e. for prime formulas $R(d_1, \ldots, d_n)$ with $d_1, \ldots, d_n \in \mathrm{D}(k)$

$$k \Vdash R(d_1, \ldots, d_n) \Rightarrow k' \Vdash R(\varphi_{kk'}(d_1), \ldots, \varphi_{kk'}(d_n)).$$

An n-ary function f is interpreted in $\mathscr{K}$ by a family $\mathrm{f} \equiv \{\mathrm{f}_k : k \in \mathrm{K}\}$, the $\mathrm{f}_k \in \mathrm{D}(k)^n \to \mathrm{D}(k)$ commuting with the restrictions for $k' \geq k$

$$\varphi_{kk'}\mathrm{f}_k(d_1, \ldots, d_n) = \mathrm{f}_{k'}(\varphi_{kk'}d_1, \ldots, \varphi_{kk'}d_n).$$

Any Kripke model for a language with equality can be transformed into a Kripke model with transition functions where equality is interpreted at each node as equality: given $(\mathrm{K}, \leq, \mathrm{D}, \Vdash)$, let

$$d^k = d/\sim_k \quad \text{for } d \in \mathrm{D}(k),$$

and put $\mathrm{D}'(k) := \{d^k : d \in \mathrm{D}(k)\}$, $\varphi_{kk'}(d^k) = d^{k'}$. We leave it to the reader to show

$$k \Vdash A(d_1, \ldots, d_n) \Leftrightarrow k \Vdash' A(d_1^k, \ldots, d_n^k).$$

5.13. *Applications to the theory of equality.* We give two examples of the use of Kripke models in the theory of equality. In the examples a, b, c and d are assumed to be distinct. The following nearly trivial model shows that $\nVdash \forall xy(x = y \vee x \neq y)$:

$1 \quad a \sim_1 b$

$0 \quad \mathrm{D}(0) = \mathrm{D}(1) = \{a, b\}$

More interesting is the following proposition, showing that the axiom of choice (equivalently, the assumption of a Skolem function) is not conservative in the presence of equality.

Proposition. *Let $\mathscr{L}$ be the language of predicate logic with equality and a single unary function symbol* f; *let $\mathscr{L}'$ be $\mathscr{L}$ without* f. *Let* **T** *be pure predicate logic with equality axioms* $\forall x(x = x)$, $\forall xyz(x = y \wedge x = z \to z = y)$. *Then* $\mathbf{T}_1 \equiv \mathbf{T} + \{\forall x(\neg x = fx), \forall xy(x = y \to fx = fy)\}$ *is not conservative over* $\mathbf{T} + \{\forall x \exists y(\neg x = y)\} \equiv \mathbf{T}_2$ *relative to* $\mathscr{L}'$.

Proof. We exhibit a Kripke model $\mathscr{K}$ for $\mathbf{T}_2$ in which a theorem of $\mathbf{T}_1$ is not valid: $F := \forall x_1 \exists y_1 \forall x_2 \exists y_2[x_1 \neq y_1 \wedge x_2 \neq y_2 \wedge (x_1 = x_2 \to y_1 = y_2)]$ is not forced at node 0; $\mathscr{K}$ is given by the picture below.

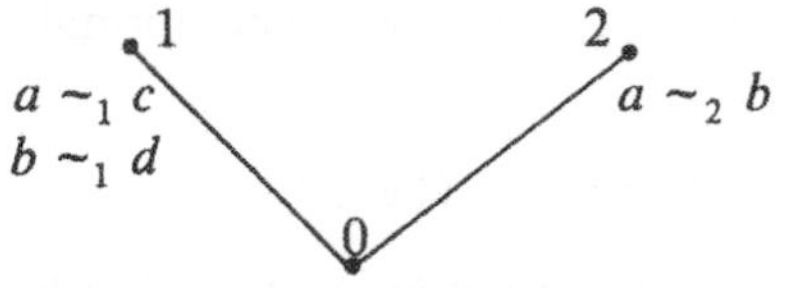

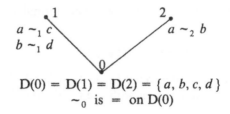

$$D(0) = D(1) = D(2) = \{a, b, c, d\}$$
$$\sim_0 \text{ is } = \text{ on } D(0)$$

We leave the verification as an exercise. □

5.14. *Kripke models for E^+-logic.* The definition in 5.11 above can also be used for E^+-logic. In the classical structures the f_i may now be partial. The strictness axioms determine the interpretation of terms built from elements of $D(k)$ with the help of the function symbols of the language via $Et := \exists x(t = x)$, $(t = s) := \exists y(t = y \wedge s = y)$ and $(f_j\vec{t} = x) := \exists \vec{y}(\vec{t} = \vec{y} \wedge f_j\vec{y} = x)$; the validity of basic relations applied to compound terms is also determined by strictness: $R_i\vec{t} := \exists \vec{y}(\vec{t} = \vec{y} \wedge R_i\vec{y})$.

The soundness theorem extends to E^+-logic without problems (exercise).

For E-logic the formulation of an appropriate version of Kripke semantics is not quite so straightforward, because of the role of the free variables and the presence of the substitution rule.

A sound interpretation is obtained by taking a Kripke model to be a partially ordered collection of classical structures with partial functions

$$\mathscr{K} \equiv \big(\mathrm{K}, \leq, \{(\mathrm{D}(k), \sim_k, \{\mathrm{R}_{i,k} : i \in I\}, \{\mathrm{f}_{j,k} : j \in J\}) : k \in \mathrm{K}\}\big),$$

where now the $D(k)$ may be possibly empty, but where $\mathbf{D} := \bigcup\{D(k) : k \in K\}$ is inhabited, and putting $k \Vdash Ed$ iff $d \in D(k)$. Strictness then enforces that for atomic formulas P with $FV(P) = \{x_1, \ldots, x_n\}$, $k \Vdash P[x_1, \ldots, x_n/d_1, \ldots, d_n]$ implies that $d_1, \ldots, d_n \in D(k)$. For formulas A

with $FV(A) \subset \{x_1, \ldots, x_n\}$ we require

$$k \Vdash A(x_1, \ldots, x_n) := \forall d_1, \ldots, d_n \in \mathbf{D}(k \Vdash A(d_1, \ldots, d_n)).$$

As shown by Unterhalt (1986) E-logic is sound for this semantics, but strong completeness fails for open theories (such as the theory of global existence $\{\boldsymbol{E}x\}$), though strong completeness holds with respect to closed theories.

Strong completeness is recovered if we let the interpretation $\mathbf{E}_k$ of $\boldsymbol{E}$ at k be a possibly empty subset of $D(k)$; strictness then must hold w.r.t. the interpretation of $\boldsymbol{E}$ only. But we shall not pursue the topic of Kripke models for E-logic here.

6. Completeness for Kripke semantics

In this section we present a classical completeness proof (of the "Henkin-type") for **IQC** and **IQCE**$^+$. We shall appeal to completeness in the next section (7.5), but the result proved there can also be obtained by purely syntactical methods (7.13). Semantics for intuitionistic logic will be treated at length in chapter 13.

N.B. The completeness proofs given rely on *classical* metamathematics.

6.1. Definition. Let Γ, Δ be sets of sentences in a given language $\mathscr{L}$. The pair (Γ, Δ) is *consistent* iff there are no finite $\Gamma_0 \subset \Gamma$, $\Delta_0 \subset \Delta$ such that $\vdash \bigwedge\Gamma_0 \to \bigvee\Delta_0$; here we take $\bigwedge\emptyset := \top$, $\bigvee\emptyset := \bot$. Γ is *consistent* iff $(\Gamma, \emptyset)$ is consistent. □

6.2. Definition. Let C be a set of constants; a set of sentences Γ in the language $\mathscr{L}$ is C-*saturated* iff

(i) Γ is consistent,
(ii) $\Gamma \vdash A \Rightarrow A \in \Gamma$,
(iii) $\Gamma \vdash A \vee B \Rightarrow \Gamma \vdash A$ or $\Gamma \vdash B$,
(iv) $\Gamma \vdash \exists x A(x) \Rightarrow$ for some $c \in \mathrm{C}$ $A(c) \in \Gamma$. □

Remark. The appeal to classical logic on the metalevel occurs in the proof of the saturation lemma 6.3 below, as well as in the proof of the principal lemma 6.5 where we argue by contradiction.

Lemma 6.3 below describes how to construct $\mathscr{L}$(C)-saturated extensions for any consistent set of $\mathscr{L}$-sentences in Γ, C a countably infinite set of new constants. A variant of the saturation construction is given in E2.6.3.

6.3. LEMMA (*saturation lemma*). *Suppose* $\Gamma \nVdash A$, Γ, A *in a language* $\mathscr{L}$; *let* $\mathrm{C} = \{c_0, c_1, c_2, \dots\}$ *be a countable set of constants not in* $\mathscr{L}$, *and let* $\mathscr{L}(\mathrm{C})$ *be* $\mathscr{L}$ *extended with* C. *Then there is a* C-*saturated* $\Gamma^\omega \supset \Gamma$, *such that* $\Gamma^\omega \nVdash A$.

PROOF. We obtain Γ^ω as $\bigcup\{\Gamma^k : k \in \mathbb{N}\}$; $\Gamma^0 = \Gamma$. $\Gamma^k \setminus \Gamma^0$ is finite. We give an inductive definition of Γ_k.

Let $g(n)$ be the first i such that c_i does not occur in $\Gamma^n \setminus \Gamma$, and let $\langle B_{i,1} \vee B_{i,2}\rangle_i$ and $\langle \exists x A_i(x)\rangle_i$ enumerate with infinite repetition all disjunctive and existential sentences of $\mathscr{L}(\mathrm{C})$. Suppose Γ^k to have been defined.

Case 1. $k = 2n$, $\Gamma^k \vdash \exists x A_n x$. We put

$$\Gamma^{2n+1} := \Gamma^{2n} \cup \{A_n(c_{g(k)})\};$$

Case 2. $k = 2n+1$, $\Gamma^k \vdash B_{n,1} \vee B_{n,2}$. Put

$$\Gamma^{2n+2} := \Gamma^{2n+1} \cup \{B_{n,i}\},$$

where i is the least of $\{1, 2\}$ such that $\Gamma^{2n+1} \cup \{B_{n,i}\} \nVdash A$.

Case 3. Cases 1 and 2 do not apply: put

$$\Gamma_{k+1} := \Gamma_k.$$

We leave it to the reader to verify that Γ^ω is saturated and that $\Gamma^\omega \nVdash A$. □

REMARK. Strictly speaking, the enumeration of the existential formulas has the form $\langle \exists x_i A_i(x_i)\rangle_i$, since the bound variable also depends on i; this dependence however is irrelevant since we agreed to regard formulas differing only in their bound variables as isomorphic.

An alternative version of the construction uses enumerations of disjunctions and existential statements (without requiring infinite repetitions), and treats at stage $n = 2k$ the first formula $\exists x A_m(x)$ not yet treated, and at stage $n = 2k + 1$ the first disjunction not yet treated. At $n = 2k$, when $\Gamma^n \vdash \exists x A_m(x)$, one adds $A_m(c_i)$ for the first $c_i \in \mathrm{C}$ not occurring in $\Gamma^n \cup \{\exists x A_m(x)\}$. □

6.4. DEFINITION (*canonical model construction*). Let $\mathrm{C}_0, \mathrm{C}_1, \mathrm{C}_2, \mathrm{C}_3, \dots$ be a countable sequence of disjoint countable sets of constants not occurring in $\mathscr{L}$; write C_n^* for $\mathrm{C}_0 \cup \mathrm{C}_1 \cup \cdots \cup \mathrm{C}_n$. Let Γ_0 be any theory (set of sentences) in $\mathscr{L}$. Then we define

$$\mathscr{K} := (\mathrm{K}, \subset, \Vdash, \mathrm{D})$$

such that

(a) K consists of all $\Gamma \supset \Gamma_0$ such that $\mathscr{L}(\Gamma) = \mathscr{L} \cup C_n^*$, Γ saturated w.r.t. C_n^* for some n;

(b) if Γ is C_n^*-saturated and $\mathscr{L}(\Gamma) = \mathscr{L} \cup C_n^*$, then $D(\Gamma) = C_n^*$;

(c) for prime P in $\mathscr{L}(D(\Gamma))$: $\Gamma \Vdash P := P \in \Gamma$. □

6.5. Lemma. *For all $\Gamma \in K$ and each sentence of $\mathscr{L}(D(\Gamma))$*

$$\Gamma \Vdash A \Leftrightarrow A \in \Gamma.$$

Proof. By induction on the complexity of A.

Case 1. For A prime the lemma holds by definition.

Case 2. For $A \equiv B \wedge C$ immediate.

Case 3. Let $A \equiv B \vee C$, then $\Gamma \Vdash B \vee C \Leftrightarrow (\Gamma \Vdash B$ or $\Gamma \Vdash C) \Leftrightarrow (B \in \Gamma$ or $C \in \Gamma) \Leftrightarrow B \vee C \in \Gamma$ (by the saturation of Γ).

Case 4. Let $A \equiv \exists x B(x)$. If $\Gamma \Vdash \exists x B(x)$, then $\Gamma \Vdash B(c)$ for some $c \in D(\Gamma)$; by the induction hypothesis $B(c) \in \Gamma$ for some $c \in D(\Gamma)$, and by the saturation properties of Γ this is equivalent to $\exists x B(x) \in \Gamma$.

Case 5. Let $A \equiv B \rightarrow C$, and suppose $\Gamma \Vdash B \rightarrow C$. Then for all saturated $\Gamma' \supset \Gamma$ we have $\Gamma' \Vdash B \Rightarrow \Gamma' \Vdash C$. Assume $\Gamma \nvdash B \rightarrow C$, then $\Gamma \cup \{B\} \nvdash C$; let Γ' be a saturated extension of $\Gamma \cup \{B\}$ such that $\Gamma' \nvdash C$ (lemma 6.3), then $\Gamma' \Vdash B$ but not $\Gamma' \Vdash C$ (induction hypothesis); this contradicts $\Gamma \Vdash B \rightarrow C$, hence $\Gamma \vdash B \rightarrow C$. The converse is trivial.

Case 6. Let $A \equiv \forall x B(x)$, and suppose $\Gamma \Vdash \forall x B(x)$, $\Gamma \nvdash \forall x B(x)$, Γ C_n^*-saturated. Now, for any $c \in C_{n+1}$, $\Gamma \nvdash B(c)$; by lemma 6.3 we can find a C_{n+1}^*-saturated $\Gamma' \supset \Gamma$ such that $\Gamma' \nvdash B(c)$, hence $\Gamma' \nVdash B(c)$ contradicting $\Gamma \Vdash \forall x B(x)$. The converse is again trivial. □

6.6. Theorem (*strong completeness for* **IQC**).

$$\Gamma \Vdash A \Rightarrow \Gamma \vdash A.$$

Proof. Suppose $\Gamma \nvdash A$, and let Γ_0 be a saturated extension of Γ, $A \notin \Gamma_0$; construct a Henkin model as above, then $\Gamma_0 \nVdash A$. This yields completeness, also for languages with equality, if equality is interpreted at each node by an equivalence relation (cf. 5.11). For models with transition functions as in 5.12 we have to take equivalence classes; the details are left to the reader. □

6.7. THEOREM. *Strong completeness holds also with respect to* **IQCE**$^+$.

PROOF. Exercise. □

The remainder of this section is devoted to certain refinements of the completeness theorem and is not needed in the sequel.

6.8. THEOREM (*transformation of a rooted Kripke model into a tree model*). *Let* $\mathscr{K} \equiv (\mathrm{K}, \leq, \Vdash)$ *be any propositional Kripke model with root (origin)* k_0, *i.e.* $\forall k \in \mathrm{K}(k_0 \leq k)$. *Then there is mapping* φ *from a Kripke model* $\mathscr{K}' \equiv (\mathrm{K}', \leq', \Vdash')$ *to* $\mathscr{K}$, *where* $(\mathrm{K}', \leq')$ *a tree with root* k_0', *such that*
(i) $\varphi(k_0') = k_0$;
(ii) *for all A and* $k' \in K'$: $\varphi(k') \Vdash A$ *iff* $k' \Vdash' A$;
(iii) $k_1' \leq' k_2' \rightarrow \varphi k_1' \leq \varphi k_2'$ (φ *is weakly order preserving*).

PROOF. K′ consists of all finite sequences $\langle k_0, \ldots, k_p \rangle$ with $k_0 < k_1 < \cdots < k_p$; $\sigma \leq' \tau$ iff σ is an initial segment of τ. We define forcing for propositional variables by

$$\langle k_0, \ldots, k_p \rangle \Vdash' P := k_p \Vdash P.$$

By formula induction one then shows

$$\langle k_0, \ldots, k_p \rangle \Vdash' A \Leftrightarrow k_p \Vdash A.$$

We illustrate the construction by an example. Consider e.g. the partially ordered set (K, ≤), as on the left in the picture below; by taking finite, < -increasing sequences of nodes, ordered by the "initial-segment-of" relation, we obtain a tree (K′, ≤′) as shown to the right.

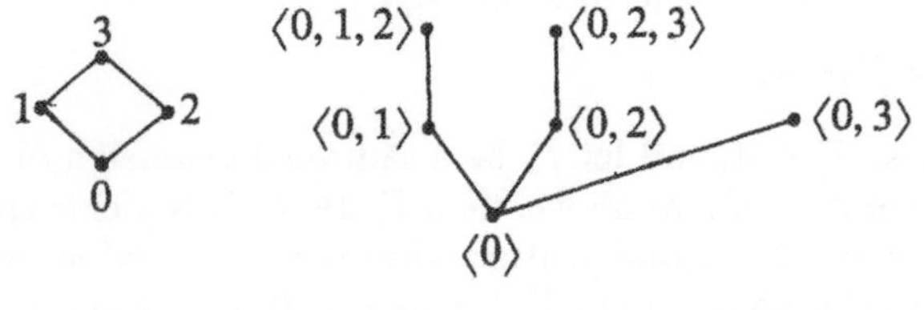

We leave the details of the proof as an exercise. □

6.9. Corollary. *For any [finite] model $\mathscr{K} \equiv (\mathrm{K}, \leq, \Vdash)$ we can construct some [finite] collection $\{\mathscr{K}_i : i \in I\}$ of tree models such that for all sentences in the appropriate language $\mathscr{K} \Vdash A$ iff for all $i \in I\, \mathscr{K}_i \Vdash A$.*

Proof. Apply the preceding theorem to the truncations of $\mathscr{K}$. □

6.10. Remark. (i) The construction permits some variations, for example
(a) We can take finite sequences $\langle k_0, \ldots, k_p \rangle$ such that $k_0 \leq k_1 \leq \cdots \leq k_p$.
(b) For finite $(\mathrm{K}, \leq)$ we can restrict attention to $\langle k_0, \ldots, k_p \rangle$ such that each k_{i+1} is an immediate successor of k_i.
(ii) The construction can also be adapted straightforwardly to predicate logic.

We shall now show that for **IPC** Kripke models on finite trees suffice for completeness.

6.11. Theorem. *Let $\mathscr{K} \equiv (\mathrm{K}, \leq, \Vdash)$ be a Kripke model for* **IPC** *with root $k_0 \in \mathrm{K}$, and suppose $k_0 \nVdash A$. Then there is a finite model $\mathscr{K}^* \equiv (\mathrm{K}^*, \leq^*, \Vdash^*)$ such that $\mathrm{K}^* \subset \mathrm{K}$, $\leq^*$ is the restriction of $\leq$ to* K, *and for all $k \in \mathrm{K}^*$ and all subformulas B of A*

$$k \Vdash^* B \Leftrightarrow k \Vdash B. \tag{1}$$

Proof. The proof proceeds by the so-called "filtration method". We successively choose nodes from K which are needed for making A invalid; the process terminates because A has only finitely many subformulas. Let S_A be the collection of subformulas of A, and put

$$S(k) := \{B \in S_A : k \Vdash B\}.$$

Initially we put

$$\mathrm{K}_0 := \{k_0\}, \qquad \mathrm{K}_{-1} := \emptyset.$$

Suppose K_n to have been defined, and let $k \in \mathrm{K}_n \setminus \mathrm{K}_{n-1}$. Let $t_{k,0}, \ldots, t_{k,n(k)}$ be a maximal subset of K such that for $i, j \leq n(k)$:

$$k \leq t_{k,i};$$

$$S(k) \neq S(t_{k,i});$$

$$k \leq k' \leq t_{k,i} \rightarrow S(k) \neq S(k') \vee S(k') \neq S(t_{k,i});$$

$$i \neq j \rightarrow S(t_{k,i}) \neq S(t_{k,j}).$$

We then put

$$\mathrm{K}_{n+1} := \mathrm{K}_n \cup \{t_{k,i} : (k \in \mathrm{K}_n \setminus \mathrm{K}_{n-1}) \wedge i \le n(k)\}.$$

Clearly, $\mathrm{K}_{n+1} = \mathrm{K}_n$ for sufficiently large n, since S_A is finite and has only finitely many subsets. Therefore we can take

$$\mathrm{K}^* := \bigcup\{\mathrm{K}_n : n \in \mathbb{N}\},$$

$$k \Vdash^* P \quad \text{iff } k \Vdash P \quad \text{for } P \text{ atomic.}$$

We prove (1) for arbitrary subformulas B by induction on the complexity of B. Consider e.g. the case $B \equiv B_1 \to B_2$. $k \Vdash^* B_1 \to B_2$ iff $\forall k' \ge^* k$ $(k' \Vdash^* B_1 \Rightarrow k' \Vdash^* B_2)$ iff $\forall k' \ge^* k(k' \Vdash B_1 \Rightarrow k' \Vdash B_2)$, hence $k \Vdash B_1 \to B_2$ implies $k \Vdash^* B_1 \to B_2$. Conversely, assume $k \Vdash^* B_1 \to B_2$ but $k \nVdash B_1 \to B_2$; then for some $k' \ge k$ $k' \Vdash B_1$ and $k' \nVdash B_2$. Suppose $k \in \mathrm{K}_i \setminus \mathrm{K}_{i-1}$. Then there must be a $t_{k,j} \in \mathrm{K}_{i+1} \setminus \mathrm{K}_i$ (why?) with $t_{k,j} \nVdash^* B_1 \to B_2$ and so $k \nVdash^* B_1 \to B_2$; contradiction. We leave the other cases to the reader. □

We obtain now as an immediate consequence:

6.12. THEOREM.
(i) *If* **IPC** $\nvdash A$, *then* A *is not valid on a Kripke model over some finite tree.*
(ii) **IPC** *is decidable.*

PROOF. (i) follows immediately from the preceding theorem together with 6.6 and 6.8.

(ii) By (i) we can enumerate all possible countermodels, that is to say we can in fact enumerate all Kripke-invalid formulas. We can also enumerate all valid (= derivable) formulas. With classical logic, every formula is either Kripke-valid or has a Kripke countermodel. Thus we can simultaneously enumerate valid and invalid formulas; each formula will occur as either valid or invalid. We conclude that **IPC** is decidable. □

REMARK. It is possible to remove the classical steps in the argument, for example via a formalization of the completeness theorem (see Smorynski 1982). For a direct syntactical proof of (ii) see section 10.4.

7. Definitional extensions

In this section we shall discuss several types of definitional extensions of first order theories:

(a) the addition of definable predicates,
(b) the addition of new variables for a definable domain,
(c) the addition of definable partial or total functions and the addition of description operators.

7.1. DEFINITION. Let $\mathbf{T}, \mathbf{T}'$ be two theories, $\mathbf{T}$ formulated in a language $\mathscr{L}$, $\mathbf{T}'$ in a language $\mathscr{L}'$, $\mathscr{L} \subset \mathscr{L}'$. $\mathbf{T}'$ is a *definitional extension* of $\mathbf{T}$, if there is a mapping φ from the formulas of $\mathscr{L}'$ to the formulas of $\mathscr{L}$ such that

(i) $\varphi(A) \equiv A$ for all formulas A of $\mathscr{L}$;
(ii) $\mathbf{T}' \vdash \varphi(A) \leftrightarrow A$ for formulas A of $\mathscr{L}'$;
(iii) $\mathbf{T}' \vdash A \Rightarrow \mathbf{T} \vdash \varphi(A)$;
(iv) φ commutes with all logical operations of $\mathscr{L}$.

In particular, $\mathbf{T}'$ is a *conservative extension* of $\mathbf{T}$, i.e.

$$\mathbf{T}' \vdash A \Rightarrow \mathbf{T} \vdash A \quad \text{for all formulas } A \text{ of } \mathscr{L}. \quad \square$$

7.2. THEOREM. *Let* $\mathbf{T}$ *be a theory in a first-order language* $\mathscr{L}$, *and let* $\mathbf{T}'$ *be an extension of* $\mathbf{T}$ *to a language* $\mathscr{L}' = \mathscr{L} \cup \{M\}$, *where* M *is an n-ary predicate constant, with an axiom*

$$A(x_1, \ldots, x_n) \leftrightarrow M(x_1, \ldots, x_n)$$

for a formula A *with* $\mathrm{FV}(A) \subset \{x_1, \ldots, x_n\}$, A *a formula of* $\mathscr{L}$. *Then* $\mathbf{T}'$ *is a definitional extension of* $\mathbf{T}$.

PROOF. Exercise. $\square$

7.3. THEOREM. *Let* $\mathbf{T}$ *be as before, and let* $\mathbf{T}'$ *have a new sort of variables (say* a, b, c*) with quantifiers* $\forall a, \exists a$; *let* A *be a definable unary predicate* $\mathbf{T}$ *such that* $\mathbf{T} \vdash \exists x A(x)$ *and let* $\mathbf{T}'$ *contain the additional axioms*

$$\forall a \exists x (a = x \wedge A(x)),$$

$$\forall x (A(x) \rightarrow \exists a (x = a)).$$

Then $\mathbf{T}'$ *is a definitional extension of* $\mathbf{T}$.

PROOF. Exercise. $\square$

7.4. *Addition of function symbols and description operator.* This is more complicated. We consider theories based on E-logic or E^+-logic with equality, and strictness for functions and predicates. We use f for function symbols and write $\vec{t}$ for a sequence of terms $t_1, \ldots, t_n$, $P\vec{t}$ for $P(t_1, \ldots, t_n)$, $\vec{s} = \vec{t}$ for $s_1 = t_1 \wedge \cdots \wedge s_n = t_n$, $f\vec{t}$ for $f(t_1, \ldots, t_n)$. Convenient axiomatizations for our purpose are the systems $\mathbf{H_4}$-**IQCE**$^+$ and $\mathbf{H_4}$-**IQCE** described in 4.7. In this section we shall use $\mathbf{L}^+$ for **IQCE**$^+$, where free variables are assumed to exist, and **L** for **IQCE**.

7.5. Theorem. *The addition of a description operator to* **L** *or* $\mathbf{L}^+$ *is a definitional extension.*

Proof. We give a semantical proof for the case of $\mathbf{L}^+$. Consider any Kripke model, say $\mathscr{K} \equiv (\mathrm{K}, \leq, \Vdash, \mathrm{D})$. We can interpret the descriptor $\boldsymbol{I}x.A(x, y)$ such that the axiom DESCR holds, as follows.

A may itself contain descriptions $\boldsymbol{I}x.B(x)$, $\boldsymbol{I}x.C(x)$ etc. We assume these to have been interpreted already such that DESCR holds, (induction hypothesis). We simply put

$$k \Vdash d = \boldsymbol{I}x.A(x, d_1, \ldots, d_n)$$

for $d_1, \ldots, d_n, d \in \mathrm{D}(k)$, iff $k \Vdash A(d, d_1, \ldots, d_n) \wedge \forall x(A(x, d_1, \ldots, d_n) \leftrightarrow x = d)$, and the validity of DESCR is immediate. This shows that the addition of DESCR is conservative, as one can easily see.

It is now not hard to construct a mapping φ showing that the addition of DESCR is in fact a definitional extension.

First we reduce arbitrary prime formulas $t = s$, $P(t_1, \ldots, t_n)$ to prime formulas of the forms $x = t$, $P(y_1, \ldots, y_n)$ by repeated use of the equivalences

$$t = s \leftrightarrow \exists y(y = t \wedge y = s),$$

$$P(t_1, \ldots, t_n) \leftrightarrow \exists y_1 \ldots y_n(y_1 = t_1 \wedge \cdots \wedge y_n = t_n \wedge P(y_1, \ldots, y_n)),$$

(y not free in t, s; y_i not free in $t_1, \ldots, t_n$). We now eliminate the descriptor from $y = t$ by repeated use of

$$y = \boldsymbol{I}x.A(x) \leftrightarrow \forall x[Ax \leftrightarrow x = y],$$

$$y = f(\vec{t}) \leftrightarrow \exists \vec{y}[z = f(\vec{y}) \wedge \vec{y} = \vec{t}].$$

By strictness and DESCR, the result is a mapping φ such that $\vdash \varphi(A) \leftrightarrow A$ for all A.

For **L** a similar argument (Unterhalt 1986) is possible but we shall not carry this out. □

7.6. COROLLARY. *Let* **T** *be any closed theory based on* **L** *or* $\mathbf{L}^+$, *and suppose* **T** *to be axiomatized by closed axioms and axiom schemas (not necessarily closed); if the axiom schemas translate under* φ *into instances or consequences of the axioms and axiom schemas, the addition of a descriptor to* **T** *is a definitional extension.* □

A typical example of such a theory **T** is **HA**.

The preceding proofs of 7.5 and 7.6 are classical, inasmuch we relied on the classically established completeness proof for intuitionistic predicate logic. Below, we shall present a straightforward syntactical proof.

7.7. NOTATION. We shall use the following abbreviation

$$\exists!\vec{x}B := \exists\vec{x}\forall\vec{z}(B[\vec{x}/\vec{z}] \leftrightarrow \vec{z} = \vec{x}). \quad \Box$$

Warning. Do not confuse $\exists!x_1x_2$ and $\exists!x_1\exists!x_2$!

7.8. LEMMA.

(i) $\exists!\vec{x}B \leftrightarrow \exists\vec{x}B \wedge \forall\vec{x}\vec{z}(B \wedge B[\vec{x}/\vec{z}] \rightarrow \vec{z} = \vec{x})$,

(ii) *Let* $\vec{x}'$ *be some permutation of* $\vec{x}$. *Then* $\exists!\vec{x}B \leftrightarrow \exists!\vec{x}'B$.

(iii) *Let* $\vec{u} \cap \mathrm{FV}(B) = \emptyset$, $\vec{x} \cap \mathrm{FV}(C) = \emptyset$, *then*

$$\exists!\vec{x}\vec{u}(B \wedge C) \leftrightarrow \exists!\vec{x}B \wedge \exists!\vec{u}C.$$

(iv) *Let* $\vec{x} \cap (\mathrm{FV}(C) \cup \mathrm{FV}(D)) = \emptyset$; *then*

$$\exists!\vec{x}B \rightarrow (\exists\vec{x}(B \wedge C) \wedge \exists\vec{x}(B \wedge D) \leftrightarrow \exists\vec{x}(B \wedge C \wedge D)).$$

PROOF. Exercise. □

7.9. DEFINITION. Let $A(\vec{x}, y)$ be a formula of **L** ($\mathbf{L}^+$) with $\vec{x}, y \in \mathrm{FV}(A)$. Then $\mathbf{L}(\varphi)$ ($\mathbf{L}^+(\varphi)$) is obtained from **L** ($\mathbf{L}^+$) by adding a function symbol φ with axiom

$$\mathrm{AX}(A, \varphi) \quad \forall\vec{x}y(\forall z(A(\vec{x}, z) \leftrightarrow y = z) \leftrightarrow \varphi\vec{x} = y)$$

and extending the axioms and rules of **L** ($\mathbf{L}^+$) to the new language. □

7.10. DEFINITION. We define a mapping * from $\mathbf{L}(\varphi)$ into **L**, and from $\mathbf{L}^+(\varphi)$ into $\mathbf{L}^+$, as follows.

For notational simplicity, let φ be unary, i.e. $\vec{x} \equiv x$. All formulas of $\mathbf{L}(\varphi)$ are supposed to be given with φ-indexing, that is to say in any formula B we assume the occurrences of φ to be indexed in such a way that indices occurring in the same prime formula occurrence are all distinct.

The definition of * is such that B^* does not depend, except for the renaming of bound variables, on the choice of indexing. To indicate an occurrence of φ with index i we write φ_i.

Below the variables $y_0, y_1, \ldots$ are always supposed to be fresh; $\vec{y}, \vec{y}', \ldots$ are sequences of fresh variables. In writing $\exists \vec{y} B$, $\exists \vec{y}' B^*$ etc. the $\vec{y}, \vec{y}'$ are tacitly assumed to contain all y_i's free in B.

We first define two auxiliary functions $\underline{\ \ }$, δ on terms, and terms plus formulas respectively. The *underlining operator* $\underline{\ \ }$ is specified by

$$\underline{t} := t \quad \text{for } \varphi\text{-free } t;$$

$$\underline{\varphi_i t} := y_i;$$

$$\underline{f(t_1, \ldots, t_n)} := f(\underline{t_1}, \ldots, \underline{t_n}) \quad \text{for function symbols } f \text{ of } \mathbf{L}.$$

The *degree* δ is given by

$$\delta(t) := 0 \quad \text{for } \varphi\text{-free } t;$$

$$\delta(\varphi_i t) := \delta(t) + 1;$$

$$\delta(f(t_1, \ldots, t_n)) := \max(\delta(t_1), \ldots, \delta(t_n)) \ (\delta(\vec{t}\,) \text{ for short});$$

$$\delta(P(t_1, \ldots, t_n)) := \delta(\vec{t}\,), \delta(\bot) := 0;$$

$$\delta(B \circ C) := \max(\delta(B), \delta(C)) \quad \text{for } \circ \in \{\wedge, \vee, \rightarrow\};$$

$$\delta(\forall x B) = \delta(\exists x B) := \delta(B).$$

In words: $\delta(B)$ is the maximum nesting depth for occurrences of φ in terms occurring in A. We now define simultaneously ε and *:

$$\varepsilon(t) := \top \quad \text{if } t \text{ is } \varphi\text{-free (e.g. } \top \equiv \forall x(x = x));$$

$$\varepsilon(\varphi_i t) := A(t, y_i)^* \wedge Et^*;$$

$$\varepsilon(f(t_1, \ldots, t_n)) := \varepsilon(t_1) \wedge \cdots \wedge \varepsilon(t_n) \ (\varepsilon(\vec{t}\,) \text{ for short});$$

$$P(t_1, \ldots, t_n)^* := \exists! \vec{y} \varepsilon(\vec{t}\,) \wedge \exists \vec{y}(\varepsilon(\vec{t}\,) \wedge P\underline{\vec{t}}), \quad \text{and}$$

$$* \text{ commutes with } \wedge, \vee, \rightarrow, \forall, \exists.$$

By induction on $\delta(B)$ one readily shows that ε and * are well-defined (note that A is φ-free, so $\delta(A(t, y_i)) = \delta(Et) = \delta(\varphi_i t) - 1$). □

7.11. LEMMA. *For all* B *in the language of* $\mathbf{L}$:
(i) B^* *is* φ*-free*;
(ii) $\mathbf{L} \vdash B^* \leftrightarrow B$ *for* φ*-free* B;
(iii) $\mathbf{L}(\varphi) \vdash (E\varphi_i t)^* \leftrightarrow \exists! y_i A(t, y_i)^* \wedge (Et)^*$;
(iv) $\mathbf{L}(\varphi) \vdash (\varphi_i t = x)^* \leftrightarrow \forall y_i(A(t, y_i)^* \leftrightarrow x = y_i) \wedge (Et)^*$,

and similarly with $\mathbf{L}^+$ *instead of* $\mathbf{L}$.

PROOF. Exercise. □

7.12. LEMMA. $\mathbf{L}^+(\varphi)$ *and* $\mathbf{L}(\varphi)$ *are definitional extensions of* $\mathbf{L}^+, \mathbf{L}$ *respectively, via the mapping* $*$.

PROOF. We have to show
(i) $\mathbf{L}(\varphi) \vdash B \leftrightarrow B^*$;
(ii) $\mathbf{L}^+(\varphi) \vdash B \Rightarrow \mathbf{L}^+ \vdash B^*$;
(iii) $\mathbf{L}(\varphi) \vdash B \Rightarrow \mathbf{L} \vdash B^*$.

To keep the notation simple, we shall assume P, f below to be unary; $\vec{y}, \vec{y}'$ are the y-variables of $\underline{t}$ and $\underline{s}$ respectively. We use $\Rightarrow$, $\Leftrightarrow$ for derivable implications and equivalences.

(i) can be proved by induction over $\delta(B)$. If $\delta(B) = 0$, B is φ-free, and we can use 7.11(ii). Now let $\delta(B) > 0$; we first show for all t with $\delta(t) \leq \delta(B)$

$$\mathbf{L}(\varphi) \vdash t = x \leftrightarrow (t = x)^*. \tag{1}$$

Case (a). t a variable; trivial.

Case (b). $t \equiv fs$. Then $\varepsilon(t) = \varepsilon(s)$, $\underline{t} = f\underline{s}$, and $fs = x \leftrightarrow \exists z(s = z \wedge fz = x)$ (STR, EAX) $\leftrightarrow \exists z((s = z)^* \wedge fz = x)$ (application of (1) with $t \equiv s$, by the induction hypothesis) $\equiv \exists z(\exists! \vec{y}\varepsilon(s) \wedge \exists \vec{y}(\varepsilon(s) \wedge \underline{s} = z) \wedge fz = x) \leftrightarrow \exists! \vec{y}\varepsilon(s) \wedge \exists \vec{y}(\varepsilon(s) \wedge f\underline{s} = x) \equiv (fs = x)^*$.

Case (c). $t \equiv \varphi_i s$. Then $\delta(s) = \delta(Es) = \delta(A(s, y_i))$, and $\delta(s) < \delta(t) \leq \delta(B)$. Thus $\varphi_i s = x \Leftrightarrow \forall y_i(A(s, y_i) \leftrightarrow x = y_i) \wedge Es$ (by $AX(A, \varphi)$) $\Leftrightarrow \forall y_i(A(s, y_i)^* \leftrightarrow x = y_i) \wedge Es^* \wedge Ex$ (induction hypothesis) $\Leftrightarrow (\varphi_i s = x)^*$ (by 7.11(iv)).

Now we can prove $B \leftrightarrow B^*$. For B prime, assume $B \equiv Pt$, $x \notin \mathrm{FV}(t)$. Then $Pt \Leftrightarrow \exists x(Px \wedge x = t)$ (by STR, EAX) $\Leftrightarrow \exists x(Px \wedge (x = t)^*)$ (by (1)) $\equiv \exists x(Px \wedge \exists! \vec{y}\varepsilon(t) \wedge \exists \vec{y}(\varepsilon(t) \wedge x = \underline{t})) \Leftrightarrow \exists! \vec{y}(\varepsilon(t) \wedge \exists! \vec{y}(\varepsilon(t) \wedge P\underline{t}) \equiv (Pt)^*$. For B not prime, we observe that $*$ commutes with all logical operators.

(ii) Since * commutes with all logical operators, all axioms and rules of **IQC** are translated into instances of themselves; we only have to look at EAX, STR, REPL, AX(A, φ).

Case (d). EAX: $(Et)^* \equiv (\exists!\vec{y}\varepsilon(t) \wedge \exists\vec{y}(\varepsilon(t) \wedge E\underline{t})) \Leftrightarrow (\exists!\vec{y}\varepsilon(t) \wedge \exists\vec{y}(\varepsilon(t) \wedge \exists x(x = \underline{t})))$(EAX) $\Leftrightarrow \exists x(\exists!\vec{y}\varepsilon(t) \wedge \exists\vec{y}(\varepsilon(t) \wedge x = \underline{t})) \equiv (\exists x(x = t))^*$.

Case (e). STR: $P(t)^* \equiv (\exists!\vec{y}\varepsilon(t) \wedge \exists\vec{y}(\varepsilon(t) \wedge P\underline{t})) \Rightarrow \exists!\vec{y}\varepsilon(t) \wedge \exists\vec{y}(\varepsilon(t) \wedge E\underline{t})$ (by STR) $\equiv (Et)^*$. One similarly proves $(E(ft))^* \to (Et)^*$. $(E(\varphi_i t))^* \to (Et)^*$ follows from 7.11(iv).

Case (f). REPL: $(Ps)^* \wedge (s = t)^* \equiv [\exists!\vec{y}'\varepsilon(s) \wedge \exists\vec{y}'(\varepsilon(s) \wedge P\underline{s}) \wedge \exists!\vec{y}\vec{y}'(\varepsilon(s) \wedge \varepsilon(t) \wedge \exists\vec{y}\vec{y}'(\varepsilon(s) \wedge \underline{s} = \underline{t})] \Rightarrow [\exists!\vec{y}'\varepsilon(s) \wedge \exists!\vec{y}\varepsilon(t) \wedge \exists\vec{y}'(\varepsilon(s) \wedge P\underline{s} \wedge \exists\vec{y}(\varepsilon(t) \wedge \underline{s} = \underline{t}))]$ (7.8(iii), (iv)) $\equiv (Pt)^*$. Quite similarly one proves

$$(Efs)^* \wedge (s = t)^* \to (fs = ft)^*$$

The remaining troublesome case is

$$\text{REPL}(\varphi_i) \equiv E(\varphi_i s) \wedge s = t \to \varphi_i s = \varphi_i t.$$

Observe that $Es \wedge s = t \to Et$ and $A(s, y_i) \wedge s = t \to A(t, y_i)$ are in fact provable in $\mathbf{L}^+(\varphi) \setminus \text{REPL}(\varphi_i)$. Now $[(E(\varphi_i s)^* \wedge (s = t)^*] \Leftrightarrow [\exists!y_i(A(s, y_i)^* \wedge (Es)^* \wedge (s = t)^*)]$ (by 7.11(iii)) $\Rightarrow [\exists!y_i(A(s, y_i)^* \wedge (Es)^* \wedge \exists!y_j A(t, y_j)^* \wedge (Et)^* \wedge \exists y_i y_j(A(s, y_i)^* \wedge A(t, y_j)^* \wedge y_i = y_j)] \Leftrightarrow [\exists!y_i\varepsilon(\varphi_i s) \wedge \exists!y_j\varepsilon(\varphi_j t) \wedge \exists y_i y_j(A(s, y_i)^* \wedge (Es)^* \wedge A(t, y_j)^* \wedge Et)^* \wedge y_i = y_j)] \equiv (\varphi_i s = \varphi_j t)^*$.

(iii) Completely similar. Use $(Ex)^* \leftrightarrow Ex$ to deal with the quantifier rules and axioms of **L**. □

7.13. *Syntactic proof of theorem 7.5.* Any proof in **T** + DESCR is actually carried out in a finite subsystem, where the descriptions $Ix.A_1(x)$, $Ix.A_2(x)$, ... used may be assumed to be such that $A_j(x)$ contains only $Ix.A_i(x)$ for $i < j$. Then, identifying each $Ix.A_i(\vec{y}, x)$ with a function $\varphi_i(\vec{y})$, we may think of the finite subsystem as being obtained by successively adding function symbols $\varphi_1, \varphi_2, \varphi_3, \ldots$ with axioms AX(A, φ_i); the AX(A, φ_i) coincide with DESCR for the corresponding description terms. □

Observe that we now also have a syntactic proof of 7.6.

7.14. *The descriptor in languages with function variables.* We now consider languages which have function variables $\alpha, \beta, \gamma, \ldots$ besides individual variables, with quantifier rules and axioms based on $\mathbf{H}_4$-**IQCE**$^+$, i.e.

$$\forall \alpha A \to A[\alpha/\beta], A[\alpha/\beta] \to \exists \alpha A;$$

$$\left.\begin{array}{l} A \to B \Rightarrow A \to \forall \beta B[\alpha/\beta]; \\ B \to A \Rightarrow \exists \beta B[\alpha/\beta] \to A. \end{array}\right\} \alpha \notin \mathrm{FV}(A), \alpha \equiv \beta \text{ or } \beta \notin \mathrm{FV}(B).$$

We consider logics **LFT** and **LFP**, based on **L**, with function variables for total and partial functions respectively. For **LFT** we add to two-sorted **L**

FAXT $\quad [\forall x E(t(x)) \to \exists \alpha \forall x(\alpha x = t(x))] \wedge \forall \alpha \forall x E(\alpha x)$

and for **LFP** we add to **L**

FAXP $\quad \exists \alpha \forall xy(t(x) = y \leftrightarrow \alpha x = y).$

In the presence of a description operator we find, taking $Iy.A(x, y)$ for $t(x)$ in FAXT and FAXP

AC!. $\quad \forall x \exists ! y A(x, y) \to \exists \alpha \forall x A(x, \alpha x)$

in **LFT**, and

APC! $\quad \exists \alpha \forall xy(\forall z(A(x, z) \leftrightarrow z = y) \leftrightarrow \alpha x = y)$

in **LFP**. Instead of adding a descriptor we may equivalently consider theories with successive addition of function symbols.

7.15. THEOREM.
(i) **LFT** + DESCR + AC! *is a definitional extension of* **LFT** + AC!
(ii) **LFP** + DESCR + APC! *is a definitional extension of* **LFP** + APC!

PROOF. (i) We have to extend the definition of $\underline{\ \ }$, δ and ε by

$$\underline{\alpha t} := \alpha \underline{t},\ \delta(\alpha t) := \delta(t),\ \delta(\forall \alpha B) = \delta(\exists \alpha B) := \delta(B),\ \varepsilon(\alpha t) := \varepsilon(t).$$

We only have to check

$$\mathbf{LFT} + \mathrm{AC!} \vdash \mathrm{FAXT}^*.$$

In **LFT** + DESCR − FAXT we can derive $Et \to \exists ! z(z = t)$, and thus

$$\mathbf{LFT} + \mathrm{AC!} - \mathrm{FAXT} \vdash (Et)^* \to (\exists ! z(z = t))^*.$$

Now $[(\forall x E(t))^*] \to [(\forall x \exists ! z(z = t))^*] \equiv [\forall x \exists ! z(\exists ! \vec{y} \varepsilon(t) \wedge \exists \vec{y}(\varepsilon(t) \wedge z = t))] \to [\exists \alpha \forall x(\exists ! \vec{y}\varepsilon(t) \wedge \exists \vec{y}(\varepsilon(t) \wedge \alpha x = t))]$ (by AC!) $\equiv [\exists \alpha \forall x (\alpha x = t)^*]$, and hence FAXT* since $(\forall \alpha \forall x E(\alpha x))^* \equiv \forall \alpha \forall x E(\alpha x)$.

(ii) We have to check

$$\mathbf{LFP} + \mathrm{APC!} \vdash \mathrm{FAXP}^*.$$

Observe **LFP** + DESCR − FAXP $\vdash t = u \leftrightarrow \forall z(t = z \to z = u)$, so

$$\mathbf{LFP} + \mathrm{APC!} - \mathrm{FAXP} \vdash (t = u)^* \leftrightarrow (\forall z(t = z \leftrightarrow z = u))^*.$$

Now $\mathrm{FAXP}^* \equiv [\exists \alpha \forall x u(\alpha x = u \leftrightarrow t = u)]^* \leftrightarrow [\exists \alpha \forall x u(\alpha x = u \leftrightarrow \forall z(t = z \leftrightarrow z = u)^*)] \equiv [\exists \alpha \forall x u(\alpha x = u \leftrightarrow \forall z(\exists ! \vec{y}\varepsilon(t) \wedge \exists \vec{y}(\varepsilon(t) \wedge t = z) \leftrightarrow z = u))]$ which is an instance of APC! □

7.16. Corollary. *Let* **T** (**T′**) *be a theory based on* **LFT** (**LFP**), *plus closed axioms and axiom schemas which translate under * into consequences of the axioms and axiom schemas. Then* **T** + DESCR + AC! (**T** + DESCR + APC!) *is a definitional extension of* **T** (**T′**). □

8. Notes

8.1. There is extensive literature on intuitionistic propositional and predicate logic. Books in which a substantial part of the contents are devoted to **IPC** and **IQC** are a.o. Dummett (1977), Fitting (1969), Gabbay (1981), Prawitz (1965), Schütte (1968); the paper van Dalen (1986) gives a survey. Further information on intuitionistic predicate and its semantics is to be found in chapters 10 and 13.

8.2. The earliest published complete formalization of **IPC** and **IQC** is found in Heyting (1930, 1930A) (some historical comments on the significance of Heyting (1930, 1930A, 1930B) are in Troelstra (1978, 1981A). Kolmogorov (1925) and Glivenko (1928) contain partial axiomatizations of minimal logic. Johannson (1937) contains a formalism equivalent to **MQC**.

8.3. *Natural deduction.* The natural deduction systems NJ and NK for intuitionistic and classical predicate logic were introduced by Gentzen (1935). Gentzen has both $\perp$ and $\neg$ as primitives. For $\neg$ Gentzen adopts the rules

$$\begin{array}{c}[A]\,(n)\\ \mathscr{D}\\ \bot\\ \hline \neg A\end{array}\,(n) \qquad \frac{A \quad \neg A}{\bot}$$

In other respects the system is identical with the one presented here. For the classical system NK Gentzen has in addition axioms $A \vee \neg A$. The versions used in this book are identical with the systems as presented in Prawitz (1965).

8.4. *E-logic and E$^+$-logic.* Heyting (1930A) contains an early, only partially successful attempt to deal with partially defined terms.

E-logic as presented here is equivalent to the system for first-order logic in Scott (1979); Scott was the first to stress the importance of a logic with partially defined terms and existence predicate, in formalizing mathematical theories based on constructive logic.

E$^+$-logic is used, for example, in Beeson (1985), and seems to be the more obvious choice of a logical basis if we have to deal with first-order axiomatizations of inhabited domains with possibly partial operations, such as the combinatory algebras studied in chapter 9.

8.5. *The "negative translation".* The translation ^{k} is found in Kolmogorov (1925) and is there applied to an incomplete formalization of intuitionistic predicate logic. The paper remained almost unobserved for a long time (presumably because it was in Russian); it is missing, for example, in the otherwise rather complete bibliography in Heyting (1934). The final section of Kolmogorov's paper makes it clear that Kolmogorov realized the possibility of applying this translation to mathematical theories. For some comments on Kolmogorov's paper, see H. Wang's "Introductory Note" to the English translation of the paper in van Heijenoort (1967).

Gödel (1933) and Gentzen (1933), independently, gave an embedding of classical arithmetic into intuitionistic arithmetic; apparently neither of them knew of Kolmogorov's paper. Gentzen's translation coincides with the translation ^{g} (the choice of ^{g} is to remind us of Gödel–Gentzen) as given here, except that the prime formulas are left unchanged. Gödel's translation has for implication $(A \to B)' := \neg(A' \wedge \neg B')$. Gentzen's paper was written in 1933, but withdrawn in the galley proof stage when he learnt of Gödel's result. The formulation of the conservative extension result for arithmetic corresponding to 3.6 is due to Bernays (in Gentzen 1933).

Kleene (1952, section 8.1) treats the Gödel and Gentzen versions of the translation at length, also for predicate logic, which requires the insertion of $\neg\neg$ in front of prime formulas.

For more historical comment see e.g. the introductory note to Gödel (1933) in his Collected Works (1986).

A number of papers in the literature contain extensions of the conservative extension result in 3.6, such as Mints and Orevkov (1967, our 3.26(ii)), Cellucci (1969, our 3.26(iii)) and Leivant (1971, our 3.20(iv)). Leivant (1985) contains another systematization of the conservative extension results, not only for predicate logic but also for mathematical theories. Our exposition in 3.11–26 is based on Leivant (1985) and (1971); 3.26(iv) is a combination of results from both papers.

8.6. *Hilbert-type systems.* The term "Hilbert type system" is used for example by Kleene (1952), and presumably originates in the side-by-side comparison of the calculi NJ, NK, LJ, LK, and LHJ, LHK ("a logistic calculus according to Hilbert and Glivenko") in Gentzen (1935). The distinctive feature of the Hilbert-type systems as understood here is that they characterize derivability without assumptions by means of axioms and rules, in contrast to sequent calculi such as LJ, LK. The name "Hilbert-type system" is historically not quite correct as a designation of this type of system, since many authors, following Frege, gave axiomatizations of this type.

It is to be noted that the deduction theorem $\Gamma, A \vdash B \Rightarrow \Gamma \vdash A \rightarrow B$ can also be obtained under the following more liberal version of $\forall$I: $\Gamma, A \vdash C \Rightarrow \Gamma, A \vdash \forall x C$ if $x \notin \mathrm{FV}(A)$ (Kleene 1952, section 22: deductions with variables held constant w.r.t. the assumption A).

8.7. *Kripke semantics.* This semantics (Kripke 1965) for **IQC** has its roots in Kripke's earlier "possible world semantics" (1963) for modal logic, in combination with the well-known embedding of **IQC** in quantificational modal logic **QS4** (cf. the introductory note to Gödel (1933A) in the Collected Works (1986)). The embedding has recently been extended to mathematical theories beyond logic; see e.g. the volume Shapiro (1985), and the papers by Flagg (1986), Flagg and Friedman (1986) and Scedrov (1986).

The proof of the completeness theorem in 6.2–6 is of the Henkin type; several authors have given such a proof, e.g. Aczel (1968), Thomason

(1968), Fitting (1969), Luckhardt (1970). Luckhardt's proof is related to the proof given by Schütte (1968, section 15) for **IPC**; see also Troelstra (1978A) and E2.6.3.

Unterhalt (1986) studied very thoroughly the completeness of E-logic and E^+-logic relative to Kripke semantics. The interpretation indicated in 5.14 for E-logic is not strongly complete as the following example of Unterhalt shows. Consider the open theory $\{Ex\}$ (the theory of global existence). If $\mathscr{K}$ is a model of $\{Ex\}$ then $\forall d \in \mathbf{D} \forall k(k \Vdash Ed)$, hence $\forall d \in \mathbf{D} \forall k(d \in \mathrm{D}(k))$, which makes D a constant function; but in Kripke models with constant domain $(\forall x R(x) \vee P) \leftrightarrow \forall x(R(x) \vee P)$ holds, (E2.5.6) while this schema is false for Kripke models with variable domains (5.9).

E-logic is slightly more general in permitting empty domains, which may be an advantage if we are working in a many-sorted theory in which some of the domains are possibly empty (e.g. type theory in the version described in 3.9.5–8).

8.8. Our sources for section 7 are Kleene (1952), section 74, and Renardel de Lavalette (1984) (also as chapter 1 of Renardel de Lavalette 1984A). The latter source also contains references to the literature. The syntactic method of proof (7.5–13) extends the syntactic treatment in Kleene (1952).

Exercises

2.1.1. Construct natural deduction trees (without short-cuts) showing the derivability of

$$(A \to B) \to (\neg B \to \neg A);$$

$$\neg\neg(A \to B) \to (A \to \neg\neg B);$$

$$\neg\neg(A \wedge B) \leftrightarrow \neg\neg A \wedge \neg\neg B, (\neg A \wedge \neg B) \leftrightarrow \neg(A \vee B);$$

$$[A \to ((B \vee (A \to C))] \to [A \to (B \vee C)];$$

$$[A \to (B \to C)] \leftrightarrow [(A \to B) \to (A \to C)];$$

$$(A \wedge B \to C) \leftrightarrow (A \to (B \to C)).$$

2.1.2. Construct natural deduction trees (without short-cuts) showing the derivability of

$$\neg\exists x A(x) \leftrightarrow \forall x \neg A(x);$$

$$\neg\forall x \neg A(x) \to \neg\neg\exists x A(x);$$

$$A \vee \forall x B(x) \to \forall x(A \vee B(x)) \quad (x \notin \mathrm{FV}(A));$$

$$\exists x A(x) \vee \exists y B(y) \leftrightarrow \exists x(A(x) \vee B(x)) \quad (x \notin \mathrm{FV}(B));$$

$$\exists x A(x) \wedge \exists y B(y) \leftrightarrow \exists x \exists y(A(x) \wedge B(y)) \quad (x \notin \mathrm{FV}(B), y \notin \mathrm{FV}(A));$$

$$(A \to \forall x B(x)) \leftrightarrow \forall x(A \to B(x)) \quad (x \notin \mathrm{FV}(A));$$

$$(\exists x A \to B) \leftrightarrow \forall x(A \to B) \quad (x \notin \mathrm{FV}(B)).$$

2.1.3. Give a deduction tree for $(A \to \neg\neg B) \to (\neg\neg A \to \neg\neg B)$ without using derived rules such as R, R′ in 1.5.

2.1.4. Prove, for A with $x \notin \mathrm{FV}(A)$:

$$(A \vee \neg A) \wedge (A \vee \forall x B(x)) \leftrightarrow \forall x((A \vee \neg A) \wedge (A \vee B(x))).$$

2.1.5. A formula is said to be *prenex* if it consists of a quantifier-free formula preceded by a string of quantifiers. A formula A *majorizes* B iff

$$\mathbf{CQC} \vdash A \leftrightarrow B, \quad \mathbf{IQC} \vdash A \to B.$$

Prove the following theorem: for each A in the language of **IQC** we can find prenex A_1, A_2 majorizing $A, \neg A$ respectively. *Hint.* First establish a lemma: let A, B be prenex or negated prenex; then there exists a prenex C majorizing $A \vee B$. This lemma can be established by induction on the number of quantifiers in $A \vee B$; at one step one uses the logical law of E2.1.4. Finally establish the theorem itself by induction on the complexity of A.

2.1.6. Show that (cf. 1.10)
(i) each formula of **IPC** constructed from a single proposition letter P is in fact logically equivalent to an $\mathrm{A}_n(P)$ ($n \in \mathbb{N}$) or $\mathrm{A}_\omega(P)$;
(ii) between the A_n implications hold as indicated in the diagram of 1.10.

2.1.7. Show that $\vee$E, $\vee$I, $\exists$E, $\exists$I can be derived from the rules for $\bot$, $\wedge$, $\to$, $\forall$ in **CQC** if we define $\vee, \exists$ as in 1.6.

2.1.8. Show that the system with axioms for equality is equivalent to the system with rules in subsection 1.8. Moreover, show that the replacement rule is derivable from its atomic cases

$$\frac{R(\vec{s}, t, \vec{s}\,') \quad t = t'}{R(\vec{s}, t', \vec{s}\,')} \qquad \frac{t = t'}{f(\vec{s}, t, \vec{s}\,') = f(\vec{s}, t', \vec{s}\,')} \qquad \frac{t = t' \quad t'' = t'}{t = t''}$$

(R a predicate symbol, f a function symbol of the language).

2.1.9. Show that $\bot_i$ is derivable from the instances of $\bot_i$ with atomic conclusion.

2.2.1. Show that for E-logic we have: if $\boldsymbol{E}^*$ is another unary predicate obeying the same laws as $\boldsymbol{E}$, then for all terms t

$$\boldsymbol{E}^*t \leftrightarrow \boldsymbol{E}t$$

is derivable. Show also

$$\exists x[Ex \wedge A(x)] \leftrightarrow \exists x A(x),$$

$$\forall x[Ex \to A(x)] \leftrightarrow \forall x A(x)$$

(Scott 1979).

2.2.2. Derive in **IQCE** $t \simeq s \wedge t' \simeq s \to t \simeq t'$; $t \simeq s \to s \simeq t$; $t = s \wedge t' = s \to t = t'$; $t = s \to s = t$.

f is *total* iff $Ex_1 \wedge \cdots \wedge Ex_n \to E(f\vec{x})$. Show that $\vec{t} = \vec{s} \to f\vec{t} = f\vec{s}$, as a schema, is equivalent to (2) of 2.6 + f is total. Prove (1) in 2.4 (Scott 1979).

2.2.3. Give deduction trees for the derivations in the proof of 2.7.

2.2.4. Let $\mathrm{FV}(A) \subset \{x_1, \ldots, x_n\}$. Show

$$\mathbf{IQCE}^+ \vdash A(x_1, \ldots, x_n) \Leftrightarrow \mathbf{IQCE} \vdash \exists x Ex \wedge Ex_1 \wedge \cdots \wedge Ex_n \to A(x_1, \ldots, x_n),$$

(Renardel de Lavalette 1984, 1984A, 1.3.3(i)).

2.2.5. Prove in **IQCE** the following properties of the descriptor:

$$Ix.Ax = Ix.Bx \leftrightarrow \exists y[\forall x(Ax \to x = y) \wedge \forall x(Bx \leftrightarrow x = y)];$$

$$Ix.Ax \simeq Ix.Bx \leftrightarrow \forall y[\forall x(Ax \leftrightarrow x = y) \wedge \forall x(Bx \leftrightarrow x = y)];$$

$$\forall x(Ax \leftrightarrow Bx) \to Ix.Ax \simeq Ix.Bx$$

(Scott 1979).

2.3.1. Prove proposition 3.8.

2.3.2. Let $B^{\perp}$ be obtained from B by replacing all prime formulas P in B, other than $\perp$, by $P \vee \perp$. Then show $\mathbf{IQC} \vdash B \leftrightarrow B^{\perp}$ and $\mathbf{IQC} \vdash B \Rightarrow \mathbf{MQC} \vdash B^{\perp}$. *Hint.* Show first by formula induction $\mathbf{MQC} \vdash \perp \to A^{\perp}$.

2.3.3. Prove *Glivenko's theorem* for propositional logic:

$$\mathbf{CPC} \vdash A \Leftrightarrow \mathbf{IPC} \vdash \neg\neg A.$$

Let DNS ("*double negation shift*") be the schema $\forall x \neg\neg A(x) \to \neg\neg \forall x A(x)$ and show

$$\mathbf{CQC} \vdash B \Leftrightarrow \mathbf{IQC} + \mathrm{DNS} \vdash \neg\neg B.$$

(Glivenko 1929, Gödel 1933).

2.3.4. Complete the proof of 3.12.

2.3.5. Complete the proof of 3.18.

2.3.6. Complete the proof of 3.20(i), (iii).

2.4.1. Prove theorem 4.3.

2.4.2. Prove theorem 4.4.

2.4.3. Complete the proofs of 4.6 and the theorem in 4.7.

2.4.4. Another Hilbert-type system is based on the following axiom schemas (Gödel 1958)

$$A \vee A \to A, \quad A \to A \wedge A, \quad A \to A \vee B, \quad A \wedge B \to A,$$

$$A \vee B \to B \vee A, \quad A \wedge B \to B \wedge A, \quad \perp \to A,$$

$$\forall x A x \to A t, \quad A t \to \exists x A x$$

and the rules $\forall I_2$, $\exists E_2$ of 4.3, $\to$ E, and

$$\frac{A \to B}{C \vee A \to C \vee B} \qquad \frac{A \to B \quad B \to C}{A \to C} \qquad \frac{A \wedge B \to C}{A \to (B \to C)} \qquad \frac{A \to (B \to C)}{A \wedge B \to C}.$$

Prove the equivalence with one of the other systems. (See e.g. Troelstra 1973, section 1.1.)

2.4.5. Show that the Hilbert-type system based on the following axioms and rules (Spector 1962) is equivalent to one of the others. The axioms are instances of

$$A \to A, \quad A \wedge B \to A, \quad A \wedge B \to B, \quad A \to A \vee B, \quad B \to A \vee B$$

$$\bot \to A, \quad \forall x A x \to A t, \quad A t \to \exists x A x$$

and the rules are $\to$ E, $\forall I_2$, $\exists I_2$ (cf. 4.3) and

$$\frac{A \to B \quad B \to C}{A \to C} \qquad \frac{A \to C \quad B \to C}{A \vee B \to C} \qquad \frac{A \to B \quad A \to C}{A \to B \wedge C}$$

$$\frac{A \wedge B \to C}{A \to (B \to C)} \qquad \frac{A \to (B \to C)}{A \wedge B \to C},$$

(see e.g. Troelstra 1973, section 1.1).

2.5.1. Construct countermodels to the following formulas

$$\neg P \vee \neg\neg P;$$

$$\neg\neg P \to (\neg P \vee P);$$

$$(P \to (Q \vee R)) \to ((P \to Q) \vee (P \to R));$$

$$[((P \to Q) \to Q) \to ((P' \to Q) \to Q)] \to [((P \to P') \to Q) \to Q].$$

It is to be noted that here as well as in the next exercise the same model may sometimes serve to refute several formulas.

2.5.2. Construct countermodels to

$$\forall x \neg\neg A x \to \neg\neg \forall x A x;$$

$$\neg\neg(\forall x \neg\neg A x \to \neg\neg \forall x A x);$$

$$\neg\neg \exists x A x \to \exists x \neg\neg A x.$$

2.5.3. Let $F_0(P_0, P_1) := ((P_0 \to P_1) \to P_0) \to P_0$, $F_{n+1}(P_0, \ldots, P_{n+2}) := F_0(P_0, F_n(P_1, \ldots, P_{n+2}))$. Construct Kripke models $\mathscr{K}_n$, $\mathscr{K}$ such that
(i) For no n is F_n valid in $\mathscr{K}$,
(ii) In $\mathscr{K}_n$ F_n is refuted and F_{n+1+k} holds for all $k \in \mathbb{N}$.

2.5.4. Let (K, $\leq$) be a finite partially ordered set such that the collection of upwards monotone sets (sets of nodes X such that $\forall k \in X \forall k' \geq k(k' \in X)$) contains at most n elements; then

$$E_n := \bigvee_{i<j\leq n} (P_i \leftrightarrow P_j)$$

holds in any Kripke model (K, $\leq$, $\Vdash$). Prove this, and deduce that there is no finite Kripke model $\mathscr{K}$ such that for all A **IPC** $\vdash A \Leftrightarrow \mathscr{K} \Vdash A$.

2.5.5. Let (K, $\leq$) be a fixed partial order such that for all Kripke models (K, $\leq$, $\Vdash$) the formula $(P \to Q) \vee (Q \to P)$ is valid. Prove: if K has a root, then (K, $\leq$) is linearly ordered; in general K satisfies $\forall xyz(x \leq y \wedge x \leq z \to y \leq z \vee z \leq y)$.

2.5.6. A Kripke model with *constant domain* is a Kripke model $\mathscr{K} \equiv$ (K, $\leq$, $\Vdash$, D) such that for all $k \in$ K D$(k) =$ D. Show, reasoning classically, that in such models, if $x \notin$ FV(A)

$$\Vdash \forall x(P \vee R(x)) \to (P \vee \forall x R(x)).$$

2.5.7. Complete the proofs of 5.3 and 5.10.

2.5.8. Show that the conditions listed at the end of 5.11 suffice to recover a Kripke model as in the definition.

2.5.9. Complete the proof of the proposition in 5.13. Reformulate the model with restrictions.

2.5.10. Extend the proof of the soundness theorem to the full language and to E$^+$-logic.

2.6.1. We call a pair of sets of sentences (Γ, Δ) *complete*, if for each sentence A, $A \in \Gamma \cup \Delta$. If (Γ, Δ) is complete and consistent, Γ has the disjunction property. Show: Each consistent (Γ, Δ) has a complete extension (Γ', Δ') with $\Gamma \subset \Gamma'$, $\Delta \subset \Delta'$, Γ' C-saturated where C is a set of new constants. Formulate a corresponding generalization of the completeness theorem.

2.6.2. Modify the construction of Γ^ω in the proof of 6.3 as indicated in the remark.

2.6.3. Yet another modification of the construction of Γ^ω in 6.3 is as follows. Let $\langle A_n \rangle_n$ be an enumeration of all sentences of $\mathscr{L}$(C), with $A_0 \equiv A$. We define simultaneously Γ^k, F_k, G_k by induction on k: put $\Gamma^0 := \Gamma$, $G_0 := A_0$, F_0 any tautology. Assuming Γ^k, F_k, G_k to have been constructed, we define $\Gamma^{k+1}, F_{k+1}, G_{k+1}$:

Case 1. $\Gamma^k \nvdash G_k \vee A_{k+1}$;

$$F_{k+1} := F_k, \Gamma^{k+1} := \Gamma^k, G_{k+1} := G_k \vee A_{k+1}.$$

Case 2. $\Gamma^k \vdash G_k \vee A_{k+1}, A_{k+1} \equiv \exists x B(x)$. Put

$$\Gamma^{k+1} := \Gamma^k \cup \{\exists x B(x), B(c_{g(c)})\},$$

$$F_{k+1} := F_k \wedge \exists x B(x) \wedge B(c_{g(c)}), \quad G_{k+1} := G_k;$$

$c_{g(c)}$ first constant not in Γ^k or G_k.

Case 3. $\Gamma^k \vdash G_k \vee A_{k+1}, A_{k+1} \equiv B_1 \vee B_2$. Put $\Gamma^{k+1} := \Gamma^k \cup \{B_1 \vee B_2, B_i\}$, $F_{k+1} := F_k \wedge B_i, G_{k+1} := G_k$, i the least of $\{1,2\}$ such that $\Gamma^0 \nvdash \neg(F_k \wedge B_i \to G_k)$.

Case 4. If 1–3 do not apply, let $\Gamma^{k+1} := \Gamma^k \cup \{A_{k+1}\}$, $G_{k+1} := G_k$, $F_{k+1} := F_k$.
Show that Γ^ω is C-saturated, and (for readers familiar with the basic notions of recursion theory) that Γ^ω is recursive in Γ.

2.6.4. Show that for theories with *decidable* equality, (i.e. $\forall xy(x = y \vee x \neq y)$ holds) we have completeness w.r.t. *normal* Kripke models where = at each node k is interpreted by equality in D(k). *Hint.* Consider quotient structures obtained by taking equivalence classes modulo $\sim_k$.

2.6.5. Carry out the completeness proof for **IQCE**$^+$ (6.7) in full.

2.6.6. Let $\{\mathcal{K}_i : i \in I\}$ be a collection of propositional Kripke models $(\mathrm{K}_i, \leq_i, \Vdash_i)$ with disjoint K_i. We put

$$\bigcup_{i \in I} \mathcal{K}_i := \Big(\bigcup_{i \in I} \mathrm{K}_i, \bigcup_{i \in I} \leq_i, \bigcup_{i \in I} \Vdash_i\Big).$$

For any Kripke model $(\mathrm{K}, \leq, \Vdash) \equiv \mathcal{K}$ let $\mathcal{K}' := (\mathrm{K} \cup \{k_0\}, \leq', \Vdash')$ where $k_0 \notin \mathrm{K}$, $\leq' \restriction \mathrm{K} = \leq$, $k \geq k_0$ for all $k \in K$, $k_0 \nVdash' P$ for P prime, $\Vdash'$ restricted to K coincides with $\Vdash$.

Let C be such that no strictly positive subformula of C is a disjunction or an existential formula (cf. 3.23). Show that if for all $i \in I$ $\mathcal{K}_i \Vdash C$, then $(\bigcup \mathcal{K}_i)' \Vdash' C$.

2.6.7. Use the preceding exercise to prove (classically) the following strong version of the disjunction property for **IPC**: if C satisfies the conditions of the preceding exercise, and **IPC** $\vdash C \to A \vee B$, then **IPC** $\vdash C \to A$ or **IPC** $\vdash C \to B$. *Hint.* Argue by contraposition, use completeness.

2.6.8. Prove Glivenko's theorem (E2.3.3) by means of Kripke models. *Hint.* Show that for a theorem A of **CPC** at each node of a finite Kripke model $\neg\neg A$ must hold (use 6.11).

2.6.9. Show that Kripke models over finite *binary* trees are sufficient for completeness for propositional formulas without $\vee$ (Segerberg 1974).
Hint. Let $(\mathrm{K}, \leq, \Vdash)$ be any finite tree model, $k \in \mathrm{K}$, $k_0 > k$, $k_1 > k$ immediate successors of k. Define $(\mathrm{K}', \leq', \Vdash')$ such that $\mathrm{K}' := \mathrm{K} \cup \{k'\}$ $(k' \notin \mathrm{K})$, $\leq' \restriction \mathrm{K} = \leq$, $k' >' k$, $\forall k'' \in \mathrm{K}((k_0 \leq k'' \text{ or } k_1 \leq k'') \Leftrightarrow k' <' k'')$ and for prime P

$$\begin{aligned} &\forall k \in \mathrm{K}(k \Vdash' P \Leftrightarrow k \Vdash P), \\ &k' \Vdash' P \Leftrightarrow (k_0 \Vdash P \text{ and } k_1 \Vdash P). \end{aligned} \tag{1}$$

Show that (1) extends to arbitrary formulas.

2.6.10. A surjection from a poset $(\mathrm{K}, \leq)$ to a poset $(\mathrm{K}', \leq')$ is called a *p-morphism* (or strongly isotone mapping) if f satisfies (i) $x \leq y \to fx \leq' fy$ and (ii) $\forall x \in \mathrm{K}' \forall y \in \mathrm{K}(fy \leq' x \to \exists z \in \mathrm{K}(y \leq z \wedge fz = x))$. Let $\mathscr{K} := (\mathrm{K}, \leq, \Vdash)$, $\mathscr{K}' := (\mathrm{K}', \leq', \Vdash')$ be two Kripke models, and f a p-morphism from $(\mathrm{K}, \leq)$ to $(\mathrm{K}', \leq')$ such that $k \Vdash A$ iff $fk \Vdash' A$ for atomic A; prove that this holds for all A.

2.6.11. If $\mathscr{A}$ is a collection of posets such that each finite tree can be obtained as the image of an element of $\mathscr{A}$ under a p-morphism (see the preceding exercise), then **IPC** is complete for Kripke models over posets from $\mathscr{A}$. In particular, show that **IPC** is complete with respect to Kripke models over the infinite binary tree.

2.6.12. Let $\mathscr{A}_i \equiv (\mathrm{K}_i, \leq_i)$ be a poset for $i < n$, and define $\mathscr{A}_0 \times \cdots \times \mathscr{A}_{n-1}$ as $(\mathrm{K}_0 \times \cdots \times \mathrm{K}_{n-1}, \leq')$ where $(k_0, \ldots, k_{n-1}) \leq' (k_0', \ldots, k_{n-1}') \Leftrightarrow \forall i < n$ $(k_i \leq_i k_i')$. If $\mathscr{A} \equiv (\mathrm{K}, \leq)$ is a poset, we let $\mathscr{A}' \equiv (\mathrm{K} \cup \{k^*\}, \leq')$ be the poset (modulo isomorphism) such that $k^* \notin \mathrm{K}$, $k \leq' k'$ iff $k \leq k'$ for all $k, k' \in \mathrm{K}$, $k^* < k$ for all $k \in \mathrm{K}$. Now define a sequence of posets as follows. $\mathscr{A}_0 :=$ the singleton poset with trivial ordering; $\mathscr{A}_1 := (\mathscr{A}_0 \times \mathscr{A}_0)'$; $\mathscr{A}_{n+1} := (\mathscr{A}_n \times \cdots \times \mathscr{A}_n)'(n + 2 \text{ factors})$. Show that **IPC** is complete for Kripke models with trees from the sequence $\langle \mathscr{A}_n \rangle_n$. (This is virtually Jaskowski's result (1936); cf. also Smorynski 1973A, or Smorynski in Troelstra (1973, 5.3.8).)

2.6.13. Give a completeness proof for **IPC** + LO with respect to linearly ordered Kripke models, where LO is the axiom schema $(A \to B) \vee (B \to A)$. *Hint.* Suppose **IPC** + LO $\nvdash F$, then a fortiori **IPC** + $\mathscr{G} \nvdash F$, where $\mathscr{G} \equiv \{(A \to B \vee (B \to A) : A \text{ and } B \text{ subformulas of } F\}$. Extract from a tree-countermodel for **IPC** + $\mathscr{G} \nvdash F$ a linear countermodel.

2.6.14. Let us call a formula *positive* if it is constructed from the function symbols and relations of the language using $\bot, \wedge, \vee, \exists, =$. A formula is *geometric* if it is of the form

$\forall \vec{x}(A \to B)$, where A and B are positive. Let A_k $(k < n)$ and B be geometric; show that if $\vdash_c \bigwedge\{A_k : k < n\} \to B$, then $\vdash_i \bigwedge\{A_k : k < n\} \to B$. *Hint.* Observe that positive formulas are forced at a node of a Kripke model iff they are classically valid at that node.

2.7.1. Prove 7.2.

2.7.2. Prove 7.3.

2.7.3. Prove 7.8, 7.11.

CHAPTER 3

ARITHMETIC

In the present chapter we treat intuitionistic/constructive arithmetic at some length. The first five sections are elementary in character, the rest mainly contains material needed in chapters 9–15.

For the reader who primarily wants an introduction to constructive and intuitionistic mathematics, a brief perusal of the sections 1, 3, and 4 suffices. The result in 3.4 explains why intuitionistic elementary arithmetic is little different in character from classical elementary arithmetic: the vast majority of well-known theorems of classical elementary number theory can be stated in Π^2_0-form and are therefore also intuitionistically provable.

Summary of the contents. Section 1 introduces the class of primitive recursive functions PRIM and lists certain of its closure properties.

Section 2 describes primitive recursive arithmetic **PRA**, a system without quantifiers, and illustrates how some basic theorems of arithmetic can be established in this system. This section is not needed for the next section on intuitionistic first-order arithmetic **HA**.

Section 4 presents a brief analysis of the informal notion of algorithm and deduces from this analysis some basic properties such as the smn-theorem and the recursion theorem. The contents of this section are needed in sections 4.3 and 4.4, and in later chapters where examples of constructive recursive mathematics are given.

Section 5 gives some examples of metamathematical properties of **HA** such as the disjunction and existence property, closure under the "independence of premiss"-rule, and closure under Markov's rule.

Sections 6 and 7 contain an introduction to elementary inductive definitions and give an outline of the formalization of elementary recursion theory in **HA** and elementary analysis **EL**. This material is needed in Chapter 9, and also to provide a rigorous backing to the proof sketch of the characterization theorem for realizability in section 4.4.

Section 8 is devoted to intuitionistic second-order logic and arithmetic with set variables, and section 9 discusses the corresponding generalization to all finite types.

Though the sections dealing with elementary recursion theory, namely 1, 4, 6 and 7 are in principle self-contained, the treatment is condensed and some preliminary acquaintance with the basics of recursion theory will be helpful.

1. Informal arithmetic and primitive recursive functions

1.1. As already noted in 1.2.1, natural numbers are generally regarded as unproblematic, and are treated as basic constructive objects all of the same kind (except from the ultra-finitist point of view mentioned in 1.2.1). Thus the "Peano-axioms" are regarded as obviously true, i.e. if S is the successor function then

$$x = y \leftrightarrow Sx = Sy,$$

$$0 \neq Sx,$$

and induction is accepted for all constructively meaningful properties of natural numbers $A(x)$:

$$A(0) \wedge \forall x(A(x) \rightarrow A(Sx)) \rightarrow \forall x A(x).$$

The justification of induction is based on the mental picture we have of the natural numbers as obtained by successively adding "abstract units"; given $A(0)$, $\forall x(A(x) \rightarrow A(Sx))$ we build parallel to the construction of $n \in \mathbb{N}$ a proof of $A(n)$:

$$\dfrac{\dfrac{A(0) \qquad A(0) \rightarrow A(1)}{A(1)} \qquad A(1) \rightarrow A(2)}{A(2) \quad \ldots}$$

etc.

Number theory uses many operations besides successor, but also for metamathematical purposes one needs at least addition and multiplication. However, it is extremely convenient to have a more ample supply; for this we shall take the class of primitive recursive functions PRIM, each primitive recursive function will be quite obviously constructively well-defined.

The remainder of this section will be devoted to the definition and properties of PRIM.

1.2. DEFINITION. The class of *primitive recursive functions*, PRIM, is generated by the following clauses

(i) Z, S, p_i^n for $i < n$, $n \geq 1$ belong to PRIM. Here Z is the *zero*-function satisfying $Z(x) = 0$, S is the *successor* function, and p_i^n is a *projection* (*-function*) determined by $p_i^n(x_0, \ldots, x_{n-1}) = x_i$.

(ii) PRIM is closed under *composition*: if f, $g_i \in$ PRIM, with $f \in \mathbb{N}^n \to \mathbb{N}$, $g_i \in \mathbb{N}^k \to \mathbb{N}$ $(1 \leq i \leq k)$, then there is an $h \in$ PRIM satisfying

$$h(\vec{x}) = f(g_1(\vec{x}), \ldots, g_n(\vec{x})).$$

(iii) PRIM is closed under *definition by recursion*, i.e. if $f \in \mathbb{N}^n \to \mathbb{N}$, $g \in \mathbb{N}^{n+2} \to \mathbb{N}$, then there is an $h \in$ PRIM such that

$$\begin{cases} h(0, \vec{x}) = f(\vec{x}) \\ h(Sy, \vec{x}) = g(h(y, \vec{x}), y, \vec{x}). \end{cases} \quad \square$$

By "PRIM is *generated* by (i), (ii) and (iii)" we mean that PRIM is the least class satisfying (i), (ii), (iii). We may think of all primitive recursive functions being numbered, as $\varphi_0, \varphi_1, \varphi_2, \ldots$ such that each φ_k is either one of Z, S, p_n^i or defined from certain φ_j's with $j < k$; properties of PRIM may now be established by induction on k (N.B. repetitions may occur).

The reader who is familiar with elementary recursion theory may skip the rest of this section. We shall now introduce some familiar arithmetical functions as elements of PRIM.

1.3. *Some primitive recursive functions.*
Constant functions are primitive recursive. Thus for fixed m the $f \in \mathbb{N}^n \to \mathbb{N}$ satisfying

$$f(\vec{x}) = m$$

is in PRIM: if $m = 0$, take $f(\vec{x}) := Zp_0^n(\vec{x})$, and for $m = Sm'$, given g with $g(\vec{x}) = m'$ for all $\vec{x}$, let $f(\vec{x}) = S(g(\vec{x}))$.

Addition is usually given by $x + 0 = x$, $x + Sy = S(x + y)$. Writing $+(y, x)$ instead of $x + y$ we can bring this in the standard form prescribed by definition 1.2:

$$+(0, x) = p_0^1(x); \quad +(Sy, x) = S(p_0^3(+(y, x), y, x)).$$

After this example of a definition in standard form we assume that the reader can provide detailed formulations, if desired.

Multiplication

$$x \cdot 0 = 0, \quad x \cdot Sy = x \cdot y + x.$$

Exponentiation

$$x^0 = 1, \quad x^{Sy} = x^y \cdot x.$$

Predecessor

$$\mathrm{prd}(0) = 0, \quad \mathrm{prd}(Sx) = x.$$

Cut-off subtraction

$$x \dot{-} 0 = x, \quad x \dot{-} Sy = \mathrm{prd}(x \dot{-} y).$$

Note that $x \dot{-} y = 0$ iff $x \leq y$, $x \dot{-} y = x - y$ otherwise, i.e. if $x \geq y$ then $y + (x \dot{-} y) = x$.

Absolute difference

$$|x - y| := (x \dot{-} y) + (y \dot{-} x).$$

Signum function sg and *inverse signum function* sg′

$$\mathrm{sg}'(x) = 1 \dot{-} x, \quad \mathrm{sg}(x) = 1 \dot{-} (1 \dot{-} x).$$

So $\mathrm{sg}(x) = 0$ iff $x = 0$, and $\mathrm{sg}(x) \leq 1$ for all x.

Maximum and *minimum*

$$\max(x, y) = x + (y \dot{-} x), \quad \min(x, y) = x \dot{-} (x \dot{-} y).$$

The properties listed below in 1.4–1.14 are presented without proof; these may be regarded as exercises, or looked up in a standard text on recursion theory.

1.4. PROPOSITION (*closure properties of* PRIM). *Let σ be a permutation of $\{1, \ldots, n\}$ and let $g \in$ PRIM. Then also $f \in$ PRIM in the following cases*
(i) $f(\vec{x}) = g(x_{\sigma(1)}, \ldots, x_{\sigma(n)})$ (*permutation of arguments*);
(ii) $f(\vec{x}) = g(x_1, \vec{x})$ (*identification of arguments*);
(iii) $f(y, \vec{x}) = g(\vec{x})$ (*dummy arguments*);
(iv) $f(z, \vec{x}) = \prod_{y<z} g(y, \vec{x})$ (*finite product*) *with*

$$\prod_{y<0} g(y, \vec{x}) = 1, \quad \prod_{y<Sz} g(y, \vec{x}) = g(z, \vec{x}) \cdot \Bigl(\prod_{y<z} g(y, \vec{x})\Bigr);$$

(v) $f(z, x) = \Sigma_{y<z} g(y, \vec{x})$ (*finite sum*) *with*

$$\sum_{y<0} g(y, \vec{x}) = 0, \quad \sum_{y<Sz} g(y, \vec{x}) = g(z, \vec{x}) + \left(\sum_{y<z} g(y, \vec{x}) \right);$$

(vi) $f(z, y, \vec{x}) = g^z(y, \vec{x})$ (*iteration*) *where*

$$g^0(y, \vec{x}) = p_0^{n+1}(y, \vec{x}), \quad g^{Sz}(y, \vec{x}) = g(g^z(y, \vec{x}), \vec{x}).$$

(vii) $f(z, \vec{x}) = \min_{y \le z}(g(y, \vec{x}) = 0)$ (*bounded minimum operator*) *where*

$$\min_{y \le z}(g(y, \vec{x}) = 0) := \sum_{u<Sz} \mathrm{sg}\left(\prod_{y<Su} g(y, \vec{x}) \right).$$

Thus $\min_{y \le z}(g(y, \vec{x}) = 0)$ *is the least* $y \le z$ *such that* $g(y, \vec{x}) = 0$, *if existing*, $z + 1$ *otherwise*. □

1.5. Definition. A relation R is said to be *primitive recursive*, iff its characteristic function χ_R, satisfying $\chi_R(\vec{x}) \le 1$, $\chi_R(\vec{x}) = 0 \leftrightarrow R(\vec{x})$, belongs to PRIM. □

Examples. $=$, $\le$ are primitive recursive since

$$\chi_=(x, y) := \mathrm{sg}(|x - y|), \ \chi_\le(x, y) := \mathrm{sg}(x \dot{-} y). \quad □$$

1.6. Proposition. *The primitive recursive relations are closed under intersection, union, complementation and bounded quantification, and under substitution: if* $R(x, \vec{y})$ *is primitive recursive, then so is* $R(f(x, \vec{y}, \vec{z}), \vec{y})$ *for any* $f \in$ PRIM. □

1.7. Definition. If $R(y, \vec{x})$ is primitive recursive, we define

$$\min_{y \le z} R(y, \vec{x}) := \min_{y \le z}(\chi_R(y, \vec{x}) = 0). \quad □$$

Examples. The following predicates and functions are primitive recursive

$x|y := \exists z \le y(x \cdot z = y)$ (*divisibility*);

$\mathrm{Prime}(x) := \forall y < x(\neg y|x \vee y = 1) \wedge x \ne 0 \wedge x \ne 1$ (x is a *prime*);

$0! = 1, (Sx)! = (x!) \cdot Sx$ (*factorial function*);

$p(Sx) := \min_{y \le p(x)!+1}[\mathrm{Prime}(y) \wedge y > p(x)], \ p(0) = 2$

($p(n)$ is the nth prime).

1.8. *Coding of p-tuples.* In arithmetic we can talk about p-tuples of natural numbers and finite sequences of natural numbers via suitable codings. For the coding of pairs we need a mapping $j \in \mathbb{N}^2 \to \mathbb{N}$ with inverses j_1, j_2 satisfying

$$j_1 j(x, y) = x, \quad j_2 j(x, y) = y.$$

It is often convenient, though not strictly necessary, that the pairing is surjective, i.e.

$$j(j_1 z, j_2 z) = z.$$

From j, j_1, j_2 we can now construct codings ν^p for p-tuples with inverses j_i^p such that

$$j_i^p \nu^p(x_1, \ldots, x_p) = x_i \quad (1 \le i \le p).$$

To achieve this, we may put for example

$$\nu^1(x_0) := x_0,$$

$$\nu^{p+1}(x_0, x_1, \ldots, x_p) := \nu^p(x_0, \ldots, x_{p-2}, j(x_{p-1}, x_p)).$$

It is obvious how the j_i^p are to be defined from j_1 and j_2. □

It is easy to see that if j is surjective, so are the ν^p for all p. There are many possible choices for j, a convenient one is

1.9. DEFINITION.

$$j(x, y) := 2^x \cdot (2y + 1) \dot{-} 1,$$

$$j_1 z := \min_{x \le z}[\exists y \le z(2^x \cdot (2y + 1) = Sz)],$$

$$j_2 z := \min_{y \le z}[\exists x \le z(2^x \cdot (2y + 1) = Sz)]. \quad □$$

Another well-known pairing function is $j'(x, y) := \frac{1}{2}(x + y)(x + y + 1) + y$.

1.10. PROPOSITION. *j as defined above is a surjective pairing with inverses j_1, j_2 which satisfies in addition monotonicity conditions*

(i) $x \le j(x, y)$, $y \le j(x, y)$,

(ii) $x < x' \to j(x, y) < j(x', y) \wedge j(y, x) < j(y, x')$,

(iii) $j(0, 0) = 0$. □

1.11. DEFINITION (*coding of finite sequences*). If j is a surjective pairing with inverses j_1, j_2, such that $j(0, 0) = 0$, and the ν^p's are the p-tuple codings defined from j, we can code arbitrary finite sequences $x_0, \ldots, x_u$ by a number $\langle x_0, \ldots, x_u \rangle$ such that

$$\langle\rangle = 0 \text{ (zero codes the empty sequence)},$$

$$\langle x_0, \ldots, x_u \rangle = j\big(u, \nu^{u+1}(x_0, \ldots, x_u)\big) + 1.$$

It does not make sense to describe this coding as primitive recursive, since it does not have a fixed number of arguments; however, it can be completely described from the following operations: the *concatenation* function $*$ such that

$$\langle x_0, \ldots, x_u \rangle * \langle x_{u+1}, \ldots, x_{u+v} \rangle = \langle x_0, \ldots, x_{u+v} \rangle,$$

the *length* function lth such that

$$\mathrm{lth}\langle\rangle = 0$$

$$\mathrm{lth}\langle x_0, \ldots, x_u \rangle = u + 1,$$

and the *decoding* function $\pi \in \mathbb{N}^2 \to \mathbb{N}$ such that

$$\pi(n, y) = x_y \text{ if } n = \langle x_0, \ldots, x_u \rangle \text{ and } y < \mathrm{lth}(n)$$

$$\pi(n, y) = 0 \text{ in all other cases.}$$

We usually write

$$(n)_y \quad \text{for } \pi(n, y). \quad \square$$

$*$, lth and π are primitive recursive in j (exercise). We put

$$n \preccurlyeq m := \exists n'(n * n' = m), \quad n \succcurlyeq m := m \preccurlyeq n,$$

$$n \prec m := n \preccurlyeq m \land n \neq m, \quad n \succ m := m \prec n.$$

1.12. Proposition.
(i) $\mathrm{lth}(n * m) = \mathrm{lth}(n) + \mathrm{lth}(m)$,
(ii) $(n * m)_y = (n)_y$ for $y < \mathrm{lth}(n)$,
$(n * m)_y = (m)_{y \dot{-} \mathrm{lth}(n)}$ for $y \geq \mathrm{lth}(n)$. $\square$

1.13. Definition (*course-of-values function*). For any function φ we define $\bar{\varphi}$ as

$$\bar{\varphi}(0, \vec{m}) = \langle\rangle, \quad \bar{\varphi}(Sn, \vec{m}) = \bar{\varphi}(n, \vec{m}) * \langle \varphi(n, \vec{m}) \rangle. \quad \square$$

1.14. Proposition (*course-of-values recursion*). *If χ and ψ are primitive recursive, so is φ given by*

$$\varphi(0, \vec{x}) = \chi(\vec{x}), \quad \varphi(Sz, \vec{x}) = \psi(\bar{\varphi}(Sz, \vec{x}), z, \vec{x}). \quad \square$$

2. Primitive recursive arithmetic PRA

2.1. The usual presentation of number-theoretic arguments may create the impression that the full force of first-order logic, and in particular the use of quantifiers, is necessary. Therefore it comes more or less as a surprise that a considerable fragment of arithmetic can be handled in a quantifier-free fragment. In this section we illustrate this by developing some arithmetic in the quantifier-free part, **PRA**, of intuitionistic arithmetic **HA**. The contents of this section are not needed later on.

The language of **PRA** contains free variables for natural numbers, constants 0 (zero), S (successor), a function symbol for each primitive recursive function, and the binary predicate = (equality). The axioms and rules are the axioms and rules of **IPC** with equality (with respect to the language of **PRA**), the axiom

$$\neg S0 = 0,$$

the induction rule

$$A(0),\ A(x) \to A(Sx) \Rightarrow A(t)$$

for quantifier-free A, and the defining equations for all primitive recursive functions.

In particular, we have constants prd, $\dot{-}$, $+$, $\cdot$ satisfying

$$\text{prd}(0) = 0, \quad \text{prd}(Sx) = x;$$

$$x \dot{-} 0 = x, \quad x \dot{-} Sy = \text{prd}(x \dot{-} y);$$

$$x + 0 = x, \quad x + Sy = S(x + y);$$

$$x \cdot 0 = 0, \quad x \cdot Sy = x \cdot y + x.$$

The equality axioms yield

$$Sx = Sy \to \text{prd}(Sx) = \text{prd}(Sy) \to x = y,$$

so we do not need $Sx = Sy \to x = y$ as an axiom.

In the absence of quantifiers, we need a substitution rule

$$\text{SUB} \qquad \frac{A(x)}{A(t)},$$

where $A(x)$ has been derived from hypotheses not containing x free. We can avoid the introduction of a substitution rule by stating all axioms as schemata, arbitrary terms replacing the variables (e.g. $t + 0 = t$, $t + St' = S(t + t')$ for addition).

In a natural deduction formulation we can take besides axioms $t = t$, and defining equations for primitive recursive functions the following rules

$$\frac{0 = S0}{\bot} \qquad \frac{t_1 = t_2 \quad A(t_1)}{A(t_2)} \qquad \frac{A(0) \quad \begin{matrix}[A(x)] \\ \vdots \\ A(Sx)\end{matrix}}{A(t)}$$

In the third (induction-) rule the assumptions indicated by $[A(x)]$ are discharged. Actually, $\neg S0 = 0$, or the corresponding natural deduction rule, is redundant since we have

2.2. PROPOSITION. *$\bot$ is definable in* **PRA** *as* $0 = S0$.

PROOF. Let φ be the primitive recursive function such that

$$\varphi(t_1, t_2, 0) = t_1, \quad \varphi(t_1, t_2, Sx) = t_2.$$

Then $0 = S0 \to t_1 = \varphi(t_1, t_2, 0) = \varphi(t_1, t_2, S0) = t_2$ and thus

$$0 = S0 \to P$$

for all prime formulas P; by an easy formula induction we then establish

$$0 = S0 \to A$$

for all quantifier-free A. In the induction $\neg B$ is to be interpreted as $B \to 0 = S0$. □

REMARKS. This proposition shows that in the absence of $\bot$ as a primitive, $0 = S0$ can play the role of $\bot$. However, if $\bot$ is present as a primitive, $0 = S0 \to \bot$ cannot be proved unless we postulate it.

Observe also, that for quite general logical reasons, the equality axioms all follow from reflexivity, symmetry, transitivity and the following special cases of substitution.

$$x_1 = x_1' \wedge x_2 = x_2' \wedge \cdots \wedge x_n = x_n' \to \varphi(x_1, \ldots, x_n) = \varphi(x_1', \ldots, x_n')$$

for all function constants φ.

Finally note that, for any term $t(\vec{x})$ in the language of **PRA**, there is a primitive recursive function φ such that $t(\vec{x}) = \varphi(\vec{x})$ holds in **PRA**.

Our next aim is to generalize the induction schema.

2.3. PROPOSITION (*special induction*). *The following schema holds in* **PRA**

$$A(0), A(Sx) \Rightarrow A(t)$$

PROOF. Immediate from the induction schema. □

2.4. PROPOSITION.
(i) $0 \dot{-} x = 0$;
(ii) $Sx \dot{-} Sy = x \dot{-} y$;
(iii) $Sx \dot{-} x = S0$;
(iv) $x \dot{-} x = 0$;
(v) $x = 0 \vee x = S(\mathrm{prd}(x))$;
(vi) $x \neq 0 \leftrightarrow x = S(\mathrm{prd}(x))$, $x = Sz \to x = S(\mathrm{prd}(x))$.

PROOF. (i) $0 \dot{-} 0 = 0$ (defining axioms for $\dot{-}$), and $0 \dot{-} Sx = \mathrm{prd}(0 \dot{-} x) = \mathrm{prd}(0) = 0$ (under the induction hypothesis $0 \dot{-} x = 0$); hence $0 \dot{-} x = 0$ by induction.

(ii) $Sx \dot{-} S0 = \mathrm{prd}(Sx \dot{-} 0) = \mathrm{prd}(Sx) = x = x \dot{-} 0$; assume $Sx \dot{-} Sy = x \dot{-} y$, then $Sx \dot{-} SSy = \mathrm{prd}(Sx \dot{-} Sy) = \mathrm{prd}(x \dot{-} y) = x \dot{-} Sy$; induction yields (ii).

(iii) $S0 \dot{-} 0 = S0$; $SSx \dot{-} Sx = Sx \dot{-} x = S0$ (with (ii) and induction hypothesis $Sx \dot{-} x = S0$). Induction gives (iii).

(iv) $x \dot{-} x = Sx \dot{-} Sx = \mathrm{prd}(Sx \dot{-} x) = \mathrm{prd}(S0) = 0$ (by (ii), (iii)).

(v) and (vi) are left to the reader; for (v) one uses induction. □

As an illustration of the process of complete formalization we exhibit a completely formalized proof of (i), based on the logical rules in section 2.1 and the axioms and rules of subsection 2.1.

(1)	$y \dot{-} 0 = 0$	(defining axiom for $\dot{-}$)
(2)	$0 \dot{-} 0 = 0$	(SUB, (1))
(3)	$y \dot{-} Sz = \mathrm{prd}(y \dot{-} z)$	(defining axiom for $\dot{-}$)
(4)	$0 \dot{-} Sz = \mathrm{prd}(0 \dot{-} z)$	(SUB, (3))
(5)	$0 \dot{-} Sx = \mathrm{prd}(0 \dot{-} x)$	(SUB, (4))
(6)	$0 \dot{-} x = 0$	(hypothesis)
(7)	$\mathrm{prd}(0) = \mathrm{prd}(0)$	(REFL)
(8)	$\mathrm{prd}(0 \dot{-} x) = \mathrm{prd}(0)$	(REPL, (6), (7))
(9)	$\mathrm{prd}(0) = 0$	(defining axiom for prd)
(10)	$\mathrm{prd}(0 \dot{-} x) = 0$	(TRANS, (8), (9))
(11)	$0 \dot{-} Sx = 0$	(TRANS, (5), (10))
(12)	$0 \dot{-} x = 0$	(ind. rule, (2), (11), elim. of (6)).

It is to be noted that the proof of $0 = S0 \to A$ in 2.2 above is not an (informal) argument in **PRA**; rather, it is a metamathematical argument showing how to generate proofs of $0 = S0 \to A$ in **PRA** for all formulas in the language of **PRA**.

2.5. Proposition. *Let t, t_0, t_1 be terms; then the following rule holds in* **PRA**

$$A(0, y), A(x, t(x, y)) \to A(Sx, y) \Rightarrow A(t_0, t_1).$$

Proof. It is enough to establish the rule in the form

$$A(0, y), A(x, \varphi(x, y)) \to A(Sx, y) \Rightarrow A(t_0, t_1),$$

for φ a primitive recursive function symbol (cf. remarks in 2.2). Assume $A(0, y)$ and $A(x, \varphi(x, y)) \to A(Sx, y)$ and define by recursion ψ such that

$$\psi(0) = t_1, \psi(Sz) = \varphi(t_0 \dot{-} Sz, \psi(z)).$$

If $Sx = t_0 \dot{-} z$, then $t_0 \dot{-} Sz = \mathrm{prd}(t_0 \dot{-} z) = \mathrm{prd}(Sx) = x$, hence, since $A(t_0 \dot{-} Sz, \varphi(t_0 \dot{-} Sz, \psi z) \leftrightarrow A(t_0 \dot{-} Sz, \psi(Sz))$ and $A(x, \varphi(x, y)) \to A(Sx, y)$,

$$Sx = t_0 \dot{-} z \to [A(t_0 \dot{-} Sz, \psi(Sz)) \to A(t_0 \dot{-} z, \psi z)].$$

Since $A(0, \psi(z))$, also

$$0 = t_0 \dot{-} z \to [A(t_0 \dot{-} Sz, \psi(Sz)) \to A(t_0 \dot{-} z, \psi z)],$$

and therefore by special induction (2.3)

$$t_0 \dot{-} z = t_0 \dot{-} z \to [A(t_0 \dot{-} Sz, \psi(Sz)) \to A(t_0 \dot{-} z, \psi z)],$$

hence

$$A(t_0 \dot{-} Sz, \psi(Sz)) \to A(t_0 \dot{-} z, \psi z)$$

and therefore

$$[A(t_0 \dot{-} z, \psi z) \to A(t_0, t_1)] \to [A(t_0 \dot{-} Sz, \psi(Sz)) \to A(t_0, t_1)].$$

Also

$$A(t_0 \dot{-} 0, \psi 0) \to A(t_0, t_1),$$

and so by induction

$$A(t_0 \dot{-} t_0, \psi(t_0)) \to A(t_0, t_1),$$

and thus $A(t_0, t_1)$. □

2.6. PROPOSITION. *The following form of double induction holds in* **PRA**:

$$A(x,0),\, A(0,y),\, A(x,y) \to A(Sx,Sy) \Rightarrow A(t_0,t_1).$$

PROOF. Assume $A(x,0)$, $A(0,y)$, $A(x,y) \to A(Sx,Sy)$ to be derivable; $y = Sz \to S(\mathrm{prd}(y)) = y$ (2.4(vi)), and so

$$y = Sz \to (A(x,\mathrm{prd}(y)) \to A(Sx,y)); \quad \text{also}$$

$$y = 0 \to (A(x,\mathrm{prd}(y)) \to A(Sx,y))$$

(since $A(Sx,0)$ by hypothesis). Therefore by 2.3

$$y = y \to (A(x,\mathrm{prd}(y)) \to A(Sx,y));$$

now detach $y = y$ and apply the preceding proposition. □

2.7. DEFINITION.

$$x \le y := (x \dot{-} y = 0),$$

$$x < y := (Sx \dot{-} y = 0). \quad \square$$

2.8. PROPOSITION.
(i) $x = y \vee x \neq y$;
(ii) $x \le y \wedge y \le x \to x = y$;
(iii) $x \le y \vee y \le x$;
(iv) $x \le y \leftrightarrow x = y \vee x < y$;
(v) $x = y \vee x < y \vee y < x$;
(vi) $\neg x \le y \leftrightarrow y < x$.

PROOF. (i) Apply double induction. $x = 0 \vee x \neq 0$ follows by special induction 2.3 on x, similarly $0 = y \vee 0 \neq y$; and if $x = y \vee x \neq y$, then $Sx = Sy \vee Sx \neq Sy$ since $Sx = Sy \leftrightarrow x = y$.

(ii) By double induction.

(iii), (iv) Double induction with respect to x and y.

(v) From (iii) and (iv).

(vi) In one direction we have to show $x \dot{-} y \neq 0 \to Sy \dot{-} x = 0$. This can be shown by double induction on x and y. $0 \dot{-} y \neq 0 \to Sy \dot{-} 0 = 0$ since $0 \dot{-} y \neq 0$ and $Sy \dot{-} 0 = 0$ are both false; $x = x \dot{-} 0 \neq 0 \to x \dot{-} 0 = S(\mathrm{prd}(x)) = x$ (2.4(vi)), hence $S0 \dot{-} x = S0 \dot{-} S(\mathrm{prd}(x)) = 0 \dot{-} \mathrm{prd}(x) = 0$ (2.4(i)). Induction step: suppose $x \dot{-} y \neq 0 \to Sy \dot{-} x = 0$ (induction hypothesis), $Sx \dot{-} Sy \neq 0$. Then $Sx \dot{-} Sy = S(\mathrm{prd}(Sx \dot{-} Sy)) =$

$S(\text{prd}(x \dot{-} y)) = x \dot{-} y$ (applying 2.4(vi), and 2.4(ii) twice), hence by our first assumption $Sy \dot{-} x = 0$, and therefore also $SSy \dot{-} Sx = 0$ (again by 2.4(ii)).

In the other direction, observe that $y < x \to y \leq x$, and also $y \neq x$; since $x \leq y$ and $y \leq x$ would imply $y = x$ by (ii), it follows that $y < x \to \neg x \leq y$. □

2.9. PROPOSITION. *For all A* **PRA** $\vdash A \vee \neg A$.

PROOF. Formula induction with 2.8(i) as basis (exercise). □

REMARK. This proposition shows that it does not matter whether we adopt **CPC** or **IPC** as our logical basis for **PRA**; **PRA** is a fragment both of intuitionistic first-order arithmetic **HA** (defined in section 3) and of classical first-order arithmetic **PA** (= $\mathbf{HA}^c$).

2.10. PROPOSITION.
(i) $x + S0 = Sx$;
(ii) $S0 + x = Sx,\ \ 0 + x = x$;
(iii) $x + (y + z) = (x + y) + z$;
(iv) $x + y = y + x$.

PROOF. (i) $x + S0 = S(x + 0) = Sx$.

(ii) Induction on x.

(iii) Induction on z. Basis step: $x + (y + 0) = x + y = (x + y) + 0$. Now assume $x + (y + z) = (x + y) + z$. Then $x + (y + Sz) = x + S(y + z) = S(x + (y + z)) = S((x + y) + z) = (x + y) + Sz$.

(iv) Induction on y. Basis step: $x + 0 = x = 0 + x$. Assume $x + y = y + x$. Then $x + Sy = S(x + y) = S(y + x) = y + Sx = y + (S0 + x) = (y + S0) + x = Sy + x$. □

2.11. PROPOSITION.
(i) $x \dot{-} y = Sz \leftrightarrow x = y + Sz$;
(ii) $x < y \wedge y < z \to x < z$;
(iii) $x \leq y \wedge y \leq z \to x \leq z$.

PROOF. (i) By double induction w.r.t. x and y. Basis step: $x \dot{-} 0 = Sz \leftrightarrow x = Sz \leftrightarrow x = 0 + Sz$ and $0 \dot{-} y = Sz \leftrightarrow 0 = y + Sz = S(y + z)$ since both sides of this equivalence are simultaneously false. For the induction step, assume $x \dot{-} y = Sz \leftrightarrow x = y + Sz$. $Sx \dot{-} Sy = Sz \leftrightarrow x \dot{-} y = Sz \leftrightarrow$

$x = y + Sz$; $x = y + Sz \to Sy + Sz = Sz + Sy = S(Sz + y) = S(y + Sz) = Sx$; $Sx = Sy + Sz \to Sx = Sz + Sy = S(Sz + y) = S(y + Sz) \to x = y + Sz$. Hence $Sx \dot{-} Sy = Sz \leftrightarrow Sx = Sy + Sz$.

(ii) Assume $x < y$, $y < z$, then $\neg y \leq x$, $\neg z \leq y$, so by 2.4(v) $y \dot{-} x = S\,\mathrm{prd}(y \dot{-} x)$, $z \dot{-} y = S\,\mathrm{prd}(z \dot{-} y)$. Let $u = \mathrm{prd}(y \dot{-} x)$, $v = \mathrm{prd}(z \dot{-} y)$, then $y = x + Su$, $z = y + Sv$ by (i), so $z = (x + Su) + Sv = x + (Su + Sv) = x + S(Su + v)$, hence $x < z$.

(iii) Use (ii) and $x \leq y \leftrightarrow x < y \vee x = y$, $y \leq z \leftrightarrow y < z \vee y = z$. □

The preceding propositions may suffice as an illustration of a quantifier-free development; for more examples we refer the reader to the exercises and the literature (see the references in 10.2).

3. Intuitionistic first-order arithmetic HA

3.1. *Description of* **HA**. Intuitionistic first-order arithmetic **HA**, also called *Heyting arithmetic*, is based on **IQC** with equality, and contains a constant 0 (zero), a unary function constant S (successor) and function symbols for all primitive recursive functions. As axioms we have, besides the equality axioms for this language,

$$\neg S0 = 0,$$

defining equations for all primitive recursive functions, such as

$$x + 0 = x, \quad x + Sy = S(x + y),$$

$$x \cdot 0 = 0, \quad x \cdot Sy = x \cdot y + x,$$

and the *induction rule*

$$A(0), A(x) \to A(Sx) \Rightarrow A(t)$$

which implies the *induction-axiom schema*

$$B(0) \wedge \forall x(B(x) \to B(Sx)) \to \forall x B(x).$$

In a natural deduction formulation we can adopt the same rules as indicated at the end of subsection 2.1.

As in the case of **PRA** we have

3.2. Proposition. *In* **HA** $\perp$ *is definable as* $0 = S0$.

Proof. Straightforward extension of the proof for **PRA**. □

In addition, it is now also possible to eliminate disjunction.

Proposition. *In* **HA** $A \vee B$ *is explicitly definable as*

$$\exists x[(x=0 \to A) \wedge (x \neq 0 \to B)].$$

Proof. It is not sufficient to show that in **HA** $(A \vee B) \leftrightarrow \exists x[(x=0 \to A) \wedge (x \neq 0 \to B)]$; we must actually show that for the defined disjunction the rules and/or axioms become derivable from the rules and axioms for the other logical operators, together with induction. Let us verify this for the natural deduction formulation. The $\vee$ I-rules become derived rules by the following deductions

$$\dfrac{\dfrac{\dfrac{A}{0=0\to A}\qquad \dfrac{\dfrac{\dfrac{(1)\;0\neq 0\quad 0=0}{\bot}}{B}\,\bot_i}{0\neq 0\to B}\,(1)}{(0=0\to A)\wedge(0\neq 0\to B)}}{\exists x[(x=0\to A)\wedge(x\neq 0\to B)]}
\qquad\qquad
\dfrac{\dfrac{\dfrac{\dfrac{\dfrac{S0\neq 0\quad S0=0\;(1)}{\bot}}{A}\,\bot_i}{S0=0\to A}\,(1)\qquad \dfrac{B}{S0\neq 0\to B}}{(S0=0\to A)\;\wedge\;(S0\neq 0)\to B)}}{\exists x[(x=0\to A)\;\wedge\;(x\neq 0\to B)]}$$

As to the $\vee$ E-rule, assume derivations

$$\begin{array}{c}[A]\\ \mathcal{D}\\ C\end{array}\qquad\qquad \begin{array}{c}[B]\\ \mathcal{D}'\\ C\end{array}$$

to be given. Then the following proof tree establishes $\vee E$ for the defined $\vee$:

(the double line stands for several deduction steps)

$$\dfrac{\exists x[(x=0\to A)\wedge(x\neq 0\to B)]\qquad \dfrac{\dfrac{\dfrac{\begin{array}{c}\dfrac{\dfrac{(3)\;(x=0\to A)\wedge(x\neq 0\to B)}{x=0\to A}\quad x=0\;(4)}{[A]}\\ \mathcal{D}\\ C\end{array}}{x=0\to C}\,(4)\qquad \dfrac{\begin{array}{c}\dfrac{\dfrac{(3)\;(x=0\to A)\wedge(x\neq 0\to B)}{x\neq 0\to B}\quad \dfrac{\dfrac{\dfrac{(1)\;x=0\quad x=Sy\;(2)}{0=Sy}}{\bot}}{x\neq 0}\,(1)}{[B]}\\ \mathcal{D}'\\ C\end{array}}{x=Sy\to C}\,(2)}{x=x\to C}\,\text{(induction)}\qquad x=x}{C}}{C}\,(3)$$

□

3.3. The further development of arithmetic itself can now be given along routine lines (cf. e.g. Kleene 1952, chapter 8). In the presence of quantifiers, many proofs from **PRA** can be shortened considerably. As an example, let us give another proof of proposition 2.5.

Assume **HA** $\vdash A(0, y)$, **HA** $\vdash A(x, \varphi(x, y)) \to A(Sx, y)$. Then **HA** $\vdash$ $\forall y A(0, y)$, **HA** $\vdash \forall y A(x, y) \to A(Sx, y)$, hence also **HA** $\vdash \forall y A(x, y) \to$ $\forall y A(Sx, y)$; now by ordinary induction applied to $B(x) := \forall y A(x, y)$ we find $\vdash \forall x B(x)$, so $\vdash \forall x \forall y A(x, y)$. □

It should be noted that here, in order to obtain the rule for quantifier-free A, we use induction over the universal formula $\forall y A(x, y)$.

Another example is provided by $\forall xy(x = y \lor x \neq y)$, which has a considerably shorter proof in **HA** (E3.3.1).

3.4. *The relationship between* **PA** $\equiv$ **HA**^{c} *and* **HA**. The theorems on the relation between **IQC** and **CQC** in 2.3.1–8 can be straightforwardly extended to **HA** and **PA** respectively:

PROPOSITION. *Let* ^{g} *be the Gödel–Gentzen translation as defined in* 2.3.4, *except that for prime formulas we now put* $(t = s)^g := t = s$. *Then*
(i) **PA** $\vdash A \Leftrightarrow$ **HA** $\vdash A^g$, **PA** $\vdash A \leftrightarrow A^g$.
(ii) *For A not containing* $\lor$ *or* $\exists$: **HA** $\vdash A$ *iff* **PA** $\vdash A$.

PROOF. The modification in the definition of ^{g} is justified by the fact that in **HA** $\vdash \neg\neg(t = s) \to t = s$.

(i) follows from the corresponding theorem for logic by the observation that for any mathematical axiom A of **PA**, A^g holds in **HA**, and (ii) is immediate from the definition of ^{g}. □

The second part of the proposition explains why so many theorems provable in **PA** can be lifted without any difficulty to **HA**: many well-known and important theorems from elementary number theory are expressible in the $\exists$-free fragment; in fact, they are often purely universal (example: the quadratic reciprocity law). The proposition does not cover theorems of the form $\forall x \exists y A(x, y)$, A quantifier-free (i.e. theorem in Π^0_2-form). However, it can in fact be shown that also

$$\mathbf{PA} \vdash \exists x A(x, y) \Rightarrow \mathbf{HA} \vdash \exists x A(x, y) \tag{1}$$

for primitive recursive A; a proof is given in the next subsection and another proof is given in 5.3. The rule (1) is directly connected with the rule

$\mathrm{MR}_{\mathrm{PR}}$, *Markov's rule for primitive recursive predicates*:

$$\mathrm{MR}_{\mathrm{PR}} \qquad \mathbf{HA} \vdash \neg\neg\exists x A(x, y) \Rightarrow \mathbf{HA} \vdash \exists x A(x, y),$$

$$A \text{ primitive recursive.}$$

For $\mathbf{HA} \vdash \neg\neg\exists x A(x, y) \leftrightarrow \neg\forall x \neg A(x, y)$, and $\neg\forall x \neg A(x, y) \equiv (\exists x A(x, y))^{g}$, so by (i) of the proposition (1) implies $\mathrm{MR}_{\mathrm{PR}}$ and vice versa.

The following is based on the digression 2.3.9–26, and may be skipped.

3.5. PROPOSITION (*corollary to 2.3.26*).

(i) *Let A be essentially isolating or* I*-wiping. Then* $\mathbf{PA} \vdash A \Rightarrow \mathbf{HA} \vdash A$.

(ii) $\mathbf{PA} \vdash \forall x \exists y A(x, y) \Rightarrow \mathbf{HA} \vdash \forall x \exists y A(x, y)$ *for quantifier-free A.*

(iii) $\mathbf{HA}$ *is closed under* $\mathrm{MR}_{\mathrm{PR}}$.

PROOF. (i) The arithmetical axioms of $\mathbf{PA}$ are M-closed under ^{g}, since they are given by spreading formulas and formula schemas; $\bot$ occurs only positively in Γ, namely in $S0 = 0 \to \bot$.

(ii) In $\mathbf{HA} \vdash A(x, y) \leftrightarrow \varphi(x, y) = 0$ for some primitive recursive function symbol φ; $\exists y(\varphi(x, y) = 0)$ is isolating, hence $\mathbf{PA} \vdash \exists y A(x, y) \Leftrightarrow \mathbf{HA} \vdash \exists y A(x, y)$.

(iii) follows by the remark above. □

As an example of a principle of $\mathbf{PA}$ not valid in $\mathbf{HA}$ we consider

3.6. *The least number principle.* In classical arithmetic one can show that $\mathbb{N}$ is well-ordered with respect to arithmetic predicates, this is expressed by the least number principle

$$\mathrm{LNP} \qquad \exists x B(x) \to \exists x [B(x) \wedge \forall y < x \neg B(y)]$$

which is classically equivalent to transfinite induction over $\mathbb{N}$:

$$\mathrm{TI}_{\omega} \qquad \forall y [\forall x < y A(x) \to A(y)] \to \forall x A(x).$$

To see this, apply TI_{ω} to $\neg B(x)$ and take the contraposition; the converse is also easy. Intuitionistically, LNP implies PEM for $\mathbf{HA}$: apply LNP to

$$A(x) := (x = 0 \wedge P) \vee (x = 1 \wedge \neg P) \vee x = 2.$$

$\exists x A(x)$ is obvious; assume there is a least y such that $A(y)$, then $y \in \{0, 1, 2\}$. $y = 2$ would imply $\neg A(0), \neg A(1)$, hence $\neg(0 = 0 \wedge P)$, $\neg(1 = 1 \wedge \neg P)$, i.e. $\neg P$ and $\neg\neg P$ which is contradictory. If $y = 0$, it follows that P; if $y = 1$, it follows that $\neg P$.

On the other hand we have the

PROPOSITION. $\neg\neg$LNP *holds in* **HA**, *that is to say*

$$\mathbf{HA} \vdash \neg\neg[\exists x A(x) \to \exists x(A(x) \wedge \forall y < x \neg A(y))].$$

First proof. We first give a direct proof. Since in intuitionistic logic $\neg\neg(P \to \neg\neg Q) \leftrightarrow (\neg\neg P \to \neg\neg Q) \leftrightarrow \neg\neg(P \to Q)$ it is sufficient to show

$$\exists x A(x) \to \neg\neg\exists x(A(x) \wedge \forall y < x \neg A(y)) \tag{1}$$

which is equivalent to

$$\forall x[A(x) \to \neg\neg\exists z(A(z) \wedge \forall y < z \neg A(y))]. \tag{2}$$

We show this by applying TI_ω, which can easily be shown to follow from the induction principle in **HA**. Assume

$$\forall u < x[A(u) \to \neg\neg\exists z(A(z) \wedge \forall y < z \neg A(y))]. \tag{3}$$

We have to show

$$A(x) \to \neg\neg\exists z(A(z) \wedge \forall y < z \neg A(y))$$

so asume

$$A(x). \tag{4}$$

If $\exists y < x A(y)$, then with (3) $\neg\neg\exists z(A(z) \wedge \forall y < z \neg A(y))$, and if $\neg\exists y < x Ay$, i.e. $\forall y < x \neg A(y)$, we have $A(x) \wedge \forall y < x \neg A(y)$, so $\neg\neg\exists x(A(x) \wedge \forall y < x \neg A(y))$. Therefore we have from (3) and (4)

$$\exists y < x A(y) \vee \forall y < x \neg A(y) \to \neg\neg\exists z(A(z) \wedge \forall y < z \neg A(y)),$$

and since from $P \vee \neg P \to \neg\neg Q$ it follows that $\neg\neg(P \vee \neg P) \to \neg\neg Q$, and hence $\neg\neg Q$, we obtain from (3)

$$A(x) \to \neg\neg\exists x(A(x) \wedge \forall y < x \neg A(y)).$$

So (2) holds. □

Another argument for $\neg\neg$LNP is given below; this depends on the work in section 2.3.

Second proof (a metamathematical argument for $\neg\neg$LNP). We use proposition 2.3.26(ii). $\neg\neg$LNP corresponds to the schema

$$\neg\neg(\exists x P(x) \to \exists x(P(x) \wedge \forall y(y < x \to \neg P(y))).$$

The only positive occurrence of $\forall$ is in $\forall y(y < x \to \neg P(y))$, and this does not contain a strictly positive occurrence of a d-formula; so the conditions for the Mints–Orevkov theorem apply. The arithmetical axioms are M-closed under ^{g} and contain $\bot$ only positively, hence from $\mathbf{PA}(P) \vdash \neg\neg\text{LNP}$ it follows that $\mathbf{HA}(P) \vdash \neg\neg\text{LNP}$, and thus $\mathbf{HA} \vdash \neg\neg\text{LNP}$. Here $\mathbf{HA}(P), \mathbf{PA}(P)$ are formulated as $\mathbf{PA}, \mathbf{HA}$ but with a unary predicate letter P added to the language. □

4. Algorithms

4.1. *The notion of an algorithm.* One possibility for a sharper delimitation of the mathematical objects (= constructions) occurring in constructive mathematics is to require all objects to be representable as algorithms; that is to say each object is represented by an algorithm operating on finite configurations in a completely deterministic way. In particular, we may think of functions from $\mathbb{N}$ to $\mathbb{N}$ as given by algorithms. And since the presentation of (description of, programme for) an algorithm is itself finitely presentable, it also makes sense to have algorithms operate on algorithms, etc.

Since we may assume finite (finitely representable) configurations to be coded in some way by natural numbers, there is no loss of generality if we restrict our attention to algorithms applicable to natural numbers, and which produce natural numbers as output.

We shall not assume algorithms to be *total* on the natural numbers; for certain numbers the computation (i.e. the process of applying the algorithm to an argument) may break off without producing a result, or keep on running without terminating. The primitive recursive functions PRIM, which we encountered in section 1 are a very important class of total algorithms on $\mathbb{N}$, but by no means exhaustive. For example, note that one can effectively enumerate all unary functions in PRIM as $f_0, f_1, f_2, \ldots$ say; then f given by $f(x) := f_x(x) + 1$ is algorithmic and total but not in PRIM.

A more general class of total algorithms on $\mathbb{N}$ is given by the following

Definition. The class TREC of *total recursive functions* is the least class of functions satisfying the closure conditions of PRIM (with TREC replacing PRIM in the clauses) and in addition:

(iv) If g is in TREC and $\forall y\, \exists x(g(x, \vec{y}) = 0)$, then so is f given by

$$f(\vec{y}) = \min_x[g(x, \vec{y}) = 0] \text{(closure under } \textit{minimalization}).$$

Here $\min_x[g(x, \vec{y}) = 0]$ is the least x for which $g(x, \vec{y}) = 0$ if such an x exists, undefined otherwise; $\min_x$ is called the *minimum operator*. □

Below we use $\simeq$ as in logic with existence predicate: $t \simeq t'$ means that t is defined iff t' is defined, and if they are defined, then their values are equal.

We also introduce the class REC of partial recursive functions by the

DEFINITION. The class of (*partial*) *recursive functions* is defined by

(i) TREC $\subset$ REC.

(ii) REC is closed under composition, that is to say if $g_1, \ldots, g_m \in$ REC, $h \in$ REC then also f defined by $f(\vec{x}) \simeq h(g_1(\vec{x}), \ldots, g_m(\vec{x}))$ belongs to REC. Here the right hand side is defined iff $g_1(\vec{x}), \ldots, g_m(\vec{x})$ are all defined and have values $y_1, \ldots, y_m$ say, and $h(y_1, \ldots, y_m)$ is defined (the functions are *strict* in the sense of E-logic and E^+-logic).

(iii) if $g \in$ TREC then f, defined by $f(\vec{y}) = \min_x[g(x, \vec{y}) = 0]$, is in REC. □

One easily sees that the functions in REC are algorithmically computable. On the other hand, the definitions of REC and TREC are concise and easy to manage, but do not give a very clear idea of the extent of the notion of recursive function. It turns out however that there are many equivalent formulations describing the same class of functions: functions computable by Turing machines or register machines, λ-definable functions, functions definable by the Gödel–Herbrand equation calculus, Post's normal systems, Markov algorithms (cf. 1.4.6); and analysis of the informal concept of an algorithm (mechanically computable function, or function computable by a human operator in a mechanical way), has provided strong grounds for identifying these informal concepts with the mathematically precise notions just mentioned. It would carry us too far to present here the details of such a concept analysis; see e.g. Kleene (1952, chapter 13).

Later on we shall not be overly precise in our use of "recursive". When we want to stress that $f \in$ REC is total we call f a total recursive function, and we use "*partial recursive function*" to keep on the safe side, or when the reader deserves a warning.

Intuitively, we think of an algorithm as an effective procedure that operates in discrete steps on a finite configuration of some kind, which is kept in a "working space" (a tape, a register, a piece of paper for a human agent). The algorithm is given by a finite set of instructions, which we call a programme.

The following features are inherent to the notion of an algorithm:

(1) A *programme* (for an algorithm) is described by means of finitely many symbols in some fixed alphabet.

(2) The execution of the programme is deterministic: each step uniquely determines the next step in the computation.

(3) At each step only finitely many symbols are stored in the memory, and so the next stage of a computation is always determined from a finite set of data.

(4) There is no upperbound to the size of the memory or the length of strings of symbols handled.

4.2. On the basis of (1)–(4) above the following assumptions seem to be justified.

(A) All $f \in$ REC are given by algorithms.

(B) Programmes and computations are coded by natural numbers.

(C) Let us call a *predicate* or *relation* algorithmic iff its characteristic function is algorithmic. Then "z codes a (terminating) computation according to programme x for arguments $\vec{y}$" is an algorithmic predicate of x, $\vec{y}$ and z. Let us denote this predicate by T ("*Kleene's T-predicate*"). There is a total algorithm U (the *result-extracting function*) which extracts the result from the code for a terminating computation. The instruction "Apply the programme with code z to $\vec{x}$" is itself algorithmic in z and $\vec{x}$.

Thus each algorithmic φ has a code, x say, and is representable as

$$\varphi(\vec{y}) \simeq U(\min_z[\chi_T(x, \vec{y}, z) = 0]),$$

or $\min_z T(x, \vec{y}, z)$ for short, where χ_T is the characteristic function of the predicate T.

(D) Let $\{x\}(y_1, \ldots, y_n)$ be the (partially defined) expression denoting the result (if existing) of applying programme x to arguments $y_1, \ldots, y_n$. The (in general partial) function defined by the algorithm with code x is usually indicated by means of the so-called "*Kleene brackets*": $\{x\}$; sometimes, if we want to indicate the number of arguments, we write $\{x\}^n$.

If we consider $\{x\}(\vec{y}, \vec{z})$, this represents for each fixed $\vec{y}$ an algorithm applied to $\vec{z}$. The code of this algorithm depends on x and $\vec{y}$; the code may be assumed to be found by an algorithm from x and $\vec{y}$, i.e. there is a *total* algorithm φ such that $\{x\}(\vec{y}, \vec{z}) \simeq \{\varphi(x, \vec{y})\}(\vec{z})$(the *smn-theorem*).

(E) Any term constructed by means of $\{\ \ \}(\ \ )$, variables and constant numbers is an algorithmic function of its free variables (the value of such a term is regarded as defined if and only if all its subterms are defined).

4.3. For the mathematical consequences of "everything is algorithmic", the properties (A)–(E) of algorithms listed in 4.2 are sufficient. For metamathematical work, one wants to know more; and the detailed analyses of the notion of an algorithm mentioned in 4.1, which motivate the identification of REC with the class of algorithmically computable functions, show that T and U in 4.2(C) and φ in 4.2(D) can in fact be chosen in PRIM.

4.4. DEFINITION. A predicate S is said to be *recursively enumerable* (abbreviation: r.e.) if, for some recursive R, $S(\vec{y}) :\equiv \exists x R(x, \vec{y})$.

If S is r.e. then S is the domain of a recursive function, namely $\min_x[\chi_R(x, \vec{y}) = 0]$ as a function of y, and conversely if $S = \mathrm{dom}(f)$ and f has code x, then $S(\vec{y}) \leftrightarrow \exists z\, T(x, \vec{y}, z)$. Thus we may also define: S is r.e. iff S is the domain of a recursive function. □

PROPOSITION. *$X = \{x : \exists y\, Txxy\}$ is r.e., but not recursive.*

PROOF. Suppose X to be recursive, then $\mathbb{N} \setminus X$ is recursive, hence for some x_0 $\mathbb{N} \setminus X = \{x : \exists y\, T(x_0, x, y)\}$. Then $x_0 \in \mathbb{N} \setminus X \leftrightarrow \exists y\, T(x_0, x_0, y) \leftrightarrow x_0 \in X$, which is contradictory. □

4.5. DEFINITION. Two sets $X, Y \subset \mathbb{N}$, $X \cap Y = \emptyset$ are said to be *recursively inseparable* if there is no recursive set Z such that $X \subset Z$, $Y \subset \mathbb{N} \setminus Z$.

PROPOSITION. *The r.e. sets*

$$A \equiv \{x : \exists y(Txxy \wedge Uy = 0)\}, \quad B \equiv \{x : \exists y(Txxy \wedge Uy = 1)\}$$

are recursively inseparable.

PROOF. Suppose $B \subset C$, $A \subset \mathbb{N} \setminus C$, C recursive; let z be a code for χ_C, then

$$x \in A \rightarrow x \notin C \leftrightarrow \exists y(Tzxy \wedge Uy = 1),$$

$$x \in B \rightarrow x \in C \leftrightarrow \exists y(Tzxy \wedge Uy = 0).$$

Now $z \notin C \leftrightarrow \exists y(Tzzy \wedge Uy = 1) \leftrightarrow z \in B$, hence $z \in C$; similarly $z \in C \rightarrow z \notin C$. Thus we have obtained a contradiction. □

4.6. PROPOSITION (*recursion theorem*).

$$\forall x\, \exists y\, \forall \vec{z}(\{x\}(y, \vec{z}) \simeq \{y\}(\vec{z}));$$

more generally, for every x there exists a (primitive) recursive φ, such that

$$\forall \vec{y}\vec{z}\big(\{x\}^{n+m+1}(\varphi(\vec{y}), \vec{y}, \vec{z}) \simeq \{\phi(\vec{y})\}^{m}(\vec{z})\big).$$

PROOF. Let ψ be such that (smn-theorem, cf. 4.2(D))

$$\{\psi(u, y', \vec{y})\}(\vec{z}) \simeq \{u\}(y', \vec{y}, \vec{z}).$$

Let v be a code for the partial recursive function $\lambda u\vec{y}\vec{z}.\{x\}(\psi(u, u, \vec{y}), \vec{y}, \vec{z})$ and take $\varphi(\vec{y}) \equiv \psi(v, v, \vec{y})$, then

$$\{\varphi(\vec{y})\}(\vec{z}) \simeq \{v\}(v, \vec{y}, \vec{z}) \simeq \{x\}(\psi(v, v, \vec{y}), \vec{y}, \vec{z})$$
$$\simeq \{x\}(\varphi(\vec{y}), \vec{y}, \vec{z}). \quad \square$$

An application is given in the following

EXAMPLE. The recursive functions are closed under a more general form of *minimalization*: if $\psi(x, \vec{y})$ is partial recursive then so is

$$\psi'(\vec{y}) \simeq \min_x[\psi(x, \vec{y}) = 0].$$

Here $\psi'(\vec{y})$ is defined and equal to x iff $\psi(x, \vec{y}) = 0$ and for all $x' < x$ $\psi(x', \vec{y})$ is defined and unequal to 0. To see this we argue as follows. Let ψ have a code z, and let φ be a total recursive function such that $\{\varphi z\}(x, \vec{y}) \simeq \{z\}(x+1, \vec{y})$ for all x and $\vec{y}$. Let $\{v\}$ be a partial recursive function such that

$$\{v\}(u, z, \vec{y}) \simeq \begin{cases} 0 & \text{if } \{z\}(0, \vec{y}) \simeq 0, \\ \{u\}(\varphi z, \vec{y}) + 1 & \text{if } \{z\}(0, y) \text{ defined and } \neq 0. \end{cases}$$

The definition as it stands does not immediately show that $\{v\} \in \mathrm{REC}$. To see this, write

$$\{v\}(u, z, \vec{y}) \simeq$$
$$\mathrm{sg}(\{z\}(0, \vec{y})) \cdot \big[U\big(j_2 \min_w[(T(z, 0, \vec{y}, j_1 w) \wedge Uj_1 w = 0)$$
$$\vee \big(T(z, 0, \vec{y}, j_1 w) \wedge Uj_1 w > 0 \wedge T(u, \varphi z, \vec{y}, j_2 w)\big)]\big) + 1\big].$$

By the recursion theorem there is a v_0 such that $\{v_0\}(z, \vec{y}) \simeq 0$ if $\{z\}(0, \vec{y}) = 0$, $\{v_0\}(z, \vec{y}) \simeq \{v_0\}(\varphi z, \vec{y}) + 1$ if $\{z\}(0, \vec{y}) > 0$. It is not hard to see that $\{v_0\}(z, \vec{y}) \simeq \min_x[\{z\}(x, \vec{y}) = 0]$. $\square$

In a similar way we can show closure under a general form of primitive recursion: given $\psi, \chi \in \mathrm{REC}$ there is a $\varphi \in \mathrm{REC}$ such that

$$\begin{cases} \varphi(0, \vec{y}) \simeq \psi(\vec{y}), \\ \varphi(x+1, \vec{y}) \simeq \chi(\varphi(x, \vec{y}), \vec{y}, x). \end{cases}$$

4.7. The following type of definition is legitimate and can be used for introducing a partial recursive function:

$$\psi(\vec{x}) \simeq \begin{cases} f_0(\vec{x}) & \text{if} \quad \exists y R_0(\vec{x}, y), \\ f_1(\vec{x}) & \text{if} \quad \exists y R_1(\vec{x}, y), \\ & \quad \cdots \\ f_n(\vec{x}) & \text{if} \quad \exists y R_n(\vec{x}, y), \\ \text{undefined otherwise,} \end{cases}$$

where $R_1, \ldots, R_n$ are recursive, $f_1, \ldots, f_n$ partial recursive, and the $\exists y R_i(\vec{x}, y)$ mutually exclusive. To spell it out: suppose $f_0, \ldots, f_n$ to have codes $e_0, \ldots, e_n$, then

$$\psi(\vec{x}) \simeq Uj_1\big(\min_y[\exists i \leq n(T(e_i, \vec{x}, j_1 y) \wedge R_i(\vec{x}, j_2 y))]\big).$$

4.8. Let us call an enumeration $\langle f_n \rangle_n$ of unary algorithmic functions *effective* if $f_n(m)$ is algorithmic in n and m. The following proposition shows that the particular coding chosen for algorithms is irrelevant, as long as the codings are effective.

PROPOSITION. *Let* $\langle f_n \rangle_n, \langle g_n \rangle_n$ *be two effective enumerations of algorithms; then there is an algorithmic permutation* θ *of* $\mathbb{N}$ *such that* $f_{\theta n} = g_n$ *(i.e.* θ *is a bijection from* $\mathbb{N}$ *to* $\mathbb{N}$*).*

PROOF. See e.g. Richman (1983). □

5. Some metamathematics of HA

5.1. *Markov's rule.* Below we shall present a simple proof (Friedman 1978 and Dragalin 1980) of closure of **HA** under the rule $\mathrm{MR}_{\mathrm{PR}}$ (Markov's rule for primitive recursive predicates):

$\mathrm{MR}_{\mathrm{PR}}$ $\quad \vdash \neg\neg\exists x Ax \Rightarrow \vdash \exists x Ax \quad (A(x) \in \mathrm{PRIM});$

another proof was given in 3.5(ii). In fact, more is true: **HA** is also closed under the rule

MR $\quad \vdash \forall x(Ax \vee \neg Ax)$ and $\vdash \neg\neg\exists x Ax \Rightarrow \vdash \exists x Ax$

(for a proof see E4.4.8).

5.2. Definition. Let A be any fixed formula of **HA**. For arbitrary formulae B we define B^A by
(i) $P^A := P \vee A$ for P prime, $\perp^A := A$,
(ii) $(B \circ C)^A := B^A \circ C^A$ for $\circ \in \{\wedge, \vee, \to\}$,
(iii) $[(Qx)B]^A := (Qx)B^A$ for $Q \in \{\forall, \exists\}$, with the proviso that where x occurs free in A, we first replace all free occurrences of x in B by a variable not occurring free in A. □

5.3. Lemma. *Let* $\Gamma^A \equiv \{B^A : B \in \Gamma\}$, *and let* $\vdash$ *be derivability in* **IQC**. *Then*
(i) $A \vdash B^A$, $\Gamma \vdash B \Rightarrow \Gamma^A \vdash B^A$;
(ii) $\mathbf{HA} \vdash B \Rightarrow \mathbf{HA} \vdash B^A$;
(iii) *for any term* t

$$\mathbf{HA} \vdash \neg\neg\exists x(t(x,y) = 0) \Rightarrow \mathbf{HA} \vdash \exists x(t(x,y) = 0).$$

Proof. (i) $A \vdash B^A$ is proved by a straightforward formula induction. The second part uses induction on the height of deduction trees of $\Gamma \vdash B$. Let us consider the induction step where the final rule applied is $\perp_i$, so the derivation is of the form

$$\begin{array}{c} [\Gamma] \\ \mathcal{D} \\ \perp \\ \hline B \end{array}$$

By the induction hypothesis and $A \vdash B^A$ we have deductions

$$\begin{array}{c} [\Gamma^A] \\ \mathcal{D}' \\ A \end{array} \qquad \begin{array}{c} [A] \\ \mathcal{D}'' \\ B^A \end{array}$$

and thus

$$\begin{array}{l}[\Gamma^A]\\ \mathscr{D}'\\ [A]\\ \mathscr{D}''\\ B^A\end{array}$$

All other steps are equally trivial.

(ii) Since by (i) the ^{A}-translation preserves derivability it suffices to show that **HA** proves the ^{A}-translation of its axioms. E.g. $(Sx = 0 \to \bot)^A \equiv (Sx = 0 \vee A) \to A$ which holds in **HA**.

(iii) Apply (ii) with $A := \exists x(t(x, y) = 0)$. □

From this lemma we immediately get

5.4. COROLLARY. **HA** *is closed under* $\mathrm{MR_{PR}}$. □

Another application of the translation ^{A} is the following.

5.5. THEOREM. **HA** *is closed under the "independence of premiss" rule*

IPR $\quad \vdash \neg A \to \exists x B \Rightarrow \vdash \exists x(\neg A \to B)$ *(x not free in A).*

PROOF. (Visser 1981, part 6, 3.1). Assume $\mathbf{HA} \vdash \neg A \to \exists x B$. Then by 5.3(ii)

$$\mathbf{HA} \vdash (\neg A)^{\neg\neg A} \to \exists x(B)^{\neg\neg A}. \tag{1}$$

Now observe that by induction on D we can prove for any C,

$$D \vdash \neg C \to (D^C \leftrightarrow D),$$

and thus also

$$\vdash D^C \to (\neg C \to D), \tag{2}$$

hence

$$\vdash (B)^{\neg\neg A} \to (\neg A \to B)$$

and so

$$\mathbf{HA} \vdash (\neg A)^{\neg\neg A} \to \exists x(\neg A \to B). \tag{3}$$

On the other hand, $(\neg A)^{\neg\neg A} \equiv (A \to \bot)^{\neg\neg A}$ which is equivalent to

$A^{\neg\neg A} \to \neg\neg A$. If we apply (2) with $D \equiv \neg A$, we find $\vdash A^{\neg\neg A} \to (\neg A \to A)$; also, by propositional logic $\vdash (\neg A \to A) \to \neg\neg A$, so $\vdash A^{\neg\neg A} \to \neg\neg A$, hence $\vdash (\neg A)^{\neg\neg A}$ and therefore with (3)

$$\mathbf{HA} \vdash \exists x(\neg A \to B). \quad \square$$

5.6. *Disjunction and existence properties.* All well-known intuitionistic formalisms containing arithmetic have the following properties for sentences

DP $\quad \vdash A \vee B \Rightarrow \vdash A \quad \text{or} \quad \vdash B$

(the *disjunction property*) and *explicit definability for numbers*

EDN $\quad \vdash \exists x A(x) \Rightarrow \vdash A(\bar{n}) \quad$ for some numeral $\bar{n}$.

This agrees well with the informal reading of $\vee$ and $\exists$: a proof of $A \vee B$ is given as a proof of A or a proof of B, i.e. from an intuitionistic argument for $A \vee B$ we should be able to extract an intuitionistic argument either for A or for B. Now, if in an intuitionistic formalism **S** we prove $\mathbf{S} \vdash A \vee B$, the formal proof is justified by an informal intuitionistic argument for $A \vee B$, from which we can then extract an intuitionistic (informal) argument for either A or B. However, it is not self-evident that this informal argument can in turn be translated into a formal proof for $\mathbf{S} \vdash A$ or $\mathbf{S} \vdash B$. This is a kind of closure property on the set of proofs represented by formal deductions in **S**, and it is certainly possible to construct systems which are intuitionistically obviously correct, though somewhat artificial, for which DP fails (see e.g. Troelstra 1973, 1.11.2).

The same, mutatis mutandis, might be said with respect to EDN; in fact, there is a close connection between the validity of DP and EDN, as shown by Friedman (cf. 10.5).

We shall now show how to prove DP and EDN (more in fact) with help of the (*Aczel-*) *slash relation.*

5.7. Definition. Γ is a set of sentences of **HA**, and deducibility $\vdash$ refers to deducibility in **HA**.

For sentences A we define

(i) $\Gamma | A \quad := \Gamma \vdash A$ for A prime or $A \equiv \bot$,
(ii) $\Gamma | A \wedge B := \Gamma | A$ and $\Gamma | B$,
(iii) $\Gamma | A \vee B := \Gamma | A$ or $\Gamma | B$,
(iv) $\Gamma | A \to B := \Gamma | A \Rightarrow \Gamma | B$ and $\Gamma \vdash A \to B$,
(v) $\Gamma | \forall x Ax := \Gamma \vdash \forall x Ax$ and $\Gamma | A\bar{n}$ for all numerals $\bar{n}$,
(vi) $\Gamma | \exists x Ax := \Gamma | A\bar{n}$ for some numeral $\bar{n}$. $\square$

LEMMA. $\Gamma|A \Rightarrow \Gamma \vdash A$.

PROOF. By induction on A. □

5.8. LEMMA. *Let $t_1, \ldots, t_n$ be closed terms, $\bar{t}_1, \ldots, \bar{t}_n$ the numerals obtained by evaluating them. Then for closed $A(t_1, \ldots, t_n)$*

$$\Gamma|A(t_1, \ldots, t_n) \Leftrightarrow \Gamma|A(\bar{t}_1, \ldots, \bar{t}_n).$$

PROOF. First observe that $\vdash t = \bar{t}$ for closed terms t. This is most easily proved as follows: assume all primitive recursive function symbols to be enumerated as $\varphi_0, \varphi_1, \varphi_2, \ldots$ such that φ_n is defined using only φ_i for $i < n$. Establish by induction on n that

$$\vdash \varphi_n(\bar{m}_1, \ldots, \bar{m}_p) = \overline{\varphi_n(m_1, \ldots, m_p)}$$

is provable for all $m_1, \ldots, m_p$, then prove by induction on the complexity of t that $\vdash t = \bar{t}$. (N.B. The proof actually shows **PRA** $\vdash t = \bar{t}$.) From **HA** $\vdash t_i = \bar{t}_i$ and the equality axioms/rules we get $\Gamma \vdash A(t_1, \ldots, t_n) \Leftrightarrow \Gamma \vdash A(\bar{t}_1, \ldots, \bar{t}_n)$.

The lemma itself is now proved by induction on A. For example, let $A \equiv B \to C$, and write $A', B', C', \ldots$ for the result of replacing in $A, B, C, \ldots$ the terms $t_1, \ldots, t_n$ by $\bar{t}_1, \ldots, \bar{t}_n$; then by the induction hypothesis,

$$\begin{aligned} \Gamma|B \to C &\Leftrightarrow (\Gamma|B \Rightarrow \Gamma|C) \quad \text{and} \quad \Gamma \vdash B \to C \\ &\Leftrightarrow ((\Gamma|B' \Rightarrow \Gamma|C') \quad \text{and} \quad \Gamma \vdash B' \to C') \\ &\Leftrightarrow \Gamma|B' \to C'. \end{aligned}$$

Or if $A \equiv \forall x B(x)$, we have $\Gamma|\forall x B(x) \Leftrightarrow (\Gamma \vdash \forall x B(x)$ and for all n $\Gamma|B(\bar{n}))$, which is equivalent to $\Gamma \vdash \forall x B'(x)$ and for all n $\Gamma|B'(\bar{n})$, and thus to $\Gamma|\forall x B'(x)$. □

5.9. THEOREM. *If $\Gamma|B$ for all $B \in \Gamma$, then*

$$\Gamma \vdash A \Rightarrow \Gamma|A.$$

PROOF. Notationally the simplest solution is a proof by induction on the length of a proof in a Hilbert-type axiomatization of **HA** (2.4.1, 3.1).

We shall check some cases and leave the others to the reader.

For formulas F with $\mathrm{FV}(F) \subset \{x, \ldots, x_n\}$ we define $\Gamma | F :=$ $\Gamma \vdash \forall x_1 \ldots x_n F$. So in order to prove the proposition we only have to prove $\Gamma | F(\overline{m}_1, \ldots, \overline{m}_n)$ for all choices of numerals $\overline{m}_1, \ldots, \overline{m}_n$.

Case 1. If $A \in \Gamma$, then $\Gamma \vdash A$ by assumption.

We first consider some logical axioms. It is no restriction to consider only closed instances of the axioms.

Case 2. $A \wedge B \to A$. By definition $\Gamma | A \wedge B \Rightarrow \Gamma | A$, and by logic $\Gamma \vdash A \wedge B \to A$ trivially, hence $\Gamma | A \wedge B \to A$.

Case 3. $A \to (B \to A \wedge B)$. We have to show $\Gamma \vdash A \to (B \to A \wedge B)$ and $\Gamma | A \Rightarrow \Gamma | B \to (A \wedge B)$. The first is obvious; as to the second, let $\Gamma | A$ and assume $\Gamma | B$; then $\Gamma | A \wedge B$, and by the lemma of 5.7 also $\Gamma \vdash A \wedge B$, and thus $\Gamma | B \to A \wedge B$. Thus we have shown $\Gamma | A \Rightarrow (\Gamma | B \to A \wedge B)$.

Case 4. To show $\Gamma | (A \to C) \to [(B \to C) \to (A \vee B \to C)]$ (1). Consider

$$\Gamma | A \to C \ (2), \qquad \Gamma | B \to C \ (3),$$
$$\Gamma | A \vee B \ (4), \qquad \Gamma | A \vee B \to C \ (5),$$
$$\Gamma | (B \to C) \to (A \vee B \to C) \ (6), \qquad \Gamma | C \ (7).$$

Call the corresponding statements with $|$ replaced by $\vdash$ (1′)–(7′). We will repeatedly use the lemma of 5.7. Since (1′) holds, we have to show (2) $\Rightarrow$ (6). So assume (2), then also (2′). In order to obtain (6), we have to show (6′) – which follows from (1′) and (2′) – and (3) $\Rightarrow$ (5). Assume (3), hence (3′) holds. In order to obtain (5), we have to shown (5′) – which follows from (3′) and (6′) – and (4) $\Rightarrow$ (7). Now, by definition (2), (3) and (4) yield (7); so on the assumption of (2) and (3) we get (5), which was what we required.

Case 5. $\Gamma | At \to \exists x Ax$. Let $\Gamma | At$, t a *closed* term. Then there is a numeral $\bar{n}$ such that $\vdash t = \bar{n}$, hence (lemma 5.8) $\Gamma | A\bar{n}$, so $\Gamma | \exists x Ax$. Then $\Gamma | At \to \exists x Ax$.

We now consider the rules

Case 6. $\Gamma | A \to B$, $\Gamma | A \Rightarrow \Gamma | B$. By the first assumption $\Gamma | A \Rightarrow \Gamma | B$, hence $\Gamma | B$.

Case 7. $\Gamma | A(x) \Rightarrow \Gamma | \forall x A(x)$; x not free in Γ. This holds by definition.

From the arithmetical axioms, we consider only the induction schema.

Case 8. Assume

$$\Gamma | A(0), \Gamma | \forall x (A(x) \to A(Sx)),$$

then in particular, for all n

$$\Gamma | A\bar{n} \to A(S\bar{n})$$

and hence by induction for all n $\Gamma | A\bar{n}$, therefore $\Gamma | \forall x A(x)$, etc. □

5.10. THEOREM. **HA** *has* DP *and* EDN.

PROOF. Because of $\mathbf{HA} \vdash A \vee B \leftrightarrow \exists x[(x = 0 \to A) \wedge (x \neq 0 \to B)]$, it suffices to consider EDN. Suppose $\mathbf{HA} \vdash \exists x Ax$; then $|\exists x Ax$, so $|A\bar{n}$ and therefore $\vdash A\bar{n}$, for some $\bar{n}$. □

REMARK. The fact that DP holds also shows that **HA** is properly contained in **PA**: let A be a Gödel-sentence (see e.g. Kleene 1952, section 42), i.e. $\mathbf{HA} \nvdash A$, $\mathbf{HA} \nvdash \neg A$; clearly $\mathbf{PA} \vdash A \vee \neg A$, but not $\mathbf{HA} \vdash A \vee \neg A$ since this would conflict with DP.

In fact the soundness theorem for the slash yields more:

5.11. PROPOSITION. *If* $\Gamma | C$ *for all* $C \in \Gamma$, *then* $\mathbf{HA} + \Gamma$ *has* DP *and* EDN.

PROOF. As for 5.10. □

In particular, we have the

5.1.2. COROLLARY. *For sentences* C, *if* $C|C$ *then*

$$\mathbf{HA} \vdash C \to \exists x Ax \Rightarrow \mathbf{HA} \vdash C \to A\bar{n} \quad \text{for some } n.$$

PROOF. The proof is trivial once the deduction theorem $\Gamma, A \vdash B \Rightarrow \Gamma \vdash A \to B$ is available for the formalism of **HA**. For Hilbert-type systems the proof of the deduction theorem is implicit in the proof showing equivalence with natural deduction formulations (cf. 2.4.2) □

It is interesting to note that the slash operator gives in fact an optimal criterion:

5.13. PROPOSITION. *Suppose* C *closed, and for all closed* $\exists x Ax$

$$\mathbf{HA} \vdash C \to \exists x Ax \Rightarrow \mathbf{HA} \vdash C \to A\bar{n} \text{ for some } n \in \mathbb{N},$$

then for all closed D *such that* $C \vdash D$ *also* $C|D$; *which implies* $C|C$.

Proof. By induction on D. We again use the deduction theorem.

Case (a). For D prime the conclusion is obvious.

Case (b). $D \equiv \exists x\, Ax$, and $C \vdash \exists x\, Ax$, then by hypothesis $\vdash C \to A\bar{n}$ for some n, i.e. $C \vdash A\bar{n}$; then by induction hypothesis $C|A\bar{n}$, and thus $C|\exists x\, Ax$. We leave the other cases to the reader. □

The preceding proposition also shows that the slash relation respects logical equivalence:

5.14. Corollary. *If $C|C$ and $\vdash C \leftrightarrow C'$, then $C'|C'$.* □

5.15. For the actual application of 5.12 it is convenient to have a syntactic criterion on C which is sufficient for $C|C$. We present one here.

Definition. The class $\mathcal{RH}$ of *Rasiowa–Harrop formulas* is defined as the class of those formulas which have no $\exists$ or $\vee$ as a s.p.p. (strictly positive part, cf. 2.3.23). □

We have the following

5.16. Proposition. *Let C be a sentence in $\mathcal{RH}$, and $\exists x\, Dx$, $A \vee B$ sentences. Then*
(i) *if $C \in \mathcal{RH}$ then $C|C$; and*
(ii) $\vdash C \to \exists x\, Dx \Rightarrow\, \vdash C \to D\bar{n}$ *for some numeral $\bar{n}$.*
$\vdash C \to A \vee B \Rightarrow\, \vdash C \to A$ *or* $\vdash C \to B$.

Proof. The proof of (i) follows, under the assumption $C \in \mathcal{RH}$, from
(a) if D is a s.p.p. of C, then $C|D$;
(b) if $A \wedge B$ is an s.p.p. of C, $C|A$ and $C|B$, then $C|A \wedge B$;
(c) if $A \to B$ is an s.p.p. of C and $C|B$ then $C|A \to B$;
(d) if $\forall x\, A$ is an s.p.p. of C and for all n $C|A[x/\bar{n}]$, then $C|\forall x\, A$;
(e) if P is prime and P an s.p.p. of C, then $C|P$.
Part (ii) readily follows from (i) with help of the deduction theorem. □

5.17. Corollary. *Since $\neg A \in \mathcal{RH}$ for all A, we have in* **HA**:
(iii) $\vdash \neg A \to \exists x\, B(x) \Rightarrow\, \vdash \neg A \to B(\bar{n})$ *for some numeral $\bar{n}$*,
(iv) $\vdash \neg A \to B \vee C \Rightarrow\, \vdash \neg A \to B$ or $\vdash \neg A \to C$. □

6. Elementary analysis and elementary inductive definitions

6.1. In this section we discuss the definition of sets by means of inductive definitions, possibly with function parameters, and show how to replace inductive definitions by explicit definitions in arithmetic or elementary analysis **EL**, an extension of **HA** with variables and quantifiers for number-theoretic functions (i.e. functions in $\mathbb{N} \to \mathbb{N}$). In the next section we shall use the theory of inductive definitions to describe, in outline, a smooth formalization of basic recursion theory. The results of this section are also used in section 9.7.

6.2. *Elementary analysis* **EL**. We first describe elementary analysis, an extension of **HA** which is in fact conservative over **HA**: if we interpret the function variables of **EL** as ranging over total recursive functions, the axioms and rules of **EL** involving functions translate into theorems and derivable rules of **HA**; the details of this proof rely on the formalization of elementary recursion theory in **HA**, which is carried out in the next section.

The language $\mathscr{L}$(**EL**) contains, in addition to the symbols of **HA**, unary function variables, denoted by metavariables $\alpha, \beta, \gamma, \delta, \ldots$, the *application operator* Ap, the *abstraction operator* λ and the *recursor* $\boldsymbol{r}$. As usual we write $\varphi(t)$, or even φt, for $\mathrm{Ap}(\varphi, t)$.

Functors (function terms) and (numerical) *terms* are defined simultaneously by

(i) the clauses for terms in **HA**;
(ii) function variables and unary function constants are functors;
(iii) if φ is a functor and t a term, then $\varphi(t)$ is a term;
(iv) if t is a term, then $\lambda x.t$ is a functor;
(v) if t and t' are terms, and φ is a functor then $\boldsymbol{r}(t, \varphi, t')$ is a term.

The *logic* of **EL** is two-sorted intuitionistic predicate logic. As non-logical axioms we have the axioms of **HA**, with induction extended to formulas of $\mathscr{L}$(**EL**),

CON $\quad (\lambda x.t)s = t[x/s] \quad (\lambda\text{-}conversion).$

axioms for the recursor (φ unary, $\varphi(x, y) := \varphi j(x, y)$)

$$\text{REC} \qquad \begin{cases} \boldsymbol{r}(t, \varphi, 0) = t, \\ \boldsymbol{r}(t, \varphi, St') = \varphi(\boldsymbol{r}(t, \varphi, t'), t'), \end{cases}$$

and the *quantifier-free axiom of choice*

$$\text{QF-AC} \qquad \forall x \exists y A(x, y) \to \exists \alpha \forall x A(x, \alpha(x))$$

$$(\alpha \notin \mathrm{FV}(A),\ A \text{ quantifier-free}).$$

6.3. Notations in **EL**. *Equality between functors* is defined:

$$\varphi = \psi := \forall x(\varphi x = \psi x).$$

Coding of n-tuples, finite sequences etcetera, can be lifted from numbers to functions via λ-abstraction:

$$\nu^p(\alpha_1, \ldots, \alpha_p) := \lambda x.\nu^p(\alpha_1 x, \ldots, \alpha_p x)$$

$$j_k^p(\alpha) := \lambda x. j_k^p(\alpha x) \quad \text{for } 1 \le k \le p,$$

$$\langle \alpha_0, \ldots, \alpha_p \rangle := \lambda x.\langle \alpha_0 x, \ldots, \alpha_p x \rangle.$$

By means of $\boldsymbol{r}$ we can introduce $\bar{\alpha}x$ satisfying

$$\bar{\alpha}0 = \langle\ \rangle,$$

$$\bar{\alpha}(x+1) = \bar{\alpha}x * \langle \alpha x \rangle,$$

i.e. $\bar{\alpha}(x+1) = \langle \alpha 0, \ldots, \alpha x \rangle$ (take $\bar{\alpha}x := r(\langle\ \rangle, \lambda x.(j_1 x * \langle \alpha(j_2 x)\rangle), x))$.

6.4. *Inductive definitions.* Inductive definitions are used frequently in mathematics for defining a set of objects as the least set satisfying certain conditions. For example, the A-generated subgroup H of a group G, for some inhabited $A \subset G$, is the least subgroup H of G containing A; we think of H as obtained by using *exclusively* the following clauses
(i) $a \in A \to a \in H$,
(ii) $a, b \in H \to a \cdot b^{-1} \in H$,
(here $\cdot$ is the group multiplication, $^{-1}$ the inverse operation in G). In other words, we put nothing in H except what is obtained by repeatedly applying (i) and (ii): H is *generated* by (i) and (ii).

Let us look at some arithmetical examples.

Example 1. The following clauses generate the graph P of the pairing function $(n, m) \mapsto 2^n 3^m$:

P1 $\quad (0,0,1) \in P$,

P2 $\quad (n, m, k) \in P \to (n+1, m, 2k) \in P$,

P3 $\quad (n, m, k) \in P \to (n, m+1, 3k) \in P$,

and $P = \{(n, m, 2^n 3^m) : n \in \mathbb{N} \wedge m \in \mathbb{N}\}$.

Example 2.

Q1 $\quad 1 \in Q$,

Q2 $\quad n \in Q \wedge m \in Q \to 2^n \cdot 3^m \in Q$.

In this case an explicit definition is not so easy to give. Q contains

$$1, 2 \cdot 3 = 6, 2^1 \cdot 3^6, 2^6 \cdot 3, 2^6 \cdot 3^6, 2^{2^6 \cdot 3} \cdot 3, \ldots$$

The inductive clauses in these examples are closure conditions on the set X to be defined, of the form

$$t_0 \in X \wedge \cdots \wedge t_{p-1} \in X \wedge C \to t \in X, \tag{1}$$

where C does not contain X, and $t_0, \ldots, t_{p-1}$, t and C may contain certain numerical parameters $\vec{y}$. Slightly more generally we may consider clauses of the form

$$\forall i < t'(t_i \in X) \wedge C \to t \in X. \tag{2}$$

Clauses of the form (2) or (1) can always be presented in a certain standard form where t is replaced by a variable, namely as

$$t = x \wedge C \wedge t_0 \in X \wedge \cdots \wedge t_{p-1} \in X \to x \in X \tag{3}$$

or

$$t = x \wedge C \wedge \forall i < t'(t_i \in X) \to x \in X \tag{4}$$

(x not free in t, t_i, t' or C). We can get rid of the parameters by writing (4) as

$$\exists \vec{y}(t = x \wedge C \wedge \forall i < t'(t_i \in X)) \to x \in X. \tag{5}$$

A finite number of clauses $A_j(X, x) \to x \in X$ may be combined into a single condition

$$A_0(X, x) \vee \cdots \vee A_{q-1}(X, x) \to x \in X \tag{6}$$

or with $A \equiv A_0 \vee \cdots \vee A_{q-1}$

$$A(X, x) \to x \in X. \tag{7}$$

We can express the fact that the inductively generated set is the *least* one satisfying the closure conditions, i.e. that elements are put in it *solely* on the basis of its closure conditions, simply by stating that any set satisfying the same closure conditions must contain the set generated by those closure conditions. Thus if P_A is generated by a number of closure conditions $B_1(P_A), \ldots, B_p(P_A)$ say, we must have for any X

$$B_1(X) \wedge \cdots \wedge B_p(X) \to (P_A \subset X);$$

or if we use the standardized version (7)

$$\forall x[A(X,x) \to x \in X] \to [P_A \subset X], \tag{8}$$

where P_A is the set *generated* by $A(X,x) \to x \in X$.

We can intuitively convince ourselves of the correctness of (8) without relying on the notion of the "least" set satisfying certain conditions as follows. Take e.g. example 2. The closure conditions may be written as

$$1 \in Q, \quad \exists nm(x = 2^n \cdot 3^m \wedge n \in Q \wedge m \in Q) \to x \in Q.$$

$1 \in Q$ holds "by an argument of depth 0", i.e. 1 is given as being in Q immediately. An element of the form $2^n \cdot 3^m$ is in Q by an "argument of depth less than $y+1$" if we have arguments of depth less than y for $n \in Q$, $m \in Q$. Thus arguments for membership in Q take the form of finite tree structure, e.g.

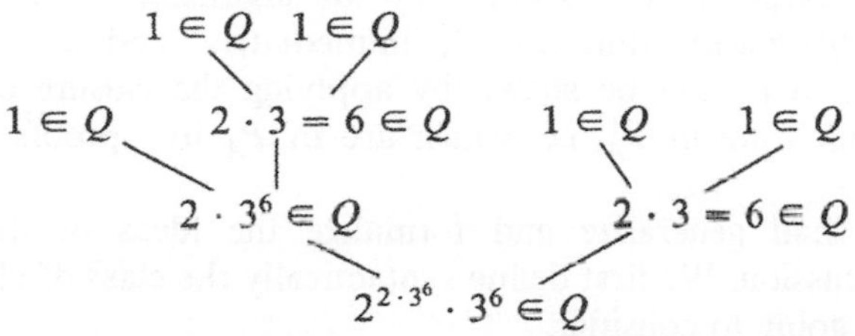

For any predicate X satisfying the same closure conditions we can, starting at terminal nodes with "$1 \in X$", build up proofs of "$t \in X$" for any node "$t \in Q$" appearing in such a tree, i.e. $Q \subset X$. Note also that the collection of $t \in Q$ appearing at the roots of such trees representing arguments of finite depth indeed satisfies the closure conditions on Q.

A similar justification of (8) also works for example 1 and more generally for any P_A generated by clauses of type (4). The same idea will be used to obtain explicit arithmetical definitions for the P_A later on.

The standard form (7) for the closure conditions of our examples gives rise to *monotone* A, i.e. A satisfying

$$A(X,x) \wedge X \subset Y \to A(Y,x).$$

In fact, any A obtained as disjunction of clauses $\exists \vec{y}(t = x \wedge C \wedge \forall i < t' (t_i \in X))$ is easily seen to be monotone.

Classically it is not difficult to show, for any A satisfying monotonicity, the existence of a least set P_A such that

$$\begin{aligned} &A(P_A, x) \to x \in P_A \\ &\forall x(A(X, x) \to x \in X) \to (P_A \subset X). \end{aligned} \tag{9}$$

To see this put

$$\Gamma_A(X) := \{x : A(X, x)\}$$

and define Γ_A^α for all ordinals α by

$$\Gamma_A^0 := \emptyset;\ \Gamma_A^{\alpha+1} := \Gamma_A(\Gamma_A^\alpha);\ \Gamma_A^\lambda := \bigcup\{\Gamma^\alpha : \alpha < \lambda\}$$

for limit ordinals λ. Then for some countable ordinal α $\Gamma_A^{\alpha+1} = \Gamma_A^\alpha$, and then Γ_A^α is the required P_A (exercise). This justification relies on the classical notion of ordinal and is thus, in its generality, not available to us.

However, in our example, and more generally for A constructed from clauses of type (4), as in (6), one has in fact already $\Gamma_A^{\omega+1} = \Gamma_A^\omega$, and $x \in \Gamma_A^{n+1}$ corresponds to "$x \in P_A$ by an argument of depth n". For $x \in \Gamma_A^1 \equiv \Gamma_A(\emptyset)$ means that $x \in P_A$ immediately, and $x \in \Gamma_A^2 \equiv \Gamma_A(\Gamma_A^1)$ means that $x \in P_A$ can be shown by applying the closure conditions to certain x which are in Γ_A^1, i.e. which are in P_A by "proofs of depth 0" etcetèra.

Next we shall generalize and formalize the ideas of the preceding informal discussion. We first define syntactically the class of closure conditions we are going to consider.

6.5. Definition. Let **H** be either **HA** or **EL** and let $\mathbf{H}(X)$ be the corresponding system with a letter X for a unary numerical predicate added. We define a class $\mathscr{P}$ of formulas in $\mathbf{H}(X)$ inductively by

(i) the formulas of **H** are in $\mathscr{P}$;
(ii) $Xt \in \mathscr{P}$, where t is a numerical term;
(iii) if $A, B \in \mathscr{P}$, then $(A \wedge B)$ and $(A \vee B) \in \mathscr{P}$;
(iv) if $A \in \mathscr{P}$, then $(\exists x A) \in \mathscr{P}$;
(v) if $A \in \mathscr{P}$, t a term, $x \notin \mathrm{FV}(t)$, then $(\forall x < t A) \in \mathscr{P}$. □

It is readily seen that the formulas $A \in \mathscr{P}$ are monotone in X. We shall now demonstrate the following

6.6. Theorem. *Each $A(X, z)$ of $\mathscr{P}$ is in $\mathbf{H}(X)$ equivalent to a formula of the form*

$$\exists x \forall y \le t[x, z](P(x, y, z) \vee (Q(x, y, z) \wedge Xt'[x, y, z]))$$

for suitable terms t, t' and formulas $P, Q \in \mathscr{L}(\mathbf{H})$. □

For the proof we need some lemmas.

6.7. Lemma. *In* $\mathbf{H}(X)$, *for given* t_0, t_1 *and* A,

$$\forall x_0 \le t_0[y] \forall x_1 \le t_1[x_0, y] A(x_0, x_1, y) \leftrightarrow \forall x \le t[y] A(s_0[x, y], s_1[x, y], y)$$

for suitable terms t, s_0, s_1 *of* $\mathscr{L}(\mathbf{H})$.

Proof. The idea is to code pairs x_0, x_1 as $x = j(x_0, x_1)$, using suitably chosen s_0, s_1 to bound the possible values of $j_1 x$ and $j_2 x$. Let

$$\psi(y) := \max\{t_1[x_0, y] : x \le t_0[y]\},$$
$$t[y] := j(t_0[y], \psi(y)).$$

(This guarantees $(\forall x \le t[y] A(j_1 x, j_2 x, y) \to \forall x_0 \le t_0[y] \forall x_1 \le t_1[x, y] A(x_0, x_1, y)$.) Now we choose s_0, s_1 such that for $i \in \{0, 1\}$

$$s_i[x, y] := \begin{cases} j_{i+1} x & \text{if } j_2 x < t_1[j_1 x, y] \wedge j_1 x < t_0[y], \\ 0 & \text{otherwise.} \end{cases}$$

So if x ranges over $\{x' : x' \le t[y]\}$, then $(s_0[x, y], s_1[x, y])$ ranges over pairs $(j_1 x, j_2 x)$ such that $j_1 x \le t_0[y]$, $j_2 x \le t_1[j_1 x, y]$. Proving the required equivalence is now routine. □

6.8. Lemma. *In* $\mathbf{H}(X)$

$$\exists xy \forall z \le t[x, y] A(x, y, z) \leftrightarrow \exists x \forall z < t[j_1 x, j_2 x] A(j_1 x, j_2 x, z),$$
$$\forall x \le t \exists y A(x, y) \leftrightarrow \exists z \forall x \le t A(x, (z)_x).$$

Proof. Straightforward. □

6.9. *Proof of theorem 6.6.* We use induction on the complexity of A.

Case 1. $A(X, z) \equiv B(z)$, X not in B. Then

$$\vdash A(X, z) \leftrightarrow \exists x \forall y \le 0(Bz \vee (0 = 1 \wedge X0)).$$

Case 2. $A(X, z) \equiv Xt'$, $x, y \notin \mathrm{FV}(t')$. Then

$$\vdash A(X, z) \leftrightarrow \exists x \forall y \le 0(0 = 1 \vee (0 = 0 \wedge Xt').$$

Case 3. Let

$$A \equiv B \wedge C,$$
$$B \leftrightarrow \exists x \forall y_0 \le t_0[x, z](P_0 \vee (Q_0 \wedge Xt'_0)),$$
$$C \leftrightarrow \exists x' \forall y_1 \le t_1[x', z](P_1 \vee (Q_1 \wedge Xt'_1))$$

(induction hypothesis). For notational simplicity we drop the variable z. Contraction of quantifiers yields

$$B \wedge C \leftrightarrow \exists x \forall y_0 \le t_0[j_1 x] \forall y_1 \le t_1[j_2 x]$$
$$[P_0(j_1 x, y_0) \vee (Q_0(j_1 x, y_0) \wedge Xt'_0[j_1 x, y_0])] \wedge$$
$$[P_1(j_2 x, y_1) \vee (Q_1(j_2 x, y_1) \wedge Xt'_1[j_2 x, y_1])].$$

Now we put

$$P(x, y, u) := (P_0(j_1 x, y) \wedge u = 0) \vee (P_1(j_2 x, y) \wedge u \ne 0),$$
$$Q(x, y, u) := (Q_0(j_1 x, y) \wedge u = 0) \vee (Q_1(j_2 x, y) \wedge u \ne 0),$$
$$t[x, u] \quad := (1 \dot{-} u) \cdot t_0[x] + sg(u) \cdot t_1[x],$$
$$t'[x, y, u] := (1 \dot{-} u) \cdot t'_0[x, y] + sg(u) \cdot t'_1[x, y].$$

Then $A \wedge B \leftrightarrow \exists x \forall u \le 1 \forall y \le t[x, u](P \vee (Q \wedge Xt')$ which is in turn provably equivalent to a formula of the class $\mathscr{P}$.

Case 4. $A \equiv B \vee C$ can be treated in a similar manner.

Case 5. $A \equiv \exists x B$, $A \equiv \forall x \le t B$ are easy. We leave these to the reader. □

6.10. We shall now show how the sets P_A defined inductively by clauses $A(X, x) \in \mathscr{P}$ can be defined explicitly in **HA** or **EL**.

Let

$$A(X, z) \equiv$$
$$\exists x \forall y \le t[x, z](P(x, y, z) \vee (Q(x, y, z) \wedge t'[x, y, z] \in X)).$$

Clearly

$$z \in \Gamma_A^1 \leftrightarrow \exists x \forall y \le t[x, z] P(x, y, z), \tag{1}$$

and

$$z \in \Gamma_A^{n+1} \leftrightarrow \exists x \forall y \le t[x, z](P \vee (Q \wedge t' \in \Gamma_A^n)).$$

With the $z \in \Gamma_A^1$ we can associate a tree of depth 1 with pairs (x', z') as labels at the nodes:

$$\begin{array}{c} (0, t'[x,0,z])\,(0, t'[x,1,z]) \ldots \quad (0, t'[x, t[x,z], z]) \\ \diagdown \quad | \quad \diagup \\ (x,z) \end{array}$$

At the bottom node we find z and the x realizing the existential quantifier in (1); there are $t[x, z] + 1$ successors. This is regarded as a *correct tree* for $z \in \Gamma_A^1$ iff $\forall y \le t[x, z] P(x, y, z)$. The labels at the top nodes are irrelevant in this case, only the number of top nodes counts; the labels assigned in the picture were chosen so as to ensure a uniform definition of "correct tree for $z \in \Gamma_A^{n+1}$" for $n > 0$.

Suppose we already know what a (correct) tree for an element of Γ_A^n is, then a (correct) tree for $z \in \Gamma_A^{n+1}$ has the form

$$\begin{array}{c} T_0 \qquad\qquad T_1 \qquad\qquad\qquad T_y \\ (x_0, t'[x,0,z])\,(x_1, t'[x,1,z]) \ldots (x_y, t'[x,y,z]) \ldots \\ \diagdown \quad | \quad \diagup \\ (x,z) \end{array}$$

(T_y a subtree with label $(x_y, t'[x, y, z])$ at its root) such that for all $i \le t[x, z]$ we have either $P(x, y, z)$ holds and T_y consists of the single node $\langle y \rangle$, or $Q(x, y, z)$ holds and T_y with label $(x_y, t'[x, y, z])$ at the bottom node is a Γ_A^n-tree for $t'[x, y, z]$. There are $t[x, z] + 1$ successors at the root.

The trees corresponding to $\bigcup\{\Gamma_A^n : n \in \mathbb{N}\}$ are all finite trees T of depth greater than 0 with finite label functions φ_T defined on their nodes such that

(i) if $n * \langle y \rangle$ is a terminal node of T and $\varphi_T n = (x, z)$, then $P(x, y, z)$ and $\varphi_T(n * \langle y \rangle) = (0, t'[x, y, z])$;
(ii) if $n * \langle y \rangle \in T$ and $\varphi_T n = (x, z)$ then $n * \langle i \rangle \in T$ iff $i \le t[x, z]$;
(iii) if $n * \langle y \rangle$ is not a terminal node of T and $\varphi_T n = (x, z)$, then $\varphi_T(n * \langle y \rangle) = (x_y, t'[x, y, z])$ for some x_y, and $Q(x, y, z)$.

(T, φ_T) is a *correct* (labeled) *tree* for z if $\varphi_T(\langle\,\rangle) = (x, z)$. It is now routine to formulate an arithmetical predicate F such that $F(x, z)$ expresses "x codes a correct tree for z".

Now we can formulate our theorem.

6.11. THEOREM. *For each $A(X, z) \in \mathscr{P}$ there is an arithmetically definable P_A such that*
(i) $\vdash \forall z(A(P_A, z) \to P_A(z))$,
(ii) $\vdash \forall z(A(Q, z) \to Q(z)) \to \forall z(P_A(z) \to Q(z))$,
for all predicates Q.

Proof. For P_A we take $\exists x F(x, z)$ as indicated above.

(i) Assume $A(P_A, z)$, i.e.

$$\exists x \forall y \leq t[x, z](P(x, y, z) \vee (Q(x, y, z) \wedge \exists u F(u, t'[x, y, z]))).$$

Suppose we can take x_0 for x. For each $y \leq t[x_0, z]$ we either know $P(x_0, y, z)$ or we know $Q(x_0, y, z)$ and $F(u_y, t'[x_0, y, z])$. We combine these data into a new tree T and a new label function $\varphi_T \equiv \varphi$

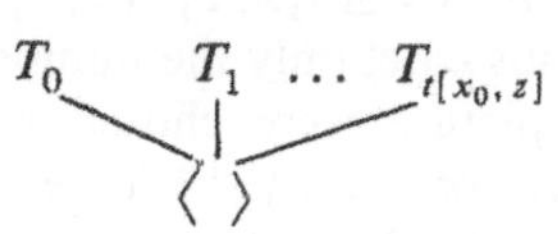

such that either (T_y, φ_{T_y}) coded by u_y consists of a single node $\langle\ \rangle$ with $\varphi_{T_y}\langle\ \rangle = \langle 0, t'[x_0, y, z]\rangle = \varphi\langle y\rangle$ and $P(x_0, y, z)$ holds, or (T_y, φ_{T_y}) is the tree given by u_y and $Q(x_0, y, z)$; the label $\varphi(\langle y\rangle * n)$ is given by the label of n in (T_y, φ_{T_y}), i.e. $\varphi(\langle y\rangle * n) = \varphi_{T_y}(n)$. Finally, $\varphi\langle\ \rangle = (x, z)$.

(ii) Suppose $\forall z(A(Q, z) \to Qz)$, i.e.

$$\exists x \forall y \leq t[x, y](P(x, y, z) \vee (Q(x, y, z) \wedge Rt'[x, y, z]) \to Rz.$$

We can show $\forall xz(F(x, z) \to Rz)$ by induction on x. Suppose

$$\forall y < x \forall z(F(u, z) \to Rz), F(x, z).$$

Then x codes a finite correct tree (T, φ); let this tree have nodes $\langle 0\rangle, \ldots, \langle t[x, z]\rangle$ of length 1, with subtrees $T_0, \ldots, T_{t[x, z]}$.

If the subtree T_y consists of a single node, $\langle y\rangle$ is terminal and $P(x, y, z)$ holds where $\varphi_T\langle\ \rangle = (x, z)$; and if $\langle y\rangle$ is not terminal, then T_y is a subtree with code $u < x$, so $\forall z(F(u, z) \to Rz)$; therefore $\varphi_T\langle y\rangle = (x_y, t'[x, y, z]) = \varphi_{T_y}\langle\ \rangle$, and $Q(x, y, z)$; as a result $Rt'[x, y, z]$, by $F(u, t'[x, y, z]) \to Rt'[x, y, z]$, and thus

$$\forall y < t[x, z](P(x, y, z) \vee (Q(x, y, z) \wedge Rt'[x, y, z]))$$

and hence Rz, etcetera. □

7. Formalization of elementary recursion theory

7.1. This section describes the formalization of elementary recursion theory in **HA** or **EL**. For reasons of technical convenience, we start with a set of initial functions and closure conditions somewhat different from the one used in 4.1; the equivalence will be intuitively obvious, even though a formal equivalence proof is long and tedious.

We shall treat partial recursive functionals in a single function argument; the treatment of partial recursive functions is obtained by dropping the function argument everywhere. Throughout this section we tacitly rely on 2.7.6 (cf. 7.4 below).

7.2. In the definition below and all following definitions, m is the code of a sequence $\langle m_0, \ldots, m_{k-1}\rangle$, and $i < k$.

DEFINITION. We inductively generate a set $\Omega(\alpha)$ of triples intended to represent the graph of partial recursive application (i.e. $(n, x, m) \in \Omega \leftrightarrow \{x\}(m_0, \ldots, m_{k-1}) \simeq n)$) by the following clauses:

$\Omega 0$ $(n, \langle 0, k, 0, n\rangle, m) \in \Omega$ (*constant function* $\lambda x.n$)

$\Omega 1$ $(m_i, \langle 0, k, 1, i\rangle, m) \in \Omega$ (*projection*)

$\Omega 2$ $(m_i + 1, \langle 0, k, 2, i\rangle, m) \in \Omega$ (*successor*)

$\Omega 3$ $(\alpha(m_i), \langle 0, k, 3, i\rangle, m) \in \Omega$ (*application*)

$\Omega 4$ if $n_0 = n_1$ then $(n_2, \langle 0, k+4, 4\rangle, n * m) \in \Omega$ and if $n_0 \neq n_1$ then $(n_3, \langle 0, k+4, 4\rangle, n * m) \in \Omega$ where $n = \langle n_0, n_1, n_2, n_3\rangle$ (*definition by cases*)

$\Omega 5$ if $(n_i, x_i, m) \in \Omega$ for $i < j$ and $(n, x_j, \langle n_0, \ldots, n_{j-1}\rangle) \in \Omega$ then $(n, \langle 1, k, \langle x_0, \ldots, x_j\rangle\rangle, m) \in \Omega$ (*substitution*)

$\Omega 6$ if $(n, x, m) \in \Omega$ then $(n, \langle 2, k+1\rangle, \langle x\rangle * m) \in \Omega$ (*reflection*)

$\Omega 7$ $(\mathrm{Sb}(x, y), \langle 0, k+2, 5\rangle, \langle x, y\rangle * m) \in \Omega$ where, if $x' := (x)_1 \dot{-} 1$, $\mathrm{Sb}(x, y) := \langle 1, x', x, \langle\langle 0, x', 0, y\rangle, \langle 0, x', 1, 0\rangle, \ldots, \langle 0, x', 1, x' \dot{-} 1\rangle\rangle\rangle$ (*smn-theorem*). □

REMARKS. (i) $\Omega 6$ simply gives an index to $\{x\}(m_0, \ldots, m_{k-1})$ as a partial recursive operation in $x, m_0, \ldots, m_{k-1}$.

(ii) $\mathrm{Sb}(x, y)$ has been defined in such a way that

$$\{x\}(y, m_0, \ldots, m_{k-1}) \simeq \{\mathrm{Sb}(x, y)\}(m_0, \ldots, m_{k-1})$$

(cf. 7.5(i) below, and the clauses $\Omega 0$, $\Omega 1$, $\Omega 5$).

7.3. LEMMA. $(n, x, m) \in \Omega \wedge (n', x, m) \in \Omega \to n = n'$.

PROOF. By induction on Ω. The form in which we have given the definition of Ω corresponds to the examples as presented in 6.4. To bring it in standard form is routine. Ω is generated by a closure condition $A_\Omega(X, x) \to x \in X$ with $A_\Omega(X, x)$ of the form

$$A_1(X, x) \vee A_2(X, x) \vee \cdots \vee A_7(X, x)$$

where $A_i(X, x)$ corresponds to clause Ωi; for example

$$A_3(X, x) := \exists k \exists i < k \exists m (x = (\alpha(m)_i, \langle 0, k, 3, i\rangle, m)),$$

$$A_6(X, x) := \exists n \exists y \exists m ((n, y, m) \in X \wedge$$
$$x = (n, \langle 2, \mathrm{lth}(m) + 1\rangle, \langle y\rangle * m)),$$

etcetera. □

The lemma justifies considering Ω as the graph of a mapping, i.e. for $\vec{m}$ of length k we put

7.4. Definition.

$$\{x\}(\alpha, m) := In.((n, x, \vec{m}) \in \Omega).$$

So $\{x\}$ represents a partial mapping from $\mathbb{N}^{\mathbb{N}} \times \mathbb{N}^k$ to $\mathbb{N}$; x is called the *index* of that mapping. □

7.5. Proposition.

(i) (*smn-theorem*) *For each* m, n *with* $0 < m < n$ *there exists a primitive recursive* S_n^m *such that*

$$\{S_n^m(x, y_0, \dots, y_{m-1})\}(\alpha, y_m, \dots, y_n) \simeq \{x\}(\alpha, y_0, \dots, y_n),$$

(ii) (*recursion theorem*). *There exists a primitive recursive function* rc *such that*

$$\{\mathrm{rc}(x)\}(\alpha, \vec{m}) \simeq \{x\}(\alpha, \mathrm{rc}(x), \vec{m}).$$

(iii) *The partial recursive functionals are closed under primitive recursion and minimalization.*

Proof. (i) For $m = 1$, $S_n^1(x, y) \equiv \mathrm{Sb}(x, y)$ satisfies (i) (cf. remark (ii) under 7.2); for $S_n^{m+1}(x, y_0, \dots, y_m)$ we can take $\mathrm{Sb}(S_n^m(x, y_0, \dots, y_{m-1}), y_m)$.

(ii) Standard from (i); cf. the argument in 4.6.

(iii) From indices x, y, z we can construct an index u such that

$$\{u\}(z, n, \vec{m}) \simeq \begin{cases} \{x\}(\vec{m}) & \text{if } n = 0, \\ \{y\}(\{z\}(n \dot{-} 1, \vec{m}), n \dot{-} 1, \vec{m}) & \text{if } n \neq 0 \end{cases}$$

(by an appeal to substitution, reflection, and definition by cases). By the recursion theorem (ii), we find $\{\mathrm{rc}(u)\}(n, \vec{m}) \simeq \{u\}(\mathrm{rc}(u), n, \vec{m})$, so

$$\{\mathrm{rc}(u)\}(0, \vec{m}) \simeq \{x\}(\vec{m}),$$

$$\{\mathrm{rc}(u)\}(n+1, \vec{m}) \simeq \{y\}(\{\mathrm{rc}(u)\}(n, \vec{m}), n, \vec{m}).$$

We leave the closure under minimalization to the reader. □

Our next aim is to obtain the

7.6. THEOREM (*normal form theorem*). *There exists a primitive recursive predicate T such that*

$$\{x\}(\alpha, m) \simeq (\min_u T(x, \alpha, m, u))_0,$$

and moreover

$$T(x, \alpha, m, u) \wedge T(x, \alpha, m, u') \to u = u'.$$

Thus, this T can be taken for Kleene's T-predicate, and $\lambda u.(u)_0$ for the result-extracting function U.

PROOF. We shall inductively define a predicate Comp_α; $\mathrm{Comp}_\alpha(u)$ expresses that u is (a code of) a finite sequence where $(u)_0$ is the output and $(u)_1$ the index of the algorithm applied; the remaining components are the numerical input (arguments) and codes of subcomputations. α is present as a parameter. Mostly we drop the subscript α. The clauses for "Comp" correspond to the clauses for Ω.

C0 $\langle n, \langle 0, k, n\rangle, m, 0\rangle \in \mathrm{Comp}$;
C1 $\langle m_i, \langle 0, k, 1, i\rangle, m, 0\rangle \in \mathrm{Comp}$;
C2 $\langle m_{i+1}, \langle 0, k, 2, i\rangle, m, 0\rangle \in \mathrm{Comp}$;
C3 $\langle \alpha(m_i), \langle 0, k, 3, i\rangle, m, 0\rangle \in \mathrm{Comp}$;
C4 $n_0 = n_1 \to \langle n_2, \langle 0, k+4, 4\rangle, n * m, 0\rangle \in \mathrm{Comp}$,
$n_0 \neq n_1 \to \langle n_3, \langle 0, k+4, 4\rangle, n * m, 0\rangle \in \mathrm{Comp}$,
where $n = \langle n_0, n_1, n_2, n_3\rangle$;
C5 If there are $u_0, \ldots, u_{t-1} \in \mathrm{Comp}$ and $u_i = \langle (u_i)_0, n_i, m, (u_i)_3\rangle$ for $i < t$, and $u_t = \langle v, n_t, \langle u_0, \ldots, u_{t-1}\rangle, (u_1)_3\rangle$ then
$\langle v, \langle 1, k, n_t, n\rangle, m, \langle u_0, \ldots, u_{t-1}\rangle\rangle \in \mathrm{Comp}$;
C6 If $v \in \mathrm{Comp}$, $v = \langle u, n, m, (v)_3\rangle$ then
$\langle u, \langle 2, k+1\rangle, \langle n\rangle * m, v\rangle \in \mathrm{Comp}$;
C7 $\langle \mathrm{Sb}(x, y), \langle 0, k+2, 5\rangle, \langle x, y\rangle * m, 0\rangle \in \mathrm{Comp}$.

We can now put

$$T(x, \alpha, m, u) := \mathrm{Comp}(u) \wedge (u)_2 = m \wedge (u_1) = x,$$

and prove by induction on Ω and Comp respectively

$$\{x\}(\alpha, \vec{m}) \simeq n \to \exists u(T(x, \alpha, m, u) \wedge (u)_0 = n),$$

$$T(x, \alpha, m, u) \to \{x\}(\alpha, x) \simeq (u)_0.$$

Note that Comp is a primitive recursive predicate (use basic facts about primitive recursive sets and the fact that coding of finite sequences is monotone, i.e. $x < n * \langle x \rangle * m$ for all x, n, m). We leave the details to the reader. □

7.7. THEOREM (*numerical form of the normal form theorem*). *There exists a primitive recursive predicate T^* such that*

$$\{x\}(\alpha, m) \simeq (\min_u T^*(x, \bar{\alpha}u, m, u))_0,$$

$$T^*(x, n, m, u) \to \mathrm{lth}(n) = u,$$

$$T^*(x, n, m, u) \wedge T(x, n', m, u') \wedge n' \preccurlyeq n \to n' = n \wedge u = u'.$$

PROOF. From Comp we define

$$\mathrm{Comp}^*(n, u) := \exists\alpha(\mathrm{Comp}(u) \wedge \bar{\alpha}u = n).$$

By induction on u one can show

$$\forall\alpha\beta\big(\bar{\alpha}u = \bar{\beta}u \to \big(\mathrm{Comp}_\alpha(u) \leftrightarrow \mathrm{Comp}_\beta(u)\big).$$

Substitution of the primitive recursive $f := \lambda x.(n)_x$ for α gives $\mathrm{Comp}_f(u) \leftrightarrow \mathrm{Comp}^*(n, u)$ and therefore Comp* is clearly primitive recursive. For T^* we can take

$$T^*(x, n, m, u) := \mathrm{Comp}^*(n, u) \wedge (u)_1 = x \wedge (u)_2 = m. \quad \square$$

7.8. NOTATION.

$$T(x, \vec{\alpha}, \vec{m}, u) := T(x, \langle\vec{\alpha}\rangle, \langle\vec{m}\rangle, u),$$

$$T^*(x, \vec{n}, \vec{m}, u) := T^*(x, n', \langle\vec{m}\rangle, u) \wedge \forall i < p\big(\mathrm{lth}(n_i) = u\big),$$

where $n' \equiv \overline{\nu_p\big(\lambda x.(n_0)_x, \ldots, \lambda x.(n_{p-1})_x\big)}(\mathrm{lth}(n_0))$, and $\vec{n} \equiv (n_0, \ldots, n_{p-1})$. We shall interpret $\{t\}(\alpha_0, \ldots, \alpha_{p-1}, m_0, \ldots)$ as $\{t\}(\nu_p(\alpha_0, \ldots, \alpha_{p-1}), m_0, \ldots)$. □

7.9. DEFINITION (*continuous function application*).

$$\alpha(\beta) \quad := \alpha\big(\bar{\beta}\big(\min_z\big(\alpha\big(\bar{\beta}z\big) \neq 0\big)\big)\big) \dot{-} 1,$$

$$(\alpha|\beta)(x) := \alpha\big(\langle x\rangle * \bar{\beta}\big(\min_z\big[\alpha(\langle x\rangle * \bar{\beta}z) \neq 0\big]\big)\big) \dot{-} 1.$$

The operators $\cdot(\cdot)$ and $\cdot | \cdot$ introduced by this definition are partial; e.g. $\neg \mathbf{E}((\lambda x.0)(\alpha))$. □

We note the following corollary of our development of elementary recursion theory:

7.10. PROPOSITION. *There are numerals $\bar{n}_0$ and $\bar{n}_1$ such that*

$$\{\bar{n}_0\}(\alpha, \beta, x) \simeq (\alpha|\beta)(x),$$

$$\{\bar{n}_1\}(\alpha, \beta) \quad \simeq \alpha(\beta).$$

PROOF. Immediate from the definition and routine facts of recursion theory. □

Let us now work in a definitional extension of **EL** with $\lambda\alpha\beta.\alpha(\beta)$ and $\lambda\alpha\beta x.(\alpha|\beta)(x)$ as primitive operations. We can then prove the following.

7.11. LEMMA.

(i) *If $\varphi[\vec{\alpha}]$ is a functor, and $\vec{\alpha}$ has length p, then for some primitive recursive f_φ*

$$\big(f_\varphi | \nu_p(\vec{\alpha})\big)(x) \simeq \varphi[\vec{\alpha}](x), \quad \text{and}$$

(ii) *if $t[\vec{\alpha}]$ is a numerical term of* **EL**, *then for some primitive recursive f_t*

$$f_t\big(\nu_p(\vec{\alpha})\big) \simeq t[\vec{\alpha}].$$

PROOF. (i) For notational simplicity let $\vec{\alpha} \equiv \alpha$. We first rewrite $\varphi[\alpha](x)$ by eliminating the notations $\varphi|\psi$, $\varphi(\psi)$ with the help of the preceding proposition. The result is a term $t[\alpha, x] \simeq \varphi[\alpha](x)$ with Kleene brackets. This can be written as $\{\bar{n}\}(\alpha, \vec{m}, x)$ for a suitable numeral $\bar{n}$, which is given by

$$\{\bar{n}\}(\alpha, \vec{m}, x) \simeq (\min_z T(\bar{n}, \bar{\alpha}(z), \langle x, \vec{m}\rangle, z))_0.$$

Now we need a functor f_φ such that

$$(f_\varphi|\beta)(x) \simeq (\min_z T(\bar{n}, \bar{\alpha}(z), \langle x, \vec{m}\rangle, z))_0,$$

which can be achieved taking f_φ such that

$$f_\varphi(\langle x\rangle * v) = \begin{cases} (z)_0 + 1 & \text{if} \\ \qquad \exists u \preccurlyeq v \exists z\, (\mathrm{lth}(u) = z \wedge T(\bar{n}, \bar{\alpha}(z), \langle x, \vec{m}\rangle, z)), \\ 0 & \text{otherwise.} \end{cases}$$

f_φ is clearly primitive recursive and

$$(\varphi|\alpha)(u) \simeq w \leftrightarrow f_\varphi\big(\langle x\rangle * \bar{\alpha}\big(\min_z\big[f_\varphi(\langle x\rangle * \bar{\alpha}z) > 0\big]\big) = w + 1,$$

which, by the definition of f_φ, means that $(f_\varphi|\alpha)(x) \simeq \{x\}(\alpha, u, m) \simeq \varphi[\alpha](x)$.

(ii) This case is similar but simpler, and left to the reader. □

7.12. DEFINITION.

$$\alpha|(\beta_0, \ldots, \beta_{n-1}) := \alpha|\nu_n(\beta_0, \ldots, \beta_{n-1}),$$

$$\alpha(\beta_0, \ldots, \beta_{n-1}) := \alpha(\nu_n(\beta_0, \ldots, \beta_{n-1})). \quad \square$$

We can now build an analogue of ordinary recursion theory with continuous function application replacing recursive function application.

7.13. THEOREM (*smn-theorem*).

(i) *There is a primitive recursive binary functional* $\wedge_n$ *such that*

$$(\alpha \wedge_n \beta_0)|(\beta_1, \ldots, \beta_n) \simeq \alpha|(\beta_0, \ldots, \beta_n).$$

(ii) *There is a primitive recursive binary functional* $\wedge'_n$ *such that*

$$(\alpha \wedge'_n \beta_0)(\beta_1, \ldots, \beta_n) \simeq \alpha(\beta_0, \ldots, \beta_n).$$

PROOF. (i) As before, we write $(\alpha|(\beta_0, \ldots, \beta_n))(x)$ as $\{\bar{n}\}((\alpha, \vec{\beta}), x)$, and thus by the normal form theorem

$$(\alpha|(\vec{\beta}))(x) \simeq \big(\min_z T\big(\bar{n}, \overline{(\alpha, \vec{\beta})}z, x, z\big)\big)_0.$$

Define $\alpha \wedge \beta_0$ by

$$(\alpha \wedge \beta_0)(0) = 0,$$

$$(\alpha \wedge \beta_0)(\langle x\rangle * n) = y + 1$$

$$\leftrightarrow \exists z \leq \mathrm{lth}(n)\Big(T\Big(\bar{n}, \overline{(\alpha, \beta_0, f_1, \ldots, f_{\mathrm{lth}(n)})}\, z, x, z\Big) \wedge (z)_0 = y\Big),$$

$$(\alpha \wedge \beta_0)(\langle x\rangle * n) = 0 \text{ otherwise},$$

where $f_i \equiv \lambda z.\, j_i^{\mathrm{lth}(n)}((n)_z)$ (for our standard coding $j_m^k(0) = 0$).

(ii) is proved similarly. □

7.14. Theorem (*recursion theorem*).

(i) *For each α there is a β such that*

$$\alpha|(\beta, \vec{\gamma}) \simeq \beta|(\vec{\gamma}).$$

(ii) *For each α there is a β such that*

$$\alpha(\beta, \vec{\gamma}) \simeq \beta(\vec{\gamma}).$$

Proof. (i) There exists a specific ε such that

$$\varepsilon|(\delta, \vec{\gamma}) \simeq \alpha|(\delta \wedge_n \delta, \vec{\gamma});$$

for $\beta \equiv \varepsilon \wedge_n \varepsilon$ we find $\beta|(\vec{\gamma}) \simeq (\varepsilon \wedge_n \varepsilon)|(\vec{\gamma}) \simeq \varepsilon|(\varepsilon, \vec{\gamma}) \simeq \alpha|(\varepsilon \wedge_n \varepsilon, \vec{\gamma}) \simeq \alpha|(\beta, \vec{\gamma})$.

(ii) Similarly. □

7.15. Notation ($\Lambda^0 x$, $\Lambda^1 x$, $\Lambda^0 \alpha$, $\Lambda^1 \alpha$).

Let t be a term in $\mathscr{L}(\mathbf{EL})$ extended with $\cdot(\cdot)$, $\cdot\,|\,\cdot$, provably total for all values of x, then $\Lambda^0 x.t$ is some term t' primitive recursive in the parameters of t minus x such that

$$\{t'\}(x) \simeq t.$$

In other words, $\Lambda^0 x.t$ is an index for t as a partial recursive function of x; the index may be assumed to depend primitive recursively on the other parameters. The following definitions are to be understood similarly.

If t is a term, $\Lambda^0 \alpha.t$ is some functor φ primitive recursive in the parameters of t minus α such that

$$\varphi(\alpha) \simeq t.$$

If φ is a functor, $\Lambda^1 x.\varphi$ is some functor ψ primitive recursive in the parameters different from x such that

$$\psi|\lambda y.x \simeq \varphi$$

and $\Lambda^1 \alpha.\varphi$ is some functor ψ' primitive recursive in the parameters of φ distinct from α such that

$$\psi'|\alpha = \varphi.$$

The choice of the terms and functors for $\Lambda^0 x.t$, $\Lambda^1 x.t$, $\Lambda^0 \alpha.\varphi$, $\Lambda^1 \alpha.\varphi$ is immaterial, but they can be defined canonically by induction on t and φ with help of the smn-theorem.

In the sequel we shall often use the abbreviations

$$\Lambda x.t := \Lambda^0 x.t, \quad \Lambda\alpha.\varphi := \Lambda^1 \alpha.\varphi. \quad \square$$

8. Intuitionistic second-order logic and arithmetic

In this and the next section we shall describe some formalisms for intuitionistic higher-order logic and arithmetic and prove some of their metamathematical properties. These sections principally serve as a reference for later developments, to be consulted when needed.

The present section is devoted primarily to **HAS**, intuitionistic second-order arithmetic with full (impredicative) comprehension (this system will be considered in E4.4.11–12, E4.5.6 and section 5.7).

8.1. *Intuitionistic second-order logic* $\mathbf{IQC}^2$. A simple version of $\mathbf{IQC}^2$ is obtained as follows. The language of $\mathbf{IQC}^2$, called $\mathscr{L}^2$ for short, is a many-sorted language obtained by adding to $\mathscr{L}(\mathbf{IQC})$ variables for n-ary relations ($n \geq 0$) $X^n, Y^n, Z^n, \ldots,$ and quantifiers for them. The superscripts ^{n} will often be omitted. If we wish to distinguish between (first-order)-quantifiers over the domain of individuals, and (second-order)-quantifiers over relations, we use $\forall^1$, $\exists^1$ and $\forall^2, \exists^2$ respectively.

Relation terms: the only relation terms are variables.

Atomic formulas: first order atomic formulas, and whenever $t_1, \ldots, t_n$ are first-order (individual) terms and X^n is a variable for an n-place relation, $X^n(t^1, \ldots, t^n)$ is an atomic formula.

N.B.: in this version a constant n-ary relation R in $\mathscr{L}(\mathbf{IQC})$ is not treated as a relation term.

Axioms and rules. As our logical basis we can now take any standard formulation of many-sorted intuitionistic logic without equality, and in addition the following *axiom of full (impredicative) comprehension*

$$\text{CA} \qquad \exists X^n \forall x_1 \ldots x_n [A(x_1, \ldots, x_n) \leftrightarrow X^n(x_1, \ldots, x_n)],$$

where A is any formula of the language not containing X^n free.

In this system equality between individuals can be defined by

$$t_1 = t_2 := \forall X^1(X(t_1) \leftrightarrow X(t_2)).$$

It is easy to see that the usual equality rules now become derivable. For example we have the following deduction of the replacement rule (where for convenience we drop the parentheses in $X(t)$)

$$\dfrac{\exists X^1 \forall x(A(x) \leftrightarrow Xx) \qquad \dfrac{\dfrac{\dfrac{(1)\ \forall x(A(x) \leftrightarrow Yx)}{A(t_2) \leftrightarrow Yt_2}}{Yt_2 \to A(t_2)} \qquad \dfrac{\dfrac{\dfrac{\forall X^1(Xt_1 \leftrightarrow Xt_2)}{Yt_1 \leftrightarrow Yt_2}}{Yt_1 \to Yt_2} \qquad \dfrac{\dfrac{\dfrac{(1)\ \forall x(A(x) \leftrightarrow Yx)}{A(t_1) \leftrightarrow Yt_1}}{A(t_1) \to Yt_1} \qquad A(t_1)}{Yt_1}}{Yt_2}}{A(t_2)}}{A(t_2)}\ \exists\text{E}(1)$$

Moreover, if we define

$$X^n = Y^n := \forall x_1 \ldots x_n (X^n(x_1, \ldots, x_n) \leftrightarrow Y^n(x_1, \ldots, x_n))$$

it is easy to verify by induction on the formula complexity that the equality rules also hold in this case.

8.2. *Other formulations of* $\mathbf{IQC}^2$. The introduction of pairing operators in the domain of individuals is an inessential change in the language. Writing (t_1, t_2) for the pair formed from t_1, t_2 we require

$$(x_1, x_2) = (y_1, y_2) \leftrightarrow x_1 = y_1 \wedge x_2 = y_2. \tag{1}$$

As soon as we have pairing operators we can also form n-tuples satisfying

$$(x_1, \ldots, x_n) = (y_1, \ldots, y_n) \leftrightarrow x_1 = y_1 \wedge \cdots \wedge x_n = y_n.$$

We may now think of $X^n(x_1, \ldots, x_n)$ as a *unary* predicate applied to the n-tuple $(x_1, \ldots, x_n)$, that is to say we only need variables for unary relations or sets, and possibly for propositions (i.e. $n = 0$). However, if we wish, we can also define quantification $\forall X^0, \exists X^0$ in terms of $\forall X^1, \exists X^1$ as follows: let c be any fixed individual constant in the language. Then

$$\forall X^0 A := \forall Y^1 A[X^0/Y^1(c)],$$

$$\exists X^0 A := \exists Y^1 A[X^0/Y^1(c)].$$

Intuitively speaking, if Y^1 ranges over all sets, $Y^1(c)$ ranges over all propositions. As to comprehension, note that

$$\exists Y^1[A \leftrightarrow Y^1 c]$$

(x, Y not free in A) holds by applying CA to $A \wedge x = x$, i.e.

$$\exists Y^1 \forall x[A \wedge x = x \leftrightarrow Y^1 x].$$

Note also that instead of requiring (1), we can also demand explicitly, in a definitional extension, the existence of $\mathbf{p}_0, \mathbf{p}_1$ decoding the pairing

$$\mathbf{p}_i(x_0, x_1) = x_i \quad (i \in \{0,1\});$$

and hence more generally we have $\mathbf{p}_i^n$ such that

$$\mathbf{p}_i^n(x_0, \ldots, x_{n-1}) = x_i \quad (i < n).$$

Finally, thinking now of $\mathbf{IQC}^2$ as a two-sorted theory with individual and set variables, we can write $t \in X$ for $X(t)$; $\in$ may be treated as a two-place relation constant for a relation between individuals and sets.

Of greater interest is the introduction of a definitional extension by means of *comprehension terms*; for every formula A in the language we introduce a second-order term (relation term) $\lambda x_1 \ldots x_n.A$ (also written as $\{(x_1, \ldots, x_n) : A\}$, with an axiom schema

$$(\lambda x_1 \ldots x_n.A(x_1, \ldots, x_n))(t_1, \ldots, t_n) \leftrightarrow A(t_1, \ldots, t_n). \tag{2}$$

In a natural deduction formulation, $\forall^2$E and $\exists^2$I now take the form

$$\forall^2\text{E}\frac{\forall X^n A(X^n)}{A(T^n)} \qquad \exists^2\text{I}\frac{A(T^n)}{\exists X^n A(X^n)},$$

where T^n is any term for an n-place relation. CA is now absorbed by these rules:

$$\frac{\forall x_1 \ldots x_n[(\lambda x_1 \ldots x_n.A)(x_1, \ldots, x_n) \leftrightarrow A] \quad \text{(axiom (2))}.}{\exists X^n \forall x_1 \ldots x_n[X(x_1, \ldots, x_n) \leftrightarrow A]}$$

In a natural deduction system it is of course more natural to replace (2) by two rules λI and λE:

$$\lambda\mathrm{I}\frac{A(t_1,\ldots,t_n)}{(\lambda x_1\ldots x_n.A(x_1,\ldots,x_n))(t_1,\ldots,t_n)},$$

$$\lambda\mathrm{E}\frac{(\lambda x_1\ldots x_n.A(x_1,\ldots,x_n))(t_1,\ldots,t_n)}{A(t_1,\ldots,t_n)}.$$

A slight variant avoids the explicit introduction of abstraction terms altogether; $\forall^2\mathrm{E}$ and $\exists^2\mathrm{I}$ now take the form

$$\forall^2\mathrm{E}'\frac{\forall X^n A(X^n)}{A^*} \qquad \exists^2\mathrm{I}'\frac{A^*}{\exists X^n A(X^n)},$$

where A^* is obtained from $A(X^n)$ by replacing all occurrences of $X^n(t_1,\ldots,t_n)$ by $B(t_1,\ldots,t_n)$ for a fixed formula $B(x_1,\ldots,x_n)$.

CONVENTION. Below we shall assume that our standard formulation of $\mathbf{IQC}^2$ has $=$ between individuals as a primitive and that abstraction terms are available, unless stated otherwise. □

8.3. *Definability of* $\vee, \wedge, \exists, \bot$. In $\mathbf{IQC}^2$ with comprehension terms we can define $\bot, \wedge, \vee, \exists^1, \exists^2$ in terms of $\to, \forall$ as follows (X not free in A, B):

$$\begin{aligned}
\bot &:= \forall X^0(X^0);\\
A\wedge B &:= \forall X^0((A\to(B\to X))\to X);\\
A\vee B &:= \forall X^0((A\to X)\wedge(B\to X)\to X);\\
\exists x A &:= \forall X^0(\forall x(A\to X)\to X);\\
\exists Y A &:= \forall X^0(\forall Y(A\to X)\to X).
\end{aligned}$$

For example, we can obtain $\vee E$ as a derived rule as follows:

$$\dfrac{\dfrac{(A\vee B)\equiv\forall X((A\to X)\wedge(B\to X)\to X)}{(A\to C)\wedge(B\to C)\to C} \qquad \dfrac{(1)\dfrac{\begin{matrix}(1)\,[A]\\ \mathscr{D}\\ C\end{matrix}}{A\to C} \quad (2)\dfrac{\begin{matrix}(2)\,[B]\\ \mathscr{D}'\\ C\end{matrix}}{B\to C}}{(A\to C)\wedge(B\to C)}}{C}$$

In the absence of comprehension terms we need $\exists^2$ as a primitive (CA is formulated with $\exists^2$). We leave the verification of the correctness of the other definitions to the reader.

8.4. *Intuitionistic second-order arithmetic* **HAS**, also called *second-order Heyting arithmetic.* This system is like **IQC**^{2}, but the language is restricted to logical operations, equality between individuals, a single individual constant 0 (zero), a single unary function constant S (successor). There are axioms

$$\forall xy(Sx = Sy \to x = y),$$

$$\forall x(0 \neq Sx),$$

and induction expressed as an axiom

$$\forall X^1(0 \in X \wedge \forall x(x \in X \to Sx \in X) \to \forall y(y \in X)).$$

Thus we think of the individual variables as ranging over $\mathbb{N}$.

Alternatively, **HAS** can be described as an extension of **HA**, obtained by adding variables for sets of natural numbers (X, Y, Z), with the corresponding quantifier-rules and axioms and full comprehension CA for $n = 1$ (either explicitly or by permitting comprehension terms); the equality rules are to hold for the full language.

It is the second version that we shall treat as the standard one; from the results mentioned in 8.6 below we see that it is in fact a definitional extension of the first version.

Instead of obtaining **HAS** from **IQC**^{2} by "identifying the domain of individuals with $\mathbb{N}$", as in the formulation above, we may also embed **HAS** in **IQC**^{2} as follows. Define

$$y \in \mathbb{N} := \forall X^1(0 \in X \wedge \forall x(x \in X \to Sx \in X) \to y \in X),$$

and interpret number variables as variables restricted to $\mathbb{N}$, and set variables as variables restricted to subsets of $\mathbb{N}$, then it is easy to see that all axioms and rules of **HAS** are derivable in **IQC**^{2} from

$$0 \neq S0, \quad \forall n,m \in \mathbb{N}(Sn = Sm \to n = m).$$

8.5. Definition. *A number structure* (X, f, c) is a structure such that

$$\forall Y^1(c \in Y \wedge \forall x(x \in Y \to fx \in Y) \to X \subset Y).$$

A *Peano structure* is a number structure satisfying in addition

$$c \neq fc, \quad \forall x, y \in X(fx = fy \to x = y). \quad \square$$

The following theorem is well-known from classical second-order logic:

8.6. THEOREM. *In* **HAS** *or* **IQC**^{2}

(i) *If* (X, f, c) *is a Peano-structure and* (X', f', c') *a number structure, there exists a unique homomorphic mapping* φ *such that*

$$\varphi c = c', \quad \varphi(fx) = f'\varphi(x) \text{ for all } x \in X.$$

(ii) *Any two Peano-structures are isomorphic.*

(iii) *In any Peano-structure* (X, c, f) *there are uniquely determined functions satisfying the equations for the primitive recursive functions (identifying* c *with* 0, f *with* S). *In particular, there are uniquely determined* + *and* $\cdot$ *such that for all* $x, y \in X$

$$x + c = x, \quad x + (fy) = f(x + y),$$

$$x \cdot c = c, \quad x \cdot (fy) = x \cdot y + x.$$

REMARK. The function φ is in fact definable in the language as a binary relation Y_φ of functional character, i.e. $\forall x \in X \exists! y \in X(Y_\varphi(x, y))$.

PROOF. The classical proof works intuitionistically as well. See E3.8.3. $\square$

8.7. *The negative translation for* **HAS**. The extension of the negative translation is completely straightforward. For simplicity, let us consider the second version of **HAS** without comprehension terms. We only have to add a clause in the atomic case:

$$X(t_1, \ldots, t_n)^g := \neg\neg X(t_1, \ldots, t_n).$$

The only axiom which needs checking is comprehension

$$\exists X \forall x_1 \ldots x_n [A \leftrightarrow X(x_1, \ldots, x_n)],$$

which translates into

$$\neg\forall X \neg\forall x_1 \ldots x_n [A^g \leftrightarrow \neg\neg X(x_1, \ldots, x_n)],$$

and this is equivalent to

$$\neg\neg\exists X \forall x_1 \ldots x_n [A^g \leftrightarrow \neg\neg X(x_1, \ldots, x_n)].$$

Observing that $\neg\neg A^g \leftrightarrow A^g$ for all A, we see that this follows from comprehension applied to A^g

$$\exists X \forall x_1 \ldots x_n [A^g \leftrightarrow X(x_1, \ldots, x_n)].$$

Cf. also the E3.8.4. As a result we have

PROPOSITION. $\mathbf{HAS}^c \vdash A \Rightarrow \mathbf{HAS} \vdash A^g$. □

8.8. *A Kripke model for* **HAS**. Let us now treat **HAS** as a two-sorted theory. Accordingly, we need to specify two domain functions D and D′ in a Kripke model for **HAS**. We consider a class of Kripke models $\mathscr{K} \equiv (\mathrm{K}, \leq, \Vdash, \mathrm{D}, \mathrm{D}')$ for **HAS** where $(\mathrm{K}, \leq)$ is any tree structure (for example, the binary tree), D is constant, namely $\mathrm{D}(k) = \mathbb{N}$ for all $k \in \mathrm{K}$, and D′ is also constant, namely $\mathrm{D}'(k)$ consists, for each k, of all families $\mathscr{S} \equiv \{S_{k'} : k' \in \mathrm{K}\}$ which satisfy

$$k_0 \leq k_1 \to S_{k_0} \subset S_{k_1}$$

$$S_{k'} \subset \mathbb{N} \quad \text{for all } k' \in \mathrm{K}.$$

Equality for individuals is interpreted by proper equality at each node, $+, \cdot, =$ are defined in the usual way on each $\mathrm{D}(k)$:

$$(k \Vdash t_1 = t_2) \Leftrightarrow t_1 = t_2.$$

Also

$$k \Vdash n \in \mathscr{S} := n \in S_k, \text{ where } \mathscr{S} \equiv \{S_{k'} : k' \in \mathrm{K}\}.$$

8.9. PROPOSITION. *A Kripke model of the class defined above is a model for* **HAS**.

PROOF. The only schema for which the verification is not entirely trivial is CA. Suppose $A(x)$ to have only x free and put

$$S_k := \{n : k \Vdash A(n)\},$$

then, for an arbitrary k', $\mathscr{S} = \{S_k : k \in K\}$ is a family belonging to $\mathrm{D}'(k')$ and for this $\mathscr{S}$ we have

$$k' \Vdash \forall x (x \in \mathscr{S} \leftrightarrow A(x)). \quad \square$$

8.10. It should be observed that for purely arithmetical sentences A we have classically

$$k \Vdash A \Leftrightarrow A \text{ is true}$$

and in particular, regarding the models from a classical point of view, $k \Vdash A \vee \neg A$. On the other hand, we do not have

$$k \Vdash \forall X \forall x [Xx \vee \neg Xx].$$

For example, let $(\mathrm{K}, \leq)$ be $\{0,1\}$ with ordering $0 < 1$, and consider $\mathcal{S} \equiv \{S_0, S_1\}$, with $S_0 = \emptyset$, $S_1 = \mathbb{N}$. Then $k \Vdash \forall x[x \in \mathcal{S} \vee \neg x \in \mathcal{S}]$ means $k \Vdash n \in \mathcal{S} \vee \neg n \in \mathcal{S}$ for each n; $0 \nVdash 0 \in \mathcal{S}$, but also $0 \nVdash \neg 0 \in \mathcal{S}$ since $1 \Vdash 0 \in \mathcal{S}$.

We shall see later that this observation can be considerably generalized and strengthened.

8.11. *The system* $\mathbf{HAS}_0$. So far we have given no thought to the legitimacy of impredicative comprehension from the constructive point of view; in **HAS** we have been restrictive on the logical side, but permissive on the definitional side, that is to say we have regarded impredicatively defined sets as well-defined. It does not obviously follow from, say, Brouwer's point of view that impredicatively defined sets are to be excluded, though in the traditional intuitionistic literature the full force of impredicative comprehension is nowhere used; nor is this the case in the writings of the Bishop- or Markov-school of constructivism.

The principal advantage of **HAS** over **HA** consists in the greater expressive power of the language, not so much in its proof-theoretic strength. We shall now describe a subsystem $\mathbf{HAS}_0$ of **HAS** which is in fact a conservative extension of **HA**.

$\mathbf{HAS}_0$ is defined completely similarly to **HAS**, except that now CA is restricted to formulas A not containing bound set variables (so-called "*arithmetical comprehension*"). Equivalently, the comprehension terms $\lambda x_1 \ldots x_n.A$ are restricted to A not containing bound set variables. Under these circumstances the induction axiom is equivalent to the arithmetical induction schema, restricted to formulas not containing bound set variables. It should be noted that in $\mathbf{HAS}_0$ the definability of $\perp, \vee, \wedge, \exists$ in terms of $\forall, \rightarrow$ no longer holds, nor can we define equality between individuals. Using classical metamathematics we give a simple proof of the following

8.12. Theorem. $\mathbf{HAS}_0$ *is conservative over* **HA**.

PROOF. We rely on the (classically established) completeness theorem for Kripke models and we shall treat $\mathbf{HAS}_0$ simply as a two-sorted first-order theory. Let $\mathscr{K} \equiv (\mathrm{K}, \leq, \Vdash, \mathrm{D})$ be an arbitrary Kripke model for **HA**; we shall show how to extend $\mathscr{K}$ to a model $\mathscr{K}' \equiv (\mathrm{K}, \leq, \Vdash', \mathrm{D}, \mathrm{D}')$ for $\mathbf{HAS}_0$, $\Vdash'$ an extension of $\Vdash$. Without loss of generality we may interpret equality in $\mathscr{K}$ at each node $k \in \mathrm{K}$ as real equality in $\mathrm{D}(k)$ (since $=$ is decidable, cf. E2.6.4), and n-ary function symbols may be interpreted at k by real functions of $\mathrm{D}(k)^n \to \mathrm{D}(k)$.

For $\mathrm{D}'(k)$ we take as elements the collection of subsets $C_{A(d_1,\ldots,d_n,x)}$ of $\mathrm{D}(k)$, with $d_1,\ldots,d_n \in \mathrm{D}(k)$, x the only variable free in $A(d_1,\ldots,d_n,x)$, $A(x_1,\ldots,x_n,x) \in \mathscr{L}(\mathbf{HA})$, and such that in the model

$$C_{A(d_1,\ldots,d_n,x)} = \{x : A(d_1,\ldots,d_n,x)\}.$$

We extend $\Vdash$ to $\Vdash'$ by stipulating

$$k \Vdash' t \in C_{A(d_1,\ldots,d_n,x)} := k \Vdash A(d_1,\ldots,d_n,t),$$

for all closed terms with parameters from $\mathrm{D}(k)$; for other prime formulas $\Vdash'$ coincides with $\Vdash$.

It is now easy to check that arithmetical induction and comprehension hold in $\mathscr{K}'$; we leave this to the reader.

This establishes our theorem, for suppose A to be such that $\mathbf{HAS}_0 \vdash A$, but $\mathbf{HA} \nvdash A$, then there should exist (by the completeness theorem) a model $\mathscr{K}$ such that $\mathscr{K} \nVdash A$. On the other hand $\mathscr{K}'$ models $\mathbf{HAS}_0$, so $\mathscr{K}' \Vdash' A$; but this contradicts the fact that $\Vdash'$ extends $\Vdash$. □

9. Higher-order logic and arithmetic

Even more so than for the preceding section, it is true that the contents of this section are not needed until much later; so the reader can safely skip this section for the time being. It is primarily for systematic reasons that we have placed this section here. We mainly concern ourselves with types obtained by finite iteration of the power-set operation. The section will be built up as follows. First, we describe one of the simplest versions of intuitionistic finite-type theory and finite-type arithmetic, together with some basic and simple metamathematical results (9.1–4). After that we shall discuss a number of possible extensions (9.5–12)

9.1. *The theories* **TYP** *and* **TYP** + EXT. **TYP** is a theory based on many-sorted intuitionistic logic. The sorts, called *types*, are given by

(i) there is a collection of *basic sorts* (*primitive types*) $\beta_0, \beta_1, \beta_2, \ldots$,

(ii) if $\sigma_1, \ldots, \sigma_n$ are types, so is $\sigma_1 \times \cdots \times \sigma_n$ (the *product type* of $\sigma_1, \ldots, \sigma_n$)

(iii) if σ is a type, then so is $[\sigma]$ (the *power type* of σ; alternative notations $\sigma \to \Omega$ or $\mathrm{P}(\sigma)$)

Intuitively, we think of the basic sorts as a collection of given domains; (ii) permits the formation of cartesian products of domains, (iii) the formation of power-domains for any given domain.

We shall assume, for the time being, the basic sorts to represent inhabited domains.

The language of **TYP** contains equality for each basic type, and a relation $\in_\sigma$ for each type σ between elements of type σ and elements of type $[\sigma]$. For each type we have a countably infinite supply of variables; for the basic types there are individual constants, function symbols and relations available.

For the products we have n-tupling ( ) with inverses $\boldsymbol{p}_i^n$ such that

$$\boldsymbol{p}_i^n(x_0, \ldots, x_{n-1}) = x_i \quad \text{for } i < n;$$

equality for power types and product types is defined by

$$t_1^{[\sigma]} = t_2^{[\sigma]} := \forall x^\sigma(x \in t_1 \leftrightarrow x \in t_2),$$

$$t_1^{\sigma_0 \times \cdots \times \sigma_{n-1}} = t_2^{\sigma_0 \times \cdots \times \sigma_{n-1}} := \bigwedge_{i<n} (\boldsymbol{p}_i^n t_1 = \boldsymbol{p}_i^n t_2).$$

Finally, we have for each σ a (*full* or *impredicative*) *comprehension axiom schema*.

$$\text{CA} \quad \exists x^{[\sigma]} \forall y^\sigma (y \in x \leftrightarrow A(y)) \quad (x \notin \mathrm{FV}(A))$$

for any formula A in the language. This completes the description of **TYP**. The additional *axiom of extensionality* states replacement for defined equality:

$$\mathrm{EXT}_\sigma \quad y^\sigma = z^\sigma \wedge y^\sigma \in w^{[\sigma]} \to z^\sigma \in w^{[\sigma]};$$

EXT is "EXT_σ for all σ".

9.2. *Elimination of the extensionality axioms.* **TYP** + EXT can be embedded into **TYP** via a suitable translation, as follows. For each type σ we define a predicate Ext_σ (abbreviating $\mathrm{Ext}_\sigma(t^\sigma)$ as $\mathrm{Ext}(t^\sigma)$):

$$\mathrm{Ext}(t^\beta) := (t^\beta = t^\beta),$$

$$\mathrm{Ext}(t^{\sigma_0 \times \cdots \times \sigma_{n-1}}) := \bigwedge_{i<n} \mathrm{Ext}(p_i^n t),$$

$$\mathrm{Ext}(t^{[\sigma]}) := \forall x^\sigma y^\sigma (\mathrm{Ext}(x^\sigma) \wedge \mathrm{Ext}(y^\sigma) \to (x \in t \leftrightarrow y \in t)).$$

We now have the following

THEOREM. *Let A_{Ext} be obtained from A by relativizing all quantifiers to* Ext, *and let* $\mathrm{FV}(A) = \{x_1^{\sigma_0}, \ldots, x_n^{\sigma_{n-1}}\}$. *Then, for A with* $\mathrm{FV}(A) = \{x_1, \ldots, x_n\}$

$$\mathbf{TYP} + \mathrm{EXT} \vdash A \Rightarrow \mathbf{TYP} \vdash \mathrm{Ext}(x_1) \wedge \cdots \wedge \mathrm{Ext}(x_n) \to A_{\mathrm{Ext}}$$

and similarly with $\mathbf{TYP}^c$ *replacing* **TYP**. □

9.3. *Extending the negative translation to* $\mathbf{TYP}^c$ + EXT. The extension in this case is not as straightforward as for **HAS** or $\mathbf{IQC}^2$. The root of the trouble is in the translation of EXT: EXT^g becomes

$$\forall x(\neg\neg x \in y \leftrightarrow \neg\neg x \in z) \wedge \neg\neg y \in w \to \neg\neg z \in w,$$

which is not obviously true. We can solve the problem, however, by first applying the embedding of 9.2, then ^{g}. By a routine argument, we prove the

LEMMA. $\mathbf{TYP}^c \vdash A \Rightarrow \mathbf{TYP} \vdash A^g$,
from which we readily obtain

THEOREM. *Let* $\mathrm{FV}(A) = \{x_1, \ldots, x_n\}$. *Then*

$$\mathbf{TYP}^c + \mathrm{EXT} \vdash A \Rightarrow$$

$$\mathbf{TYP} \vdash \mathrm{Ext}(x_1)^g \wedge \cdots \wedge \mathrm{Ext}(x_n)^g \to (A_{\mathrm{Ext}})^g. \quad \square$$

9.4. *Heyting's arithmetic in higher types* **HAH**. In its simplest version this theory is based on **TYP** with only a single basic type 0 (the type of the natural numbers); at type 0 we have the language and axioms of **HA** available, and induction holds for all formulas of the language.

In the next few sections we briefly explore certain (definitional) extensions of **TYP** with a more flexible language.

9.5. *Intuitionistic higher-order logic with existence predicate and description* *operator.* The logical basis is **IQCE** with description operator (cf. section 2.2). The collection of types is given by the clauses (i)–(iii), moreover we shall also permit the empty cartesian product yielding a type denoted by $\mathbb{1}$. We permit constants for each type; atomic formulas are Et^σ, $t_1^\sigma = t_2^\sigma$, $t_1^\sigma \in t_2^{[\sigma]}$.

The axioms and schemas now become

$$t_1^\sigma \in t_2^{[\sigma]} \to Et_1 \wedge Et_2 \ (\textit{strictness of } \in),$$

$$E(x_0, \ldots, x_{n-1}) \leftrightarrow Ex_0 \wedge \cdots \wedge Ex_{n-1},$$

$$Ez \leftrightarrow \bigwedge_{i<n} Ep_i^n z \ \ (z \in \sigma_0 \times \cdots \times \sigma_{n-1}),$$

$$\forall x_0 \cdots x_{n-1}\left[(x_0, \ldots, x_{n-1}) = Iz.\left(\bigwedge_{i<n} p_i^n z = x_i\right)\right],$$

$$(\textit{product} \text{ and } \textit{projection axioms};\ z \in \sigma_0 \times \cdots \times \sigma_{n-1}),$$

$$E\left(Iy^{[\sigma]}.\forall x^\sigma[x \in y \leftrightarrow A(x)]\right)\ (\textit{comprehension}).$$

REMARKS.

(i) The axioms for product and projection can be replaced by

$$Ez \to (p_0^n z, \ldots, p_{n-1}^n z) = z, \text{ where } z \in \sigma_0 \times \cdots \times \sigma_{n-1},$$

$$(x_0, \ldots, x_{n-1}) = (y_0, \ldots, y_{n-1}) \leftrightarrow \bigwedge_{i<n} x_i = y_i.$$

The verification of this equivalence is left as an exercise for the reader.

(ii) If in the product and projection axioms we take the empty product $\mathbb{1}$, and interpret the empty conjunction as denoting truth $\top$, we find that $(\ ) \in \mathbb{1}$, and also that this is the unique element in $\mathbb{1}$. For example, taking the formulation above, we see $(\ ) = (\ ) \leftrightarrow \top$, so $E(\ )$; also $Ez \to (\ ) = z$. Let $\Omega := [\mathbb{1}] \equiv \mathrm{P}(\mathbb{1})$; we discover that Ω is isomorphic to the collection of propositions; see 9.9 below.

(iii) Each type σ corresponds to an existing object σ^* of type $[\sigma]$ (intuitively speaking, σ^* is the set of *all* objects of type σ), namely

$$\sigma^* := Iy.\forall x^\sigma[x \in y \leftrightarrow x = x].$$

We can now enlarge our collection of types by introducing the generalized types.

9.6. DEFINITION (*generalized types*).
(i) $\{x^\sigma : A(x)\} := Iy.\forall x^\sigma[x \in y \leftrightarrow A(x)]$,
(ii) $t_0^{[\sigma_0]} \times \cdots \times t_{n-1}^{[\sigma_{n-1}]} := \{z : \bigwedge_{i<n} p_i^n z \in t_i^{[\sigma]}\}$,
(iii) $\mathrm{P}(t^{[\sigma]}) := \{x^{[\sigma]} : \forall y(y \in x \leftrightarrow y \in t^{[\sigma]})\}$. (Alternative notation $[t^{[\sigma]}]$.)

Let us use X, Y for sets defined by means of (i), (ii), (iii); these sets include the σ^* mentioned above. We can define in the usual way (assuming all variables and terms to have been provided with the appropriate types)

$$\forall x \in X.A(x) \quad := \forall x(x \in X \rightarrow A(x)),$$

$$\exists x \in X.A(x) \quad := \exists x(x \in X \wedge A(x)),$$

$$Ix \in X.A(x) \quad := Ix[x \in X \wedge A(x)],$$

$$\{x \in X : A(x)\} := \{x : x \in X \wedge A(x)\}. \quad \square$$

We can easily establish the following

9.7. THEOREM. *Assume the types to be such that all formulas below are well-formed, then we have*
(i) $\{x \in t : A(x)\} \in \mathrm{P}(t)$;
(ii) $s \in \{x \in t : A(x)\} \leftrightarrow s \in t \wedge A(s)$;
(iii) $\{x \in t : A(x)\} = \{x \in t : B(x)\} \leftrightarrow \forall x \in t(A(x) \leftrightarrow B(x))$;
(iv) $\bigwedge_{i<n} s_i \in t_i \rightarrow (s_0, \ldots, s_{n-1}) \in t_0 \times \cdots \times t_{n-1} \wedge$
$\bigwedge_{i<n} p_i^n(s_0, \ldots, s_{n-1}) = s_i$;
(v) $z \in t_0 \times \cdots \times t_{n-1} \rightarrow \bigwedge_{i<n} p_i^n z \in t_i \wedge z = (p_0^n z, \ldots, p_{n-1}^n z)$;
(vi) $(x_0, \ldots, x_{n-1}) = (y_0, \ldots, y_{n-1}) \leftrightarrow \bigwedge_{i<n} x_i = y_i$.

PROOF. Straightforward, and left as an exercise. $\square$

9.8. Next we note that the axioms and schemata in 9.5 also hold for general types. Take for example the first axiom

$$t_1^\sigma \in t_2^{[\sigma]} \rightarrow Et_1 \wedge Et_2.$$

Here $t_1 \in \sigma^*$, $t_2 \in [\sigma]^*$. More generally we have for *any* t such that $t_1 \in t$, $t_2 \in \mathrm{P}(t)$ are well-formed

$$t_1 \in t \wedge t_2 \in \mathrm{P}(t) \wedge t_1 \in t_2 \rightarrow Et_1 \wedge Et_2,$$

and similarly for the other axioms.

More precisely, we can enlarge our collection of type symbols and extend

the axioms and schemata of type theory to them, so as to obtain a definitional extension of the theory. We only have to add to the type-forming operations the following clause:
(iv) If σ is a type, $t \in \mathrm{P}(\sigma)$, then so is $\sigma \upharpoonright t$.
Intuitively, $\sigma \upharpoonright t$ represents the set of all type-σ objects x such that $x \in t$. We shall assume $\sigma \upharpoonright t$ to be part of σ, so that $t_1^\sigma = t_2^{\sigma \upharpoonright t}$ makes sense. We add to the axioms

$$t_1 \in t^{P(\sigma)} \rightarrow t_1 = Iy^{\sigma \upharpoonright t}.[t_1 = y],$$

$$E\left(t_1^{\sigma \upharpoonright t}\right) \rightarrow t_1^\sigma \in t.$$

The reader should work out for himself a variant where terms of types $t_1^{\sigma \upharpoonright t}$ and t_1^σ are kept syntactically distinct, with the help of an embedding operator for type $\sigma \upharpoonright t$ into σ. The final result is the following

THEOREM. *The extension of type theory with existence predicate and description operator to the generalized types is definitional.*

PROOF. Left as an exercise. □

9.9. *The isomorphism between propositions and* $\Omega \equiv \mathrm{P}(\mathbf{1})$. We return briefly to remark (ii) in 9.5. Since the only element of $\mathbf{1}$ is the empty sequence, for which we shall write 0, we have

$$\{x \in \mathbf{1} : A(x)\} = \{0 : A(0)\},$$

and also

$$\{0 : A(0)\} = \{0 : B(0)\} \leftrightarrow (A(0) \leftrightarrow B(0)).$$

To each $x \in \mathrm{P}(\mathbf{1})$ there corresponds a proposition $0 \in x$, and to each proposition A we have a corresponding element $\{0 : A\} \in \mathrm{P}(\mathbf{1})$. Clearly

$$\{0 : 0 \in x\} = x, \quad x = y \leftrightarrow \{0 : 0 \in x\} = \{0 : 0 \in y\}.$$

9.10. *Relations and functional types.* n-ary relations are represented as elements of the power type of a product type. If X and Y are generalized types, we can define the *set of functions* $X \rightarrow Y$ (or Y^X) as

$$f \in Y^X := \{f \in \mathrm{P}(X \times Y) : \forall u \in x(Iv.[(u, v) \in f] \in Y)\};$$

the function notation is available since we have the description operator,

$$f(u) := Iv \in y.[(u, v) \in f].$$

9.11. *The use of variable types.* One of the disadvantages of type theory in the form considered above, in comparison with set theory, is the fact that all objects are supposed to be located in a specific type: all possible kinds of power-sets may contain topological spaces for example, or the elements of groups and fields may be located in any type.

This disadvantage of the typed language is rather easily removed by admitting variables for types in the language, e.g. $\alpha, \beta \ldots$ (there are no quantifiers over type variables). This permits us to formulate statements for arbitrary types in the theory. It is easy to see that such an extension with variable types is conservative.

9.12. *Other formalisms of type theory.* There are many formalisms of type theory possible which are intermediate in character between the system **TYP** + EXT and the system of 9.5. For example, one may consider systems without partially defined terms, but with a liberal supply of types; if comprehension types are present, one then has to provide for the possibility of empty types (cf. Boileau and Joyal 1981).

10. Notes

10.1. *Primitive recursive functions.* The class of primitive recursive functions is introduced in Gödel (1931) (under the name "rekursive Funktionen"). Even after the discovery of the general recursive functions the primitive recursive functions have retained their interest as an important subclass and have been the subject of many investigations.

10.2. *Primitive recursive arithmetic.* Skolem (1923) develops primitive recursive arithmetic in an informal setting; Skolem's work is extensively discussed in Hilbert and Bernays (1934), section 7, as an example of finitistic reasoning. A neat formalism for **PRA** is presented by Curry (1941), who also gives a proof that the addition of propositional logic to the purely equational part of the theory is conservative. The latter result is also found in Goodstein (1945) (written in 1941). An extensive and detailed treatment of primitive recursive arithmetic is to be found in the monograph Goodstein (1957).

The strong version of induction in primitive recursive arithmetic stated in 2.5, as well as its equivalence with the ordinary induction rule, is due to Skolem (1939), with a correction by Peter (1940).

Further references to work on recursive arithmetic are given in notes at the end of Goodstein's monograph, and in the introduction to the translation of Skolem's paper in van Heijenoort (1967).

PRA as a system is mathematically more powerful than one would at first expect, see e.g. Mints (1976), Sieg (1985).

10.3. *Intuitionistic first-order arithmetic.* The system **HA** is not singled out as a subsystem in Heyting (1930A, 1930B). A system H′ which is virtually the same as **HA** appears in Gödel (1933); our version of **HA** coincides with version C mentioned in Kleene (1945), section 4.

In Kleene (1952) quite a lot of theorems of elementary arithmetic are proved in **HA**.

The natural deduction version of the rule of induction is found in Prawitz (1971) and Jervell (1971).

10.4. *Algorithms.* As noted in 4.3, it is not the precise formalization in arithmetic which matters for "algorithmic mathematics" such as CRM, but rather a few essential properties such as the existence of a universal binary algorithm for all unary algorithms, and a decidable computation predicate (such as Kleene's T). A nice exposition based on these principles has been given by Richman (1983). For an example of an axiomatic approach to the notion of computation see e.g. Fenstad (1980).

10.5. *Disjunction and explicit definability properties.* These are perhaps the most widely known and investigated metamathematical properties of formal systems based on intuitionistic logic. The general form of the explicit definability property is

ED $\quad \vdash \exists a A(a) \Rightarrow \vdash A(t)$ for a suitable t,

where a and t are of the same sort.

Gödel (1932) states the DP for **IPC** without proof; a proof for **IQC** is indicated in Gentzen (1935). The explicit definability property ED was presumably known to Gentzen (cf. Heyting 1934, page 18) and is certainly implicit in Gentzen (1935), since the same method which yields DP for **IPC** also gives ED for **IQC**.

Kleene (1945, end of section 8), and in greater detail in Kleene (1952, theorem 62(b)) gives a proof of EDN and DP for **HA**, using a variant of realizability. In Kleene (1962) the Kleene slash was introduced ($|^*$ of E3.5.3), and used to obtain DP and EDN under implication for suitable premisses.

The Kleene slash is closely related to the Aczel slash $|$ (Aczel 1968), on which our exposition in section 5 is based (cf. E3.5.3); both yield a characterization of the premisses for which relativized EDN and DP hold in **IQC**, **IPC** and **HA** (Kleene 1962).

The class $\mathcal{RH}$ appears in Rasiowa (1954, 1955) for **IQC** and in Harrop (1960) also for **HA**. More details are given in the introduction of Harrop (1956) and in the footnotes to Kleene (1962).

The Kleene slash has been generalized to stronger systems, see e.g. Friedman (1973, 1977), and has also model-theoretic counterparts (cf. Smorynski 1973, de Jongh and Smorynski 1976, Aczel 1968, van Dalen 1984 and the discussion in 13.7.7). It is a powerful tool, see e.g. Friedman and Scedrov (1983) for a recent application.

The fact that all the known proofs establish EDN and DP simultaneously is no mere coincidence: Friedman (1975) showed that for systems containing a certain minimum of arithmetic, DP implies EDN. For more information about generalizations of DP and ED see also 10.5.3–5.

Many authors have treated ED and DP as necessary (and sometimes also as sufficient) conditions for a system to be "constructive". This view seems to be ill-founded: for example, if a formal system **S** proves $A \vee B$ and is constructively well motivated, the formal proof in **S** corresponds to an informal constructive proof of $A \vee B$ and hence there ought to be an informal constructive proof of either A or B; but it does not follow that this informal proof of A or B should have a formal counterpart in **S**; that only follows if **S** has suitable closure properties. In Troelstra (1973A) an example, due to Kreisel, is presented of an intuitionistically correct extension of **HA** which does not satisfy EDN.

ED and DP are also not sufficient for determining the constructive meaning of $\vee$ and $\exists$ relative to the other logical operators, as may be seen from the fact that both **HA** + IP and **HA** + CT_0 + MP satisfy EDN and DP, but are inconsistent when taken together (Troelstra 1973A). Here IP is the schema

$$\text{IP} \quad (\neg A \to \exists x B) \to \exists x(\neg A \to B)\ (x \notin \mathrm{FV}(A)),$$

CT_0 and MP are defined in 4.3.2 and 4.5.1 respectively. Another example has been given in J.R. Moschovakis (1981).

10.6. *Markov's rule.* The particularly elegant proof presented here is due to Friedman (1978) and, independently, to Dragalin (1980). The earliest proof is in Kreisel (1958), remark 6.1. Before the Friedman–Dragalin argument became available proofs of Markov's rule were based on a variety of more complicated methods (see e.g. Troelstra 1973, section 3.8, Smorynski 1973, Girard 1973). For a generalization of Friedman's argument, see Leivant (1985).

10.7. *The formalization of elementary recursion theory.* The formalization as outlined here is based on three elements: 1^0 Kleene's method of indexing, introduced in Kleene (1958), 2^0 the use of (elementary) inductive definitions, and 3^0 E-logic with terms constructed with Kleene brackets, relying on the fact that the introduction of symbols for provably partial functions with the appropriate characterizing axioms produces definitional extensions (2.7.6). Kleene's monograph (1969) treats the formalization in great detail, but does not make use of 2^0 or 3^0; the treatment of Kleene brackets in Kleene (1969) is ad hoc. Section 6 is an expansion of Troelstra 1973, section 1.4.

10.8. *Higher-order logic.* The relativization of theories with extensionality to theories without extensionality has been used by many authors: see e.g. Luckhardt (1973) and the references given there.

Kreisel (1968A) showed how to extend the Gödel–Gentzen negative translation to second-order logic with set variables, Myhill (1974) extended the translation to simple type theory with set variables.

For the type theory with E-logic in 9.5–8 we followed Scott (1979). The device of variable types is used, for example, by Girard (1971).

Exercises

3.1.1. Write down the "defining equations" according to the official definition, for the functions in 1.3 (demonstrated in 1.3 for the cases of a constant function and addition).

3.1.2. Prove 1.4, 1.6.

3.1.3. Prove 1.10, 1.12, and the assertion at the end of 1.11.

3.1.4. Prove 1.14.

3.2.1. Prove in **PRA** $x \dot{-} (x \dot{-} y) = y \dot{-} (y \dot{-} x)$ and $(x + y) \dot{-} y = x$ (Goodstein 1957, examples II, 2.4.21, and formula 2.6.11).

3.2.2. Define the "integer-square-root" function R by $R0 = 0$, $R(Sx) = Rx + (1 \dot{-} ((SRx)^2 \dot{-} Sx))$(so Rx is the maximal y such that $y^2 \le x$) and prove $Sx \dot{-} (SRx)^2 = 0$, $(Rx)^2 \dot{-} x = 0$ (Goodstein 1957, 1.6, examples II 2.8.3).

3.2.3. Introduce a (non-surjective) pairing function P with inverses P_1, P_2 given by $P(x, y) = (x + y)^2 + x$, $P_1 z = z \dot{-} (Rz)^2$, $P_2 z = Rz \dot{-} P_1 z$, where R is taken from E3.2.2. Prove $P_1 P(x, y) = x$, $P_2 P(x, y) = y$ (Goodstein 1957, chapter 7).

3.2.4. Use the existence of pairing with inverses to reduce definition of φ_i $(1 \le i \le n)$ from ψ_i, χ_i $(1 \le i \le n)$ by

$$\begin{cases} \varphi_i(0, \vec{x}) = \psi_i(\vec{x}), \\ \varphi_i(Sz, \vec{x}) = \chi_i(z, \varphi_1(z, \vec{x}), \ldots, \varphi_n(z, \vec{x}), \vec{x}) \end{cases}$$

(*definition by simultaneous recursion*) to ordinary recursion.

3.2.5. Generalize proposition 2.5 to

$$A(0, \vec{y}), \quad A(x, \varphi_1(x, \vec{y}), \ldots, \varphi_n(x, \vec{y})) \to A(Sx, \vec{y}) \Rightarrow A(t_0, \vec{t}),$$

and give a corresponding generalization of proposition 2.6.

3.2.6. Prove 2.9.

3.2.7. Give a formalization of the proofs of 2.4(ii)–(iv) in a linear natural deduction style.

3.2.8. Show that $0 \ne S0$ is not provable from the other axioms in **PRA** if $\perp$ is treated as a primitive symbol.

3.3.1. Show that in **HA** proofs for
(i) $x = y \vee x \ne y$, and
(ii) $x = y \vee x < y \vee y < x$
can be given which are simpler than the proofs of these facts presented in section 2 in the context of **PRA**.

3.3.2. Prove in **HA** 1.10, 1.12.

3.3.3. Give a formalized proof of the strong induction rule (2.5) in **HA**.

3.4.1. Let φ, χ, θ be recursive. Show that then also φ given by

$$\begin{cases} \varphi(0, \vec{x}) = \psi(\vec{x}), \\ \varphi(Sz, \vec{x}) = \chi(z, \vec{x}, \varphi(z, \theta(z, \vec{x}))) \end{cases}$$

is recursive.

3.4.2. Let $\langle f_i \rangle_i$ effectively enumerate all partial algorithmic functions, and let J be any finite subset of $\mathbb{N}$ with $f_j = f$ for all $j \in J$. Then for some $i \notin J$ $f_i = f$, and i can be found by an algorithm (see e.g. Richman 1983).
Hint. Let θ be a total function such that $f_{\theta m} = f$ if $f_m(m)$ is defined, $f_{\theta m}$ totally undefined otherwise; then show that for some m

$$\theta m \notin J \text{ iff } f_m(m) \text{ defined.}$$

If $\theta(m) \notin J$, $f_m(m)$ is defined, $f_{\theta m} = f$; take $i = \theta m$.
If $\theta(m) \in J$, $f_{\theta m} = f$ totally undefined; then consider θ' such that $f_{\theta' m} = \lambda x.1$ if $f_m(m)$ is defined, otherwise $f_{\theta' m}$ everywhere undefined, etcetera.

3.5.1. Prove the lemma in 5.7.

3.5.2. Check the remaining cases in the proof of 5.9.

3.5.3. The "*Kleene slash*" is formulated similarly to the Aczel slash: (i), (ii) as before, and
(iii) $\Gamma|^* A \vee B := (\Gamma|^* A$ and $\vdash A)$ or $(\Gamma|^* B$ and $\vdash B)$,
(iv) $\Gamma|^* A \to B := (\Gamma|^* A$ and $\vdash A) \Rightarrow \Gamma|^* B$,
(v) $\Gamma|^* \forall x B(x) := \Gamma|^* A(\bar{n})$ for all n,
(vi) $\Gamma|^* \exists x B(x) := \Gamma|^* B(\bar{n})$ and $\Gamma \vdash B(\bar{n})$ for some n.
Show

$$(\Gamma|^* A \text{ and } \vdash A) \Leftrightarrow \Gamma|A.$$

3.5.4. Complete the proof of 5.12.

3.5.5. Give the proof of 5.16 in full.

3.6.1. Bring the closure conditions for examples 1 and 2 in 6.4 in the standard form indicated by (7) via (5), (6).

3.6.2. For the predicate Q of 6.4 write down a formula $A(x, z)$ expressing that $z \in Q$ by a tree of depth x. Show that $\exists x A(x, z)$ satisfies the closure conditions and is minimal. Do the same for example 1.

3.6.3. Show, arguing classically, that the definition of P_A as the first Γ_A^α for which $\Gamma_A^\alpha = \Gamma_A^{\alpha+1}$ satisfies (9) in 6.4.

3.6.4. Complete the proof of 6.7, and the proof of 6.6 (in 6.9) (Troelstra 1973, 1.4.3, 1.4.4).

3.6.5. Write down explicitly an arithmetical predicate $F(x, z)$ expressing "x is a correct tree for z" (Troelstra 1973, 1.4.5).

3.6.6. Fill in the details of the proof of 6.11 (Troelstra 1973, 1.4.5).

3.7.1. Bring the other inductive clauses in the definition of Ω (lemma 7.3) also in standard form.

3.7.2. Prove closure under minimalization for the partial recursive functionals (7.5(iii)). *Hint.* Let ψ be recursive such that $\{\psi(z_0)\}(x, \vec{y}) \simeq \{z_0\}(Sx, \vec{y})$, and apply the recursion theorem to the partial recursive $\{z\}$ such that $\{z\}(x_0, x_1, \vec{y}) \simeq 0$ if $\{x_1\}(0, \vec{y}) \simeq 0$, and $\simeq \{x_0\}(\psi(x_1)) + 1$ if $\{x_1\}(0, \vec{y}) > 0$.

3.7.3. Let the primitive recursive functionals be generated by Ω0–Ω5 and

$\Omega 8$ if $(n, m, m) \in \Omega$ then $(n, \langle 3, k+1, x, y\rangle, \langle 0\rangle * m) \in \Omega$;
if $(n, \langle 3, k+1, x, y\rangle, \langle z\rangle * m) \in \Omega$ and $(n', y, \langle n, z\rangle * m) \in \Omega$
then $(n', \langle 3, k+1, x, y\rangle, \langle Sz\rangle * m) \in \Omega$.

Show that there is an index for a primitive recursive function representing Sb(x, y); show also that the primitive recursive functionals are total.

3.7.4. Prove the second fact in 7.6:

$$T(x, \alpha, m, u) \wedge T(x, \alpha, m, u') \rightarrow u = u'.$$

3.7.5. Check the second and third property of T^* in 7.7.

3.7.6. Prove 7.11(ii), 7.13(ii), 7.14(ii).

3.8.1. Check in detail that the addition of propositional variables with full comprehension for propositions is conservative over the system of **IQC**^{2} with propositional variables deleted (8.2).

3.8.2. Complete the proof of the definability of $\perp, \wedge, \vee, \exists$ in 8.3. (Prawitz 1965, chapter V, theorem 1).

3.8.3. Prove theorem 8.6.
Hint. (i) Let us suppose, for simplicity, to be working in **HAS** and choose $(\mathbb{N}, 0, S)$ for (X, c, f). Now define

$$Y_\varphi(u, v) :\equiv \forall Z^2\big((0, c') \in Z^2 \wedge \forall xy\big((x, y) \in Z^2 \to (Sx, fy) \in Z^2\big) \to (u, v) \in Z^2\big).$$

With induction we can show that Y_φ is a relation with functional character.

(ii) (In $\mathbf{IQC}^2$) relying on (i) we let $\lambda y.x + y$ be the isomorphism from the Peano-structure (X, c, f) onto (X, x, f) (for any $x \in X$), and we let $\lambda y.x \cdot y)$ be the isomorphism from (X, c, f) onto $(X, c, \lambda y.x + y)$ etcetera (Henkin 1960).

3.8.4. Show that the negative translation can be applied directly to **HAS** formulated with comprehension terms.

3.8.5. Show that $\mathbf{HAS}^c$ and **HAS** prove the same Π^0_2-formulas, e.g. by applying 2.3.16–18 to **HAS** as embedded in $\mathbf{IQC}^2$.

3.8.6. Show in detail that $\mathscr{K}'$ as constructed from $\mathscr{K}$ in the proof of 8.9 indeed validates comprehension (de Jongh and Smorynski 1976).

3.8.7. Consider the following conservative extension **H** of **HAS**: we add constants $C_{B,V}$ for each formula B of **H** not containing free set variables, and each set $V \subset \mathbb{N}$, and we add axioms

$$(x_1, \ldots, x_n) \in C_{B,V} \leftrightarrow B(x_1, \ldots, x_n),$$

where $\mathrm{FV}(B) = \{x_1, \ldots, x_n\}$. We now define the following extension of the slash (cf. 5.7) for sentences of **H**, as follows

(i) $|(t = s) :\equiv\ \vdash t = s$;
$|(t_1, \ldots, t_n) \in C_{B,V} :\equiv (|t_1|, \ldots, |t_n|) \in V$ (here $|t|$ is the value of t in $\mathbb{N}$);
(ii) the clauses for $A \wedge B$, $A \vee B$, $A \to B$, $\forall xA$ and $\exists xA$ are as before;
(iii) $|\forall XA(X) :\equiv$ for all $C_{B,V}$ $|A(C_{B,V})$;
(iv) $|\exists XA(X) :\equiv$ for some $C_{B,V}$ $|A(C_{B,V})$
(deducibility refers to deducibility in **H**).

Prove the soundness theorem in the following form

$$\mathbf{HAS} \vdash A \Rightarrow |A$$

for all sentences A.

Derive from this the numerical existence property and the (closed) set-existence property (*explicit definability for sets*) in the form

$$\mathbf{HAS} \vdash \exists X A \Rightarrow \mathbf{HAS} \vdash A[X/\lambda x.B],$$

where $\mathrm{FV}(B) = \{x\}$, A not containing free set variables (Friedman 1973; Troelstra 1973, 3.1.19–23).

3.9.1. Prove the theorem in 9.2 (Myhill 1974, Luckhardt 1973, chapter 2).

3.9.2. Prove the lemma and theorem in 9.3 (Myhill 1974).

3.9.3. Prove the equivalence in remark (i) of 9.5.

3.9.4. Prove theorem 9.7 (Scott 1979).

3.9.5. Give the details of the theorem in 9.8; devise a formulation where types σ and $\sigma \upharpoonright t$ are treated as disjoint, and t^{σ} and $t^{\sigma \upharpoonright t}$ are regarded as syntactically distinct (Scott 1979).

3.9.6. Show that the following yields a formal system equivalent to **TYP** + EXT: Ω itself is regarded as a type; formulas are terms of type Ω, and if t_1, t_2 are terms of type σ, $t_1 = t_2$ is a term of type Ω. If t is of type Ω, we regard $(\ ) \in t$ as synonymous with t itself ($(\ )$ is the empty sequence); $\top$ is a term of type Ω. We have the axioms

$$A \Rightarrow A; \quad \top \Rightarrow x = x; \quad A \wedge x = t \Rightarrow A[x/t];$$

$$A \Rightarrow \vec{x} \in \{(\vec{x}) : A\}, \quad \vec{x} \in \{(\vec{x}) : A\} \Rightarrow A,$$

where $\Rightarrow A$ is short for $\top \Rightarrow A$. We also have the rules

$$\frac{A \Rightarrow B,\ B \Rightarrow C}{A \Rightarrow C} \quad \frac{A \Rightarrow B,\ A \Rightarrow C}{A \Rightarrow B \wedge C} \quad \frac{A \Rightarrow B \wedge C}{A \Rightarrow B} \quad \frac{A \Rightarrow B \wedge C}{A \Rightarrow C}$$

$$\frac{A \Rightarrow B}{A[x/t] \Rightarrow B[x/t]} \quad \frac{A \wedge (\vec{x}) \in t_1 \Rightarrow (\vec{x}) \in t_2 \quad A \wedge (\vec{x}) \in t_2 \Rightarrow (\vec{x}) \in t_1}{A \Rightarrow t_1 = t_2}$$

and we define

$$\forall x A := (\{x : A\} = \{x : \top\}),$$

$$A \to B := (A \wedge B = A),$$

$$A \vee B := \forall \omega^{\Omega}((A \to \omega) \wedge (B \to \omega) \to \omega = \top),$$

$$\bot := \forall \omega(\omega = \top),$$

$$\exists x A := \forall \omega^{\Omega}((\forall x A \to \omega) \to \omega = \top)$$

(Boileau and Joyal 1981).

CHAPTER 4

NON-CLASSICAL AXIOMS

The present chapter is devoted to a discussion of several principles (axioms) which play an important role in various forms of constructive mathematics. The most interesting ones among such principles are those which clearly express that we have an interpretation of logic in mind which differs from the classical reading; in other words, principles which are formally incompatible with classical logic. The main examples of such principles are various (intuitionistic) forms of Church's thesis (sections 3, 4), (partially) expressing the idea that "all operations in mathematics are algorithmic" (corresponding to "Constructive Recursive Mathematics" CRM) and continuity principles for choice sequences, originating in the adoption of non-predeterminate sequences as legitimate objects of intuitionistic mathematics (section 6 and part of section 7).

Besides these "non-classical" principles there are also some principles, which are valid on a classical reading of the logical operators, but which in a constructive setting are in need of a different justification, and which, moreover, are mathematically interesting in combination with the non-classical principles mentioned above. The principal examples are axioms of countable choice (section 2), Markov's principle (section 5), the compactness of the Cantor discontinuum (section 7), bar induction (section 8), Kripke's schema (section 9). Section 9 also discusses the uniformity principle, which is classically false.

The reader can proceed in at least two different ways. The sections may be read consecutively, giving an overview of important principles of constructive mathematics. It is also possible to postpone reading the various sections until needed for mathematical applications. Sections 8 and 9 of the present chapter hardly play a role in chapters 5–8 on constructive mathematics.

1. Preliminaries

1.1. Notations and conventions. Unless stated otherwise, variables ranging over the natural numbers are represented by ordinary lower case letters. Greek lower case letters $\alpha, \beta, \gamma, \delta$ are metavariables for variables ranging over $\mathbb{N} \to \mathbb{N}$, roman capital letters X, Y, Z for set- or relation variables. We define

$$\alpha = \beta := \forall x(\alpha x = \beta x), \quad \alpha \leq \beta := \forall x(\alpha x \leq \beta x).$$

For initial segments we use

$$\bar{\alpha}0 := \langle\ \rangle = 0, \quad \bar{\alpha}(x+1) := \bar{\alpha}x * \langle \alpha x \rangle,$$

$$\alpha \in n := (\bar{\alpha}(\mathrm{lth}(n)) = n) := \exists x(\bar{\alpha}x = n) \ (\alpha \text{ has initial segment } n).$$

Concatenation can be extended to concatenation of a finite sequence with an infinite one:

$$(n * \alpha)(x) = \begin{cases} (n)_x & \text{if } x < \mathrm{lth}(n), \\ \alpha(x \dot{-} \mathrm{lth}(n)) & \text{if } x \geq \mathrm{lth}(n). \end{cases}$$

We may think of $n * \alpha$ as "n followed by α". For predicates of sequences we shall assume extensionality:

$$A(\alpha) \wedge \alpha = \beta \to A(\beta).$$

We shall use notations such as $t \in X$, $X(t)$ (or even Xt, if no confusion is to be feared) interchangeably. □

1.2. Definition. We define a *tree* T as an inhabited, decidable set of finite sequences of natural numbers closed under predecessor; so T is a tree iff

$$\langle\ \rangle \in T, \quad \forall n(n \in T \vee n \notin T),$$

$$\forall nm(n \in T \wedge m \prec n \to m \in T).$$

A *spread* is a tree in which each node has at least one successor:

$$\forall n \in T \exists x(n * \langle x \rangle \in T).$$

A *finitary spread* or *fan* is a finitely branching spread, i.e. a spread satisfying

$$\forall n \in T \exists z \forall x(n * \langle x \rangle \in T \to x \leq z).$$

A sequence α is a *branch* of the tree T iff all initial segments belong to T:

$$\alpha \in T := \forall x(\bar{\alpha}x \in T).$$

For quantifiers over finite sequences ranging over a tree T we shall often write $\forall n \in T$ or $\forall n_T$, $\exists n \in T$ or $\exists n_T$; similarly we use $\forall \alpha_T, \exists \alpha_T$. □

1.3. DEFINITION. The *universal tree* or *universal spread* T_U consists of all finite sequences of natural numbers; so modulo coding $T_U = \mathbb{N}$. The infinite branches of T_U constitute $\mathbb{N}^{\mathbb{N}}$.

The *binary tree* consisting of all 01-sequences will be denoted by T_{01}. T' is a *subtree* of T iff $T' \subset T$ and T' is a tree. □

1.4. LEMMA. *Let $T' \subset T$, T' and T spreads. Then there is a mapping $\Gamma \in T \to T'$ such that*
(i) $\text{lth}(\Gamma n) = \text{lth}(n)$ *for all* $n \in T$,
(ii) $\Gamma n = n$ *for* $n \in T'$.
N.B. Γ can be extended to infinite branches by $\Gamma\alpha := \lambda x.(\Gamma\bar{\alpha}(x+1))_x$; then $\Gamma(\Gamma\alpha) = \alpha$ for $\alpha \in T$, $\Gamma\alpha = \alpha$ for $\alpha \in T'$.

PROOF. Exercise. □

1.5. DEFINITION. A *topology* on a set X can be defined, as usual, as a family $\mathcal{T}$ of subsets of X, containing $\emptyset$, X and which is closed under finite intersections and arbitrary unions. A *basis* $\mathcal{B}$ for a topology $\mathcal{T}$ on X is a family $\mathcal{B} \subset \mathcal{T}$ such that

$$\forall Y \in \mathcal{T}(Y = \bigcup\{X : X \in \mathcal{B} \wedge X \subset Y\}). \quad \square$$

Observe that for a basis $\mathcal{B}$:

$$\text{if } U \in \mathcal{B},\ U' \in \mathcal{B},\ x \in U \cap U' \quad \text{then } \exists U'' \in \mathcal{B}(x \in U'' \subset U \cap U');$$
$$\text{if } x \in X \text{ then } \exists U \in \mathcal{B}(x \in U).$$

A spread T can be given a topology by taking the infinite branches as points, and as basis the sets

$$V_n := \{\alpha : \alpha \in n \wedge \alpha \in T\} \quad \text{for each } n \in T.$$

Thus T_{01} as a topological space is $2^{\mathbb{N}}$, the Cantor space. T_U is $\mathbb{N}^{\mathbb{N}}$, the Baire space, with the product topology obtained from the discrete topology on $\mathbb{N}$, as in the classical theory. (For more information about constructive topology see chapter 7.)

A *topological space* is a pair $(X, \mathscr{T})$ consisting of a set X with a topology $\mathscr{T}$ on X; the elements of $\mathscr{T}$ are the *open* sets of the space and the elements of X are the *points* of the space. We define the *interior* of a set $Y \subset X$ as

$$\mathrm{Int}(Y) := \bigcup\{V \in \mathscr{T} : V \subset Y\}.$$

A *continuous* mapping from a space Γ to a space Γ' is a mapping for which the inverse image of an open set is always open, or in terms of bases: the inverse image of a basis element is a union of basis elements. A *homeomorphism* from a topological space Γ to a space Γ' is a continuous bijection with continuous inverse.

Thus, on spelling it out, continuity for $\Phi \in \mathbb{N}^{\mathbb{N}} \to \mathbb{N}$ and $\Psi \in \mathbb{N}^{\mathbb{N}} \to \mathbb{N}^{\mathbb{N}}$ is equivalent to

$$\forall\alpha\exists x\forall\beta \in \bar{\alpha}x(\Phi\alpha = \Phi\beta), \quad \text{and}$$

$$\forall x\forall\alpha\exists y\forall\beta \in \bar{\alpha}y((\Psi\alpha)(x) = (\Psi\beta)(x)),$$

respectively. The second line states that $\lambda\alpha.(\Psi\alpha)(x)$ is continuous for all x.

□

1.6. *Axiomatic basis.* Most of the discussion is informal, i.e. no particular formalism is presupposed. The principles of arithmetic, in particular induction, will be tacitly used; for functions we shall assume function comprehension

$$\mathrm{AC}_{00}! \qquad \forall x\exists! yA(x, y) \to \exists\alpha\forall xA(x, \alpha x),$$

i.e. if $A(x, y)$ is a binary relation with functional character

$$\forall xyy'(A(x, y) \wedge A(x, y') \to y = y') \wedge \forall x\exists yA(x, y),$$

then there is a function with its graph given by A.

As to the existence of sets, as a rule we shall accept comprehension in the form

$$\exists X\forall x[x \in X \leftrightarrow A(x)] \quad (X \text{ not free in } A)$$

for A constructed from prime formulas, propositional operations, quantifiers over natural numbers and quantifiers over number-theoretic functions. By the existence of pairing functions this implies the corresponding principle for n-place relations:

$$\exists X^n\forall\vec{x}[\vec{x} \in X^n \leftrightarrow A(\vec{x})].$$

We also accept a corresponding principle for sets of number-theoretic functions:

$$\exists X \forall \alpha [\alpha \in X \leftrightarrow A(\alpha)]$$

under the same restrictions as above. The use of *impredicative* comprehension, where also set- and relation quantifiers are permitted in the construction of A, will be noted explicitly (cf. also section 3.8).

2. Choice axioms

2.1. *Axioms of countable choice.* The simplest axiom (schema) of countable choice is AC-NN, also frequently denoted by AC_{00}

AC-NN $\quad \forall n \exists m A(n, m) \to \exists \alpha \forall n A(n, \alpha n).$

Informally, this schema may be justified as follows: a proof of the premiss should provide us with a method to find, for each $n \in \mathbb{N}$, an $m \in \mathbb{N}$ such that $A(n, m)$; such a method is nothing else but the description of a function assigning the required m to n.

In the same manner one can justify the more general countable choice axiom AC-N ($\equiv AC_0$) for arbitrary domains D:

AC-N $\quad \forall n \exists d \in \mathrm{D} A(n, d) \to \exists \varphi \in \mathbb{N} \to \mathrm{D} \forall n A(n, \varphi n).$

The principle

AC-NF $\quad \forall n \exists \alpha A(n, \alpha) \to \exists \beta \forall n A(n, (\beta)_n)$

(also denoted by AC_{01}), where $(\beta)_n := \lambda m.\beta(j(n, m))$, is a special case of AC-N, since a mapping $\varphi \in \mathbb{N} \to \mathbb{N}^{\mathbb{N}}$ may be coded by a $\beta \in \mathbb{N} \to \mathbb{N}$ such that $\beta(j(n, m)) = \varphi(n)(m)$.

It might be tempting to assume the more general axiom AC for any two domains D, D′:

$$\forall x \in \mathrm{D} \exists y \in \mathrm{D}' A(x, y) \to \exists \varphi \in \mathrm{D} \to \mathrm{D}' \forall x \in \mathrm{D} A(x, \varphi x). \quad (1)$$

However, we assume that domains of quantification always come equipped with a notion of equality on them, and although a proof of the premiss of (1) must contain a method for transforming any *proof* of $x \in \mathrm{D}$ into a proof of $\exists y \in \mathrm{D}' A(x, y)$, which implies that we must have a method for extracting, from the proof that $x \in \mathrm{D}$, a $y \in \mathrm{D}'$ such that $A(x, y)$, there is no guarantee that such a method respects the equality $=_{\mathrm{D}}$ on D. For example, consider the true statement

$$\forall x \in \mathbb{R} \exists n \in \mathbb{N} (x < n). \quad (2)$$

There is no continuous function assigning n to x, because such a function would have to be constant; on the other hand we have no hope of giving an example of a discontinuous function from $\mathbb{R}$ to $\mathbb{R}$ (cf. the discussion in 1.3.6). Also, free use of set comprehension and AC permits us to derive PEM (exercise 4.2.1).

Note however, that if we represent reals by fundamental sequences of rationals with a fixed rate of convergence, i.e. sequences $\langle r_n \rangle_n$ such that e.g. $\forall nm(|r_n - r_{n+m}| < 2^{-n})$, then we can assign to each $\langle r_n \rangle_n$ an $m \in \mathbb{N}$ by means of a function φ such that for the real x determined by $\langle r_n \rangle_n$ we have $x < m$ (take $\varphi(\langle r_n \rangle_n)$ to be the least natural number k such that $r_0 + 1 \leq k$).

The situation of example (2), where AC fails, may be described in more general terms as follows.

$=_{\mathrm{D}}$ is a defined notion, i.e. there is an underlying set $(\mathrm{D}'', =_{\mathrm{D}''})$ such that D consists of equivalence classes of elements of D″ modulo some equivalence relation $\approx$. Then the method φ implicit in $\forall x \in \mathrm{D}\, \exists y \in \mathrm{D}'$ perhaps respects $=_{\mathrm{D}''}$, but not necessarily $=_{\mathrm{D}}$. In example (2) above D″ is the collection of fundamental sequences.

Another at first sight surprising aspect is the following. Suppose that $(\mathrm{D}, =_{\mathrm{D}})$ is obtained as a subset of $(\mathrm{D}'', =_{\mathrm{D}''})$ and that $=_{\mathrm{D}}$ is the restriction of $=_{\mathrm{D}''}$ to D. Now one might expect that if AC-D″ holds, then AC-D also holds where the φ in AC is the restriction of a function defined on D″. However, the *method* φ given by the truth of the premiss of AC can act not only on the element $d \in \mathrm{D} \subset \mathrm{D}''$ as such, but also use information from the proof that $d \in \mathrm{D}''$ actually belongs to D. As a result, φ is not necessarily the restriction to D of a partial function from D″ to D′. In this situation even

$$\text{AC-DD}'! \qquad \forall x \in \mathrm{D}\, \exists! y \in \mathrm{D}' A(x, y) \rightarrow \exists \varphi \in \mathrm{D} \rightarrow \mathrm{D}'\, \forall x \in \mathrm{D}\, A(x, \varphi x)$$

may fail. We shall encounter this situation in 4.11.

So why is AC-N acceptable? Primarily because $n \in \mathbb{N}$ is immediately given to us, it is not in need of a proof. No extra information is needed to tell us that n, as an object constructed in a canonical way from units is a natural number.

It should be noted on the other hand that AC-NN is only justified in case our notion of function is as *wide as possible*, i.e. embraces every possible method for assigning numbers to numbers.

2.2. *The presentation axiom.* Reflection on the reasons of the failure of the general axiom of choice, described above, suggests a principle called the *presentation axiom*: each set D is the surjective image of an underlying set

D″ such that AC-D″D′ holds for arbitrary D′. We shall not make use of this principle in our constructive mathematics. For arithmetic it can be justified by recursive realizability (end of 4.12); it also crops up in connection with constructive set theory **CZF** (11.8.25).

2.3. *Axioms of dependent choice.* Let x and y range over some domain D, then the *axiom of dependent choices* is the principle

DC-D $\quad \forall x \exists y A(x, y) \to$

$$\forall x \exists \varphi \in \mathbb{N} \to \mathrm{D}[\varphi 0 = x \wedge \forall n A(\varphi n, \varphi(n+1))].$$

Informally, we may justify DC-D as follows: let $x \in \mathrm{D}$; put $\varphi 0 = x$; find y such that $A(\varphi 0, y)$, put $\varphi 1 = y$; find z such that $A(\varphi 1, z)$, put $\varphi 2 = z$ etc. For DC-$\mathbb{N}$ (resp. DC-$\mathbb{N}^{\mathbb{N}}$) one often writes DC_0 or DC-N (resp. DC_1).

An even more general principle is the *axiom of relativized dependent choices* over D

RDC-D $\quad \forall x \in \mathrm{D}[Ax \to \exists y \in \mathrm{D}(B(x, y) \wedge Ay)] \to$

$\quad \forall x \in \mathrm{D}[Ax \to$

$$\exists \varphi \in \mathbb{N} \to \mathrm{D}(\varphi 0 = x \wedge \forall n B(\varphi n, \varphi(n+1)))].$$

Some logical relationships are given in the exercises.

2.4. Remarks (*on terminology*). In the literature one frequently encounters the distinction between "operation" and "function", as well as the terms "extensional" and "intensional" applied to operations and functions.

Given two domains D, D′, the term *function* is reserved for an assignment φ of values in D′ to the elements of D which respects equality, i.e. $x =_{\mathrm{D}} y \to \varphi x =_{\mathrm{D}'} \varphi y$ for all $x, y \in \mathrm{D}$; an *operation* need not respect $=_{\mathrm{D}}$ but perhaps only a narrower relation of equality between the objects in D (cf. also the discussion of the presentation axiom above).

Informally, a notion of equality is said to be *extensional*, whenever the equality is defined as in the corresponding classical set-theoretic setting. A function is *extensional* if it respects extensional equality on its domain (in other words, if its domain is supposed to be provided with extensional equality). For example, $\alpha = \beta := \forall x(\alpha x = \beta x)$ is the usual extensional equality between functions, and an extensional functional Φ on $\mathbb{N} \to \mathbb{N}$ with values in $\mathbb{N}$ is an operation on $\mathbb{N} \to \mathbb{N}$ which respects extensional equality on $\mathbb{N} \to \mathbb{N}$.

We use, informally, the terms *non-extensional* or *intensional* if the equality or function/operation is not extensional in the sense just described. A non-extensional function does not respect extensional equality but a narrower equality relation, e.g. when all functions in $\mathbb{N} \to \mathbb{N}$ are assumed to be given by algorithms and a functional Φ on $\mathbb{N} \to \mathbb{N}$ respects equality between algorithms only.

Of course, if algorithms describing functions α, β are given by code numbers, one might avoid talking about "non-extensional equality" between α and β by working with the code numbers of α and β; in other words, we might make a notational distinction between the algorithms and the functions described by them. However, in metamathematical researches it is not always convenient to do so (cf. chapter 9).

In specific cases the precise sense of "non-extensional", "extensional" will always be clear from the context.

3. Church's thesis

3.1. *Church's thesis* CT. If we think of all our mathematical objects (predicates, operators etc.) as completely determined, fully described, we are in the realm of "lawlike" constructive mathematics. In particular, this means that elements of $\mathbb{N} \to \mathbb{N}$ are *lawlike*, i.e. completely determined by a law given in advance.

The idea of lawlikeness may be strengthened further by identifying "lawlike" with "recursive". The identification can be motivated in two different ways:

foundationally, by regarding the identification of lawlikeness with recursiveness as intuitively justified, on the basis of an analysis of the notion of lawlikeness;

pragmatically, by treating "recursive" as a mathematically precise concept which can replace the informal notion of lawlikeness in mathematical developments.

The difference between these is one of emphasis: on the foundational approach, the arguments in favour of "lawlike = recursive" carry much more weight than on the pragmatical approach.

The identification of "lawlike" with "recursive" for functions can be expressed by an axiom ("Church's Thesis")

$$\text{CT} \qquad \forall \alpha \exists x \forall y \exists z (Txyz \wedge Uz = \alpha y).$$

For a striking consequence of CT see E4.7.7: Cantor space and Baire space are homeomorphic.

3.2. *The arithmetical form of Church's thesis* CT_0. Taken by itself, CT may be regarded as a definition of the notion of "number-theoretic function", by identifying $\mathbb{N}^{\mathbb{N}}$ with the collection of all total recursive functions. This changes if we combine CT with AC-NN; then we obtain the following schema

$$\mathrm{CT}_0 \qquad \forall n \exists m A(n,m) \to \exists k \forall n \exists m [A(n, Um) \wedge Tknm].$$

CT_0 is called the *arithmetical form of Church's thesis*, since only numerical quantifiers appear in it.

The acceptation of CT_0 is equivalent to the assumption that the function implicit in an intuitionistic proof of a $\forall n \exists m$-statement is always recursive. Recall, on the other hand, that AC-NN was regarded as plausible precisely because of the fact that *any* method for assigning numbers to numbers was regarded as a function. AC-NN may still be regarded as rather plausible if we think of our mathematical universe as consisting of lawlike objects only. Note, that if we adopt CT_0 it means that we also have to accept the identification of lawlike with recursive. If we feel convinced by the arguments in favour of an identification of "lawlike" with "recursive", we have no reason to doubt the laws of intuitionistic predicate logic, even if we adopt CT_0. If, on the other hand, we do not regard the identification of "lawlike" and "recursive" as justified, CT_0 imposes a reading of quantifier-combinations "$\forall n \exists m$" stronger than what is required by the BHK-interpretation. This leaves us with a problem: are the laws of intuitionistic predicate logic still valid, if we read all combinations $\forall n \exists m$ in the stronger sense dictated by CT_0? (N.B. This is not obvious: if in implications $\forall n \exists m A(n, m) \to \forall n' \exists m' B(n', m')$ the meaning of both premiss and conclusion is changed, the new interpretation of the implication is not necessarily comparable with the original one.)

For the time being, however, we shall tacitly assume that adopting CT_0 is compatible with intuitionistic predicate logic; in the next section a metamathematical justification for this fact is discussed.

In view of $(A \vee B) \leftrightarrow \exists z((z = 0 \to A) \wedge (z \neq 0 \to B))$ we can formulate a version of CT_0 for disjunctions

$$\mathrm{CT}_0^{\vee} \qquad \forall x(Ax \vee Bx) \to \exists \alpha \in \mathrm{TREC}\, \forall x((\alpha x = 0 \to A) \wedge (\alpha x \neq 0 \to B)).$$

3.3. CT does not conflict with classical logic unless certain function-existence principles are present, such as "countable (function-) comprehension"

AC-NN! $\forall n \exists ! m A(n, m) \to \exists \alpha \forall n A(n, \alpha n)$;

classically AC-NN! establishes the existence of non-recursive functions, for example, an α satisfying

$$\alpha n = \begin{cases} 1 & \text{if } \exists m Tnnm, \\ 0 & \text{if } \neg\exists m Tnnm. \end{cases}$$

Since CT_0 implicitly contains AC-NN, we have an almost immediate conflict with classical logic:

3.4. PROPOSITION.

$$\mathbf{HA} + \mathrm{CT}_0 \vdash \neg\forall x[\neg\exists y Txxy \vee \neg\neg\exists y Txxy].$$

PROOF. Assume $\forall x[\neg\exists y Txxy \vee \neg\neg\exists y Txxy]$ and apply $\mathrm{CT}_0^{\vee}$; we find a recursive α such that

$$\forall x((\alpha x = 0 \to \neg\exists y Txxy) \wedge (\alpha x \neq 0 \to \neg\neg\exists y Txxy));$$

hence $\forall x(\alpha x = 0 \leftrightarrow \neg\exists y Txxy)$, and this would make $\exists y Txxy$ recursive. Contradiction, by 3.4.4. □

COROLLARY 1. *Assuming* CT_0, $\neg\forall$-PEM *holds.*

PROOF. Let the α in $\forall\alpha(\neg\exists x(\alpha x = 0) \vee \neg\neg\exists x(\alpha x = 0))$ range over the characteristic functions φ_x of $Txxy$ as predicate in y. □

COROLLARY 2. *Assuming* CT_0, *the following schema of predicate logic*

$$\forall x \neg\neg A(x) \to \neg\neg\forall x A(x)$$

is not generally valid.

PROOF. Take $A(x) := \neg\exists y Txxy \vee \neg\neg\exists y Txxy$. □

REMARK. $\forall$-PEM is implied by $\exists$-PEM, hence $\neg\forall$-PEM $\to \neg\exists$-PEM, where

$\exists$-PEM $\forall\alpha(\exists x(\alpha x = 0) \vee \neg\exists x(\alpha x = 0))$.

Bishop (1967) called PEM "the principle of omniscience", and $\exists$-PEM "the principle of limited omniscience". Many well-known theorems of classical analysis do not require full PEM, but only $\exists$-PEM.

3.5. PROPOSITION. *Assuming* CT_0, *we can show* $\neg\forall x A(x, \alpha, \beta)$ *for suitable* α, β, *where*

$$A(x, \alpha, \beta) := \neg(\exists y(\alpha(x, y) = 0) \wedge \exists y(\beta(x, y) = 0)) \\ \rightarrow \neg\exists y(\alpha(x, y) = 0) \vee \neg\exists y(\beta(x, y) = 0).$$

PROOF. Let $X := \exists y(\alpha(x, y) = 0)$, $Y := \exists y(\beta(x, y) = 0)$ represent two disjoint, r.e., recursively inseparable sets (e.g. by taking for α, β the characteristic functions of $Txxy \wedge Uy = 0$, $Txxy \wedge Uy = 1$ respectively; cf. proposition 3.4.5). Then the premiss of $A(x, \alpha, \beta)$ holds for all x, so if $\forall x A(x, \alpha, \beta)$, then

$$\forall x(\neg\exists y(\alpha(x, y) = 0) \vee \neg\exists y(\beta(x, y) = 0));$$

with $\mathrm{CT}_0^{\vee}$ we find a recursive γ such that

$$\gamma(x) = 0 \rightarrow \neg\exists y(\alpha(x, y) = 0),$$
$$\gamma(x) \neq 0 \rightarrow \neg\exists y(\beta(x, y) = 0),$$

and this contradicts the recursive inseparability of X and Y. □

3.6. COROLLARY. *The logical schema*

$$\neg(A \wedge B) \rightarrow (\neg A \vee \neg B)$$

has been refuted by the preceding proposition in its universally quantified form

$$\neg\forall x[\neg(A(x) \wedge B(x)) \rightarrow (\neg A(x) \vee \neg B(x))],$$

with $A(x) \equiv \exists y(\alpha(x, y) = 0)$, $B(x) \equiv \exists y(\beta(x, y) = 0)$; *a fortiori we have refuted it in the form*

$$\neg\mathrm{SEP} \qquad \neg\forall\alpha\beta[\neg(\exists y(\alpha y = 0) \wedge \exists y(\beta y = 0)) \\ \rightarrow (\neg\exists y(\alpha y = 0) \vee \neg\exists y(\beta y = 0))].$$

4. Realizability

4.1. Kleene (1945) proposed realizability as a systematic method of making the constructive content of arithmetical sentences explicit. Kleene's starting point was that an existential statement $B \equiv \exists x A(x_1, \ldots, x_n, x)$ is an "incomplete communication": one can only assert B, if x can be produced (depending on the parameters $x_1, \ldots, x_n$). Similarly, one must be able to decide between the disjuncts in order to assert a disjunction.

If one now also assumes that all existential and disjunction statements have to be fulfilled *recursively* in the parameters, one arrives at Kleene's notion of recursive realizability.

In the form in which it is given by Kleene, realizability is comparable to a classical truth definition: it interprets sentences of arithmetic in the intended structure $\mathcal{N} \equiv (\mathbb{N}, 0, S, +, \cdot, \ldots)$ of the natural numbers. Let us use $n, m, \ldots$ for arbitrary elements of $\mathbb{N}$, and let $\bar{n}, \bar{m}, \ldots$ be the corresponding numerals in the language of arithmetic. "n realizes A" is defined by induction on the complexity of A; n is a number *hereditarily coding* all information concerning the explicit realization of existence quantifiers and disjunctions occurring in A. The clauses are

(i) n realizes $(t = s)$ iff "$t = s$" holds in $\mathcal{N}$;
(so if $t = s$ holds, any number realizes it; the truth of a prime sentence gives no interesting information on the realization of $\exists$ or $\vee$);

(ii) n realizes $A \wedge B$ iff $j_1 n$ realizes A and $j_2 n$ realizes B;

(iii) n realizes $A \vee B$ iff $j_1 n = 0$ and $j_2 n$ realizes A, or $j_1 n \neq 0$ and $j_2 n$ realizes B;

(iv) n realizes $A \to B$ iff for each m realizing A, $\{n\}(m)$ is defined and realizes B;

(v) n realizes $\exists x A x$ iff $j_2 n$ realizes $A[x/\overline{j_1 n}]$;

(vi) n realizes $\forall x A(x)$ iff for all m, $\{n\}(m)$ is defined and realizes $A(\bar{m})$.

For arbitrary formulas A we may define "n realizes A" as "n realizes the universal closure of A".

It is to be noted that if we read in the BHK-interpretation "realizing number" for "proof" and "partial recursive operation" for "construction", the realizability interpretation appears as a variant of the BHK-interpretation.

Instead of working with the semantical version given above, it is more advantageous to reformulate realizability as a syntactical mapping assigning a formula $x \mathbf{r} A$ of **HA** to each formula A of **HA**, where $x \notin \mathrm{FV}(A)$ of **HA**.

4.2. DEFINITION. $x \mathbf{r} A$ is defined inductively by the following clauses, where $x \notin \mathrm{FV}(A)$.

(i) $x \mathbf{r} [t = s] := [t = s]$,

(ii) $x \mathbf{r} (A \wedge B) := j_1 x \mathbf{r} A \wedge j_2 x \mathbf{r} B$,

(iii) $x \mathbf{r} (A \vee B) := (j_1 x = 0 \wedge j_2 x \mathbf{r} A) \vee (j_1 x \neq 0 \wedge j_2 x \mathbf{r} B)$,

(iv) $x \mathbf{r} (A \to B) := \forall y (y \mathbf{r} A \to \exists u (Txyu \wedge Uu \mathbf{r} B)$ $(u \notin \mathrm{FV}(B))$,

(v) $x \mathbf{r} \forall y A := \forall y \exists u (Txyu \wedge Uu \mathbf{r} A)$, $(u \notin \mathrm{FV}(A))$,

(vi) $x \mathbf{r} \exists y A := j_2 x \mathbf{r} A[y/j_1 x]$.

In more informal notation with "Kleene brackets" (iv) and (v) may be rendered as:

$$x\,\mathbf{r}\,(A \to B) := \forall y(y\,\mathbf{r}\,A \to E\{x\}(y) \wedge \{x\}(y)\,\mathbf{r}\,B),$$

$$x\,\mathbf{r}\,\forall yA \quad := \forall y(E\{x\}(y) \wedge \{x\}(y)\,\mathbf{r}\,A).$$

REMARKS.

(i) Clause (iii) is equivalent to $x\,\mathbf{r}\,(A \vee B) := (j_1x = 0 \to j_2x\,\mathbf{r}\,A) \wedge (j_1x \neq 0 \to j_2x\,\mathbf{r}\,B)$.

(ii) $x\,\mathbf{r}\,\neg A \leftrightarrow \forall y(y\,\mathbf{r}\,A \to \{x\}(y)\,\mathbf{r}\,(0 = 1))$, hence $x\,\mathbf{r}\,\neg A \leftrightarrow \forall y(\neg\, y\,\mathbf{r}\,A)$. Thus a negation is realized by some number iff it is realized by all numbers.

(iii) $x\,\mathbf{r}\,\neg\neg A \leftrightarrow \forall y(\neg\, y\,\mathbf{r}\,\neg A) \leftrightarrow \neg\exists y(y\,\mathbf{r}\,\neg A) \leftrightarrow \neg\exists y\forall z(\neg\, z\,\mathbf{r}\,A) \leftrightarrow \neg\neg\exists z(z\,\mathbf{r}\,A)$. □

4.3. *Characterization of realizability.* We shall now sketch an axiomatic characterization of realizability.

In doing so, we shall freely use the Kleene bracket notation, and E-logic on an informal level. A detailed formal treatment can be given in a definitional extension of **HA** with Kleene brackets and the existence predicate E, in which elementary recursion theory can be formalized (see sections 2.7 and 3.7). Some remarks on a more precise treatment are made in 4.13.

In our characterization, an important role is played by the almost negative formulas.

4.4. DEFINITION. A formula A of $\mathscr{L}$(**HA**) *is almost negative* or *essentially* $\exists$*-free* iff A does not contain $\vee$, and $\exists$ only immediately in front of prime formulas.

In other words, A is almost negative iff it is constructed from prime formulas and existentially quantified prime formulas by means of $\wedge, \to, \forall$. □

For an essentially $\exists$-free formula φ there is no non-trivial information on the realization of $\exists, \vee$: φ is true iff realizable, and if φ is realizable then it can be realized by a canonical term depending on φ only. This is the content of the following

4.5. PROPOSITION. *Let* $A(\vec{x})$ *be essentially* $\exists$*-free*, $\mathrm{FV}(A) \subset \vec{x}$. *Then there is a term* ψ_A *with Kleene brackets containing at most* $\vec{x}$ *free such that*

(i) $\mathbf{HA} \vdash \exists y(y\,\mathbf{r}\,A) \to A$,

(ii) $\mathbf{HA} \vdash A \to E\psi_A \wedge \psi_A\,\mathbf{r}\,A$.

Proof. ψ_A is constructed by induction on the complexity of A.

(a) $\psi_{t=s} := 0$,
(b) $\psi_{\exists y(t=s)} := j(\min_y[t=s], 0)$,
(c) $\psi_{A \wedge B} := j(\psi_A, \psi_B)$,
(d) $\psi_{\forall y A} := \Lambda y.\psi_{A(y)}$,
(e) $\psi_{A \to B} := \Lambda y.\psi_B \quad (y \notin \mathrm{FV}(B))$.

In (d), (e) we used the notation $\Lambda y.t$ for a term t with Kleene brackets to indicate a canonically chosen index for the partial recursive function $\lambda y.t$. For the proof one establishes (i) and (ii) simultaneously by induction on the complexity of A. We leave this as an exercise. □

Because of the properties (i) and (ii) the almost negative formulas are sometimes called "self-realizing". We note the following

4.6. Lemma. *For all* A:
(i) $x \mathbin{\mathbf{r}} A$ *is equivalent to an almost negative formula*;
(ii) $(t \mathbin{\mathbf{r}} A)[x/s] \equiv t[x/s] \mathbin{\mathbf{r}} A[x/s]$.

Proof. By induction on A. Consider e.g. the clause for implication: $x \mathbin{\mathbf{r}} (B \to C) \leftrightarrow \forall y(y \mathbin{\mathbf{r}} B \to \exists z(Txyz \wedge Uz \mathbin{\mathbf{r}} C)) \leftrightarrow \forall y(y \mathbin{\mathbf{r}} B \to \exists z Txyz \wedge \forall u(Txyu \to Uu \mathbin{\mathbf{r}} C))$; and the latter formula is equivalent to an almost negative formula since by induction hypothesis $y \mathbin{\mathbf{r}} B$ and $Uu \mathbin{\mathbf{r}} C$ are equivalent to almost negative formulas, and $\exists z Txyz$ can be expressed as $\exists z(\chi_T(x, y, z) = 0)$, where χ_T is the characteristic function of T. □

Remark. From this lemma we can conclude that the properties (i) and (ii) of 4.5 characterize almost negative formulas up to logical equivalence. For this we have to show that $t = x$, for a term t with Kleene brackets, can be expressed in **HA** as $\exists! y A(y)$, where $A(y)$ is a conjunction of prime formulas in **HA**; hence $t = x$ can also be expressed as $\exists! y(t_1 = t_2)$, for suitable t_1, t_2. Then $\psi_A \mathbin{\mathbf{r}} A \wedge E\psi_A$ is logically equivalent to a formula

$$\exists x \exists y(t_1 = t_2) \wedge \forall yx(t_1 = t_2 \to x \mathbin{\mathbf{r}} A).$$

We leave the details as an exercise. □

4.7. Proposition (*idempotency of realizability*).

$$\mathbf{HA} \vdash \exists x(x \mathbin{\mathbf{r}} \exists y(y \mathbin{\mathbf{r}} A)) \leftrightarrow \exists y(y \mathbin{\mathbf{r}} A).$$

PROOF. $\leftarrow$: $y \mathbf{r} A$ is almost negative, so (proposition 4.5) $y \mathbf{r} A \to (\psi_{(y\mathbf{r}A)} \mathbf{r} (y \mathbf{r} A)) \wedge E\psi_{(y\mathbf{r}A)}$, hence $y \mathbf{r} A \to (j(y, \psi_{(y\mathbf{r}A)}) \mathbf{r} \exists y(y \mathbf{r} A)) \wedge Ej(y, \psi_{(y\mathbf{r}A)})$ and so $\exists x(x \mathbf{r} \exists y(y \mathbf{r} A))$. $\to$ is even easier: if $x \mathbf{r} \exists y(y \mathbf{r} A)$, we have $j_2 x \mathbf{r} (j_1 x \mathbf{r} A)$, and by 4.5(i) $j_1 x \mathbf{r} A$ i.e. $\exists x(x \mathbf{r} A)$. □

4.8. DEFINITION. The *extended Church's thesis* is the following schema, where A is almost negative:

$$\mathrm{ECT}_0 \qquad \forall x(Ax \to \exists y Bxy) \to \exists z \forall x(Ax \to \exists u(Tzxu \wedge B(x, Uu))). \quad \square$$

4.9. LEMMA. *For any instance* F *of* ECT_0 *there is a term* t *such that* $\mathbf{HA} \vdash Et \wedge t \mathbf{r} F$.

PROOF. Let $F \equiv \forall x[A \to \exists y By] \to \exists z \forall x[A \to \exists u(Tzxu \wedge B(Uu))]$. Assume $u \mathbf{r} \forall x[A \to \exists y By]$. We put

$$t := \{\{u\}(x)\}(\psi_A).$$

Then

$$\forall x[A \to Et \wedge t \mathbf{r} \exists y By]$$

and thus

$$\forall x[A \to Et \wedge j_2 t \mathbf{r} B[y/j_1 t]].$$

Take $t_1 := \Lambda x. j_1 t$, $t_2 := \min_v T(t_1, x, v)$. Then $\forall x[A \to j_2 t \mathbf{r} B[y/t_1 x]]$ and thus for

$$t^* := j(t_1, \Lambda x \Lambda w. j(t_2, j(0, j_2 t)))$$

we find $t^* \mathbf{r} \exists z \forall x[A \to \exists u(Tzxu \wedge B(Uu))]$. Therefore $\Lambda u. t^* \mathbf{r} F$. □

Note that $\Lambda u.\, t^*$ depends on A only!

4.10. THEOREM (*characterization of realizability*).
(i) $\mathbf{HA} + \mathrm{ECT}_0 \vdash A \leftrightarrow \exists x(x \mathbf{r} A)$,
(ii) $\mathbf{HA} + \mathrm{ECT}_0 \vdash A \Leftrightarrow \mathbf{HA} \vdash \exists x(x \mathbf{r} A)$.

PROOF. (i) is readily proved by induction on the complexity of A, with help of lemma 4.6; we leave the details as an exercise.

(ii) $\Leftarrow$ follows from (i). $\Rightarrow$ is proved by induction on the length of deductions in $\mathbf{HA} + \mathrm{ECT}_0$ ("$\Rightarrow$" is the *soundness theorem* for realizability).

The proof is quite straightforward if we base ourselves, for example, on a Hilbert-type formalization of **HA** (such as the formalization obtained by taking H_1-**IQC** of 2.4.1).

As the basis of the induction, we have to indicate terms realizing the axioms.

Logical axioms.

$$\Lambda x. j_1 x \,\mathbf{r}\, (A \wedge B \to A); \; \Lambda x. j_2 x \,\mathbf{r}\, (A \wedge B \to B),$$

$$\Lambda x \Lambda y. j(x, y) \,\mathbf{r}\, (A \to (B \to A \wedge B)),$$

$$\Lambda x. j(0, x) \,\mathbf{r}\, (A \to A \vee B); \; \Lambda x. j(1, x) \,\mathbf{r}\, (B \to A \vee B),$$

$$\Lambda x \Lambda y \Lambda z. [(1 \dot{-} \mathrm{sg}(j_1 z))(\{x\}(j_2 z)) + \mathrm{sg}(j_1 z)(\{y\}(j_2 z))] \\ \mathbf{r}\, ((A \to C) \to [(B \to C) \to (A \vee B \to C)]),$$

$$\Lambda u. j(t, u) \,\mathbf{r}\, (A(t) \to \exists x A x),$$

$$\Lambda u \Lambda v. (\{\{u\}(j_1 v)\}(j_2 v)) \,\mathbf{r}\, (\forall x (A \to B) \to (\exists x A \to B)) \quad (x \text{ not free in } B),$$

$$\Lambda u \Lambda v. u \,\mathbf{r}\, (A \to (B \to A)),$$

$$\Lambda u \Lambda v \Lambda w. \{\{u\}(w)\}(\{v\}(w)) \\ \mathbf{r}\, ([A \to (B \to C)] \to [(A \to B) \to (A \to C)]),$$

$$\Lambda u. \{u\}(t) \,\mathbf{r}\, (\forall x A x \to A t),$$

$$\Lambda u \Lambda v \Lambda x. \{\{u\}(x)\}(v) \,\mathbf{r}\, (\forall x (B \to A) \to (B \to \forall x A)) \quad (x \text{ not free in } B).$$

Let us pause for a moment to check e.g. the case of $(A \to C) \to [(B \to C) \to (A \vee B \to C)]$. Suppose $x \,\mathbf{r}\, (A \to C)$, $y \,\mathbf{r}\, (B \to C)$, $z \,\mathbf{r}\, (A \vee B)$. Then if $j_1 z = 0$, $j_2 z \,\mathbf{r}\, A$, so $\{x\}(j_2 z) \,\mathbf{r}\, C$; and if $j_1 z \neq 0$, then $j_2 z \,\mathbf{r}\, B$, and $\{y\}(j_2 z) \,\mathbf{r}\, C$. It is now clear that, if t is the term for this axiom schema indicated above, then $\{\{\{t\}(x)\}(y)\}(z)$ always realizes C. We leave the verification of other cases to the reader.

Non-logical axioms. Equations are all realized by 0; an implication between prime formulas such as $x = y \to Sx = Sy$ by $\Lambda u.0$ etc. So the only nontrivial case is induction. Suppose $u\,\mathbf{r}\,(A(0) \wedge \forall x(A(x) \rightarrow A(Sx)))$, then

$$j_1 u\,\mathbf{r}\,A(0),$$

$$j_2 u\,\mathbf{r}\,\forall x(A(x) \rightarrow A(Sx)).$$

Using the fact that the partial recursive functions are closed under (primitive) recursion we define an index $t(u)$ such that

$$\{t(u)\}(0) \simeq j_1 u,$$

$$\{t(u)\}(Sx) \simeq \{\{j_2 u\}(x)\}(\{t(u)\}(x)).$$

By induction on x one then shows $\{t(u)\}(x)\,\mathbf{r}\,A(x)$, so

$$\Lambda u.t(u)\,\mathbf{r}\,(A(0) \wedge \forall x(A(x) \rightarrow A(Sx)) \rightarrow \forall x A(x)).$$

ECT_0 has already been treated in lemma 4.9.

Rules. Suppose $t\,\mathbf{r}\,(A \rightarrow B)$, $t'\,\mathbf{r}\,A$. Then $\{t\}(t')$ is also defined and realizes B. If we have shown $t\,\mathbf{r}\,A(x)$, then clearly $\Lambda x.t\,\mathbf{r}\,\forall x A(x)$. So the property of being realizable is transmitted from the premisses to the conclusion. □

Remark. (ii) can be refined to: for each sentence A such that $\mathbf{HA} \vdash A$ there is a numeral $\bar{n}$ such that $\mathbf{HA} \vdash \bar{n}\,\mathbf{r}\,A$ (E4.4.3).

4.11. ECT_0 is stronger than CT_0, i.e. $\mathbf{HA} + \mathrm{CT}_0 \nvdash F$ for some instance F of ECT_0 (cf. 10.3). On the other hand ECT_0 cannot be strengthened to ECT^*, the schema obtained by dropping the condition "A almost negative" for ECT_0. In fact, even the following special case

$$\mathrm{ECT}^*! \quad \forall x[Ax \rightarrow \exists! y B(x, y)]$$
$$\rightarrow \exists z \forall x[Ax \rightarrow E\{z\}(x) \wedge B(x, \{z\}(x))]$$

conflicts with **HA**. To see this, take

$$A(x) \quad := \exists u Txxu \vee \neg\exists u Txxu,$$

$$B(x, y) := (\exists u Txxu \wedge y = 0) \vee (\neg\exists u Txxu \wedge y = 1).$$

We leave it to the reader to show that this particular instance of $\mathrm{ECT}^*!$ is incompatible with **HA** (E4.4.4).

Observe that if the premiss of $\mathrm{ECT}^*!$ holds, then B describes the graph of a function with domain $\mathrm{D} \equiv \{x : Ax\} \subset \mathbb{N}$. We cannot always assume such functions to be partial recursive. This shows that AC-D$\mathbb{N}$! contradicts the assumption that all partial functions from $\mathbb{N}$ to $\mathbb{N}$ are partial recursive ("partial Church's thesis"). Cf. also our remarks in 2.1.

4.12. Realizability provides a systematic method for deciphering the constructive meaning of logically compound formulas, but it does not give any information about the constructive meaning of $\exists$-free statements.

As an example, consider functions from $\mathbb{N}^{\mathbb{N}} \to \mathbb{N}$. By CT_0, all functions given by a graph predicate A, i.e. a predicate such that $\forall x \exists ! y A(x, y)$, are total recursive. Functions from $\mathbb{N}^{\mathbb{N}} \to \mathbb{N}$ can be given by a graph predicate A, i.e. a binary predicate $A(x, y)$ such that

$$\forall x \in V_1 (\exists y A(x, y) \wedge \forall x' \in V_1 (x \approx_1 x \wedge A(x', y') \to y = y') \quad (1)$$

where $x \in V_1 := \forall u \exists v Txuv$ (i.e. x is the code of a total recursive function) and $x \approx_1 x' := \forall y(\{x\}(y) = \{x'\}(y))$ (that is to say x and x' represent extensionally the same function).

Now $\forall x \in V_1\, \exists y A(x, y)$ is of the form $\forall x(\forall u \exists v Txuv \to \exists y A(x, y))$; by ECT_0 it follows that for some z $\forall x \in V_1(E\{z\}(x) \wedge A(x, \{z\}(x)))$. Moreover, by (1) z should also satisfy

$$x \in V_1 \wedge x' \in V_1 \wedge x \approx_1 x' \to \{z\}(x) = \{z\}(x').$$

Thus we know that under the realizability interpretation functionals in $\mathbb{N}^{\mathbb{N}} \to \mathbb{N}$ can be interpreted by partial recursive operations.

Essentially the same method of constructively re-interpreting the meaning of logically complex statements has been proposed by Shanin as the basis for CRM (cf. Shanin 1958, 1958A, Kleene 1960). It turns out that almost the whole mathematical practice of CRM can be formalized in $\mathbf{HA} + \mathrm{ECT}_0 + \mathrm{MR}_{\mathrm{PR}}$, where $\mathrm{MR}_{\mathrm{PR}}$ is the simplest form of Markov's principle:

$\mathrm{MR}_{\mathrm{PR}}$ $\qquad \neg\neg\exists x A x \to \exists x A x$ (A primitive recursive),

a principle adopted by the Russian constructivist school. $\mathrm{MR}_{\mathrm{PR}}$ will be discussed in the next section.

On the other hand, $\exists x(x\, \mathbf{r}\, A)$ as we have defined it here, is not accepted as the *fundamental* interpretation of the constructive meaning of "A is true", in CRM, for the following reason. Shanin (1958) regards the constructive meaning of negative formulas as immediate. Nevertheless, realizability transforms such formulas into more complex formulas (albeit logically equivalent ones; but this can be established only after the semantics has been correctly explained): $x\, \mathbf{r}\, A$ is more complex than A, also for negative A. Therefore Shanin gives a method for "deciphering the constructive meaning of statements" in which $\exists$-free statements are not transformed into more complex statements. Afterwards one can establish equivalence with realizability relative to **HA**.

Shanin's method is much more complex to describe than Kleene's realizability; and as follows from the equivalence of these interpretations relative to **HA**, the difference in semantic framework has no consequences for the practice of constructive mathematics. For these reasons we have restricted attention to realizability.

ECT_0 justifies certain choice axioms such as AC_0 and DC discussed in section 2, but also the presentation axiom relative to **HA**: if we assume ECT_0, then any arithmetical $\{x : A(x)\}$ is a surjective image of $A' \equiv \{(x, y) : y \mathbf{r} A(x)\}$ and choice over the domain A' is a consequence of ECT_0, since A' is given by an almost negative predicate.

4.13. *The formal treatment of realizability.* As noted above, though the realizability definition itself can be given entirely in **HA**, the treatment of the soundness and characterization results is made much easier if we assume Kleene brackets to be present in the arithmetical language, and accordingly we use a logic with existence predicate (E-logic or E^+-logic, cf. section 2.2). Then "Et" is a new prime formula, and realizability can now be specified by the following clauses

(i) $t \mathbf{r} P := Et \wedge P$ for P prime,
(ii) $t \mathbf{r} (A \wedge B) := (j_1 t \mathbf{r} A) \wedge (j_2 t \mathbf{r} B)$,
(iii) $t \mathbf{r} (A \to B) := Et \wedge \forall x (x \mathbf{r} A \to tx \mathbf{r} B)$,
(iv) $t \mathbf{r} \forall x A := \forall x (tx \mathbf{r} A)$,
(v) $t \mathbf{r} \exists x A := j_2 t \mathbf{r} A(j_1 t)$.

Since $\vee$ is definable we do not need a clause for $\vee$; if we wish to treat $\vee$ as a primitive, we can add

$$t \mathbf{r} A \vee B := (j_1 t = 0 \wedge j_2 t \mathbf{r} A) \vee (j_1 t \neq 0 \wedge j_2 t \mathbf{r} B).$$

We now show by induction on the complexity of A that $(t \mathbf{r} A) \to Et$ is provable. The details of the formal treatment are similar to the treatment in the abstract setting of section 9.5.

5. Markov's principle

5.1. In CRM, the following principle is accepted:

MP $\quad \forall x (Ax \vee \neg Ax) \wedge \neg\neg \exists x Ax \to \exists x Ax$

(Markov's principle, also often denoted by "M" in the literature), or in a form with function variables

$$\neg\neg\exists x(\alpha x = 0) \to \exists x(\alpha x = 0).$$

Of course, in the setting of CRM, α is an *algorithmic* (recursive) function, and A an *algorithmically* decidable predicate. Thus MP may be paraphrased as: when there is an algorithm for testing A, and by "indirect means" we know that we cannot avoid encountering x such that Ax, then in fact we *can* find x such that Ax: the testing algorithm provides the method for finding x.

One is tempted to object to MP, that $\neg\neg\exists x(\alpha x = 0)$ gives us no *effective* method of actually producing an $n \in \mathbb{N}$ such that $\alpha n = 0$; however, this objection does not help in the case one accepts a proof of $\neg\neg\exists x(\alpha x = 0)$, together with an algorithm for α as a method for finding x (so to speak "by definition"). The acceptability of MP within the context of general constructivism is therefore not all that easy to decide.

N.B. In the presence of AC-NN!, the principle with function variables is equivalent to MP for predicates, but the two forms are not equivalent in the absence of any form of function comprehension. Thus for the lawless sequences of 6.2(a), discussed at length in chapter 12, the functional form of MP is refutable, while MP for predicates not containing choice parameters free is consistent with the theory **LS** of lawless sequences.

Since the general schema of the BHK-interpretation still leaves us a lot of freedom, one might seek a solution in the form of a modification of the BHK-interpretation which so to speak makes MP automatically true. In fact, as shown by Gödel (1958) it is possible to give an interpretation of logic in the context of $\mathscr{L}(\mathbf{HA})$ which makes MP true without having to assume it beforehand. That is to say, for a suitable translation ^{D} we have

$$\mathbf{HA} + \mathrm{CT}_0 + \mathrm{MP} + \mathrm{IP}_0 \vdash A \Rightarrow \mathbf{HA} \vdash A^{\mathrm{D}},$$

where $A^{\mathrm{D}} \equiv A$ for Π^2_0-formulas, and IP_0 is the schema

$$\mathrm{IP}_0 \qquad \forall x(A \vee \neg A) \wedge (\forall xA \to \exists yB) \to \exists y(\forall xA \to B)$$

($y \notin \mathrm{FV}(A)$). The translation ^{D} may be extended to other languages and stronger systems; cf. 10.4.

5.2. In the presence of CT, MP is equivalent to its special case

$$\mathrm{MP}_{\mathrm{PR}} \qquad \neg\neg\exists xA(x) \to \exists xA(x) \ (A \text{ primitive recursive});$$

in fact this schema is easily seen to be equivalent to a single axiom

$$\forall xy(\neg\neg\exists z Txyz \to \exists z Txyx)$$

(E4.5.2).

5.3. *Markov's principle and Post's theorem.* Post's theorem, well-known from recursion theory, states that if a set and its complement are both recursively enumerable, then the set is recursive. The proof of this fact uses $\mathrm{MP}_{\mathrm{PR}}$: if z_0, z_1 are codes for enumerating X and its complement, then $\forall x \neg\neg\exists y(Tz_0xy \vee Tz_1xy)$ hence, by $\mathrm{MP}_{\mathrm{PR}}$, $\forall x \exists y(Tz_0xy \vee Tz_1xy)$, and this gives in turn

$$\forall x \exists! z\left[(z = 0 \wedge \exists y T(z_0, x, y)) \vee (z = 1 \wedge \exists y T(z_1, x, y))\right]$$

from which one readily obtains a characteristic function for X. An abstract version of Post's theorem (for the case of an inhabited set with inhabited complement) in which the recursiveness has been dropped is

$$\text{PT} \quad \forall xy(\beta x \neq \gamma y) \wedge \forall x \neg\neg\exists y(\beta y = x \vee \gamma y = x) \to \forall x \exists y((\beta y = x) \vee (\gamma y = x)).$$

PT is in fact equivalent to MP (exercise).

An interesting mathematical consequence of MP is given by the following

5.4. Proposition. MP *is equivalent to* $\forall x \in \mathbb{R}(x \neq 0 \to x \# 0)$ *where* $\mathbb{R}$ *is the set of reals as defined in* 1.2.3, *namely as the set of equivalence classes of fundamental sequences, and* $x \# 0$ *means* $\exists k(|x| > 2^{-k})$.

Proof. The detailed treatment of the theory of the real numbers is reserved for the next chapter, but that need not bother us here. Let the fundamental sequence $\langle r_n \rangle_n$ determine the real x; without loss of generality we may assume $\forall n(|r_n - x| < 2^{-n})$. If $x = 0$, this means $\forall n(|r_n| < 2^{-n})$, and therefore $x \neq 0$ is equivalent to $\neg\forall n(|r_n| < 2^{-n})$ and hence to $\neg\neg\exists n(|r_n| \geq 2^{-n})$. With MP we get $|r_n| \geq 2^{-n}$ for some n, and since $|r_n - x| < 2^{-n}$, i.e. $|r_n - x| < 2^{-n} - 2^{-m}$ for some m, it follows that $|x| > 2^{-m}$. Conversely, suppose $\neg\neg\exists x(\alpha x = 0)$. We construct $\langle r_n \rangle_n$ as

$$r_n = \begin{cases} 0 & \text{if } \forall m \leq n(\alpha m \neq 0), \\ 2^{-m} & \text{if } m \leq n,\ \alpha m = 0,\ \forall m' < m(\alpha m' \neq 0); \end{cases}$$

then $\langle r_n \rangle_n$ is a fundamental sequence determining a real number x, and $x \neq 0$. Assuming $\forall x \in \mathbb{R}(x \neq 0 \to x \# 0)$ we then also have $x \# 0$, i.e. for some k $|x| > 2^{-k}$. Then clearly $\exists m < k(\alpha m = 0)$ (otherwise $|x| \leq 2^{-k}$). □

5.5. *Realizability in the presence of MP.* Assuming $\mathrm{MP_{PR}}$, the notions of "almost negative formula" and "negative formula" coincide modulo logical equivalence, since by $\mathrm{MP_{PR}}$ $\exists x(t = s) \leftrightarrow \neg\forall x \neg(t = s)$. This fact permits a characterization of realizability by a variant of $\mathrm{ECT_0}$ which is in certain respects to be preferred.

PROPOSITION. *Assuming* $\mathrm{MP_{PR}}$, $\mathrm{ECT_0}$ *is equivalent to the schema*

$$\mathrm{ECT}_0' \qquad \forall x[\neg Ax \to \exists y B(x, y)] \\ \to \exists z \forall x[\neg Ax \to \exists u(Tzxu \wedge B(x, Uu))].$$

PROOF. Exercise. □

ECT_0' is more convenient to handle than $\mathrm{ECT_0}$ since we do not need syntactical restrictions on $A(x)$. This advantage becomes particularly clear when one wishes to extend the characterization of realizability to second-order arithmetic. Especially in combination with MP (or $\mathrm{MP_{PR}}$) the schema $\mathrm{ECT_0}$ (or ECT_0') yields interesting mathematical consequences not obtainable by $\mathrm{CT_0}$ alone. Among those consequences we mention in particular the

THEOREM. *Assuming* $\mathrm{ECT_0}$ + MP, *all mappings from complete separable metric spaces into metric spaces are continuous.*

PROOF. See 6.4.12 for the case of mappings from $\mathbb{R}$ into $\mathbb{R}$, and 7.2.11 for the general case. □

6. Choice sequences and continuity axioms

6.1. In 1917 L.E.J. Brouwer introduced choice sequences in intuitionistic mathematics, in order to capture more satisfactorily the intuition of the continuum.

Choice sequences have given traditional intuitionistic mathematics its characteristic flavour. The idea of a choice sequence is that of a sequence as a process, *not necessarily predetermined* by some law or algorithm, successively producing elements of a sequence. The possibility of using choice sequences mathematically is given by the fact that in practice all operations mapping sequences to other sequences or to natural numbers are continuous, that is to say can be carried out by operating on (computing from) initial segments.

In the introduction (1.3.6) we observed that we were unable to give an example of a total discontinuous function from $\mathbb{R}$ to $\mathbb{R}$, which suggests that, constructively, all total functions from $\mathbb{R}$ to $\mathbb{R}$ are continuous. The absence of an example of a total discontinuous function is not a proof of the conjecture however.

Permitting fundamental sequences to be given by choice sequences results in a wider notion of real number, for which the statement "all total functions from $\mathbb{R}$ to $\mathbb{R}$ are continuous" can be proved by means of plausible axioms for choice sequences (cf. theorem 6.3.4). This should not greatly surprise the reader, since the condition "f is total on $\mathbb{R}$" becomes much stronger if we extend $\mathbb{R}$ by permitting also reals defined via fundamental sequences given by choice sequences. In other words, the meaning of "f is total on $\mathbb{R}$", when permitting choice sequences with intuitionistic logic in the definition of reals, is very different from its classical meaning.

6.2. *Varieties of choice sequences.* In this section we are, deliberately, not very specific about the way one should think about choice sequences. We illustrate the ample possibilities and pitfalls by means of two special examples.

(a) *Lawless sequences.* A lawless sequence is to be viewed as a process (say α) of choosing values $\alpha 0, \alpha 1, \alpha 2, \ldots \in \mathbb{N}$ such that at any stage in the generation of α we know only finitely many values of α. If we generate a lawless sequence, then we have *a priori* decided not to make general restrictions on future values at any stage. Thus at each stage we can only choose finitely many new values (or exactly one) of the sequence; we cannot, for example, decide that all values to be chosen after stage t are to be even.

A good "picture" (intuitive model) of a lawless sequence is that of the sequence of casts of an (abstract) die, permitting perhaps finitely many deliberate placings of the die in the beginning. We can throw the die as often as we like, but at any stage we only know the result of finitely many casts.

The very idea of a lawless sequence implies that we can never assert that its value will coincide with the values of some lawlike sequence, since this would result in an overall restriction. Let a be a variable for lawlike sequences. In order to be able to assert $\exists a(\alpha = a)$, we must *know* at some stage that $\alpha = a$ for some lawlike a, which is clearly impossible. Thus we have

$$\forall\alpha\neg\exists a(\alpha = a).$$

One should contrast this with the behaviour of the

(b) *Hesitant sequences*. The name "hesitant sequences" is ad hoc, and the notion is discussed here only for purposes of illustration. A hesitant sequence (say β) is a process of generating values $\beta 0, \beta 1, \beta 2, \ldots \in \mathbb{N}$ such that at any stage we either decide that henceforth we are going to conform to a law in determining future values, or, if we have not already decided to conform to a law at an earlier stage, we freely choose a new value of β. Thus at any stage in the construction of β we either know only finitely many values of β, *or* we know that all future values of β will be chosen according to a law.

Therefore the lawlike sequences are in some sense contained in the hesitant sequences: they correspond to those hesitant sequences for which already at stage 0 one decides to conform to a law. Also, for α ranging over hesitant sequences

$$\forall\alpha\neg\neg\exists a(\alpha = a).$$

For to have a proof of $\neg\exists a(\beta = a)$ for some hesitant β would mean that at some stage we would be sure that β would never conform to a law; but the very idea of a hesitant sequence prevents this, since we always may decide later on to become law-abiding, and therefore $\neg\neg\exists a(\beta = a)$.

On the other hand, we cannot expect $\forall\alpha\exists a(\alpha = a)$ to hold for hesitant sequences, since we have no method for assigning a law to each hesitant α; the decision whether or not to conform to a law may stay open indefinitely (hence the name "hesitant").

As shown by the conflicting properties $\neg\neg\exists a(\beta = a)$, $\neg\exists a(\alpha = a)$, the universes of lawless and hesitant sequences are disjoint, notwithstanding the fact that at first sight hesitant sequences may seem to be very similar to lawless ones.

In what follows we shall assume choice sequences to embrace all possibilities from lawless to lawlike. The lawless sequences will be treated in considerable detail in Chapter 12.

6.3. *Weak or local continuity* WC–N. Suppose we are given some functional $\Phi \in \mathbb{N}^{\mathbb{N}} \to \mathbb{N}$ defined for *all* choice sequences. Then the fact that Φ

is total, i.e. everywhere defined, and that for an arbitrary choice sequence we cannot, *in general*, assume more than an initial segment to be known at any given moment leads to the supposition that Φ must be continuous on $\mathbb{N}^{\mathbb{N}}$, that is to say

$$\forall \alpha \exists x \forall \beta \in \bar{\alpha}x (\Phi\beta = \Phi\alpha).$$

Combining this with a choice principle

$$\forall \alpha \exists x A(\alpha, x) \to \exists \Phi \in \mathbb{N}^{\mathbb{N}} \to \mathbb{N}\, \forall \alpha A(\alpha, \Phi\alpha)$$

one is led to the schema of *weak* or *local continuity* for numbers

$$\text{WC-N} \qquad \forall \alpha \exists x A(\alpha, x) \to \forall \alpha \exists x \exists y \forall \beta \in \bar{\alpha}x\, A(\beta, y).$$

The term "local" refers to the fact that the schema only states the existence of a suitable y for a whole neighbourhood of α for each α individually. Via the equivalence $(A \vee B) \leftrightarrow \exists x[(x = 0 \wedge A) \vee (x \neq 0 \wedge B)]$ we obtain a version of WC-N for disjunctions:

$$\text{WC-N}^{\vee} \qquad \forall \alpha (A\alpha \vee B\alpha) \to \forall \alpha \exists x (\forall \beta \in \bar{\alpha}x\, A(\beta) \vee \forall \beta \in \bar{\alpha}x\, B(\beta)).$$

Weak continuity conflicts with classical logic, in much the same way as in the case of CT_0, as the following two propositions show.

6.4. PROPOSITION. WC-N$^{\vee}$ *refutes* $\forall$-PEM.

PROOF. Assume $\forall$-PEM, i.e.

$$\forall \alpha (\forall x (\alpha x = 0) \vee \neg \forall x (\alpha x = 0)),$$

then by WC-N$^{\vee}$

$$\forall \alpha \exists y (\forall \beta \in \bar{\alpha}y \forall x (\beta x = 0) \vee \forall \beta \in \bar{\alpha}y \neg \forall x (\beta x = 0)).$$

Now specialize α to $\lambda x.0$, then for some y, with $n = (\overline{\lambda x.0})(y)$

$$\forall \beta \in n \forall x (\beta x = 0) \vee \forall \beta \in n \neg \forall x (\beta x = 0).$$

The first disjunct is false, as may be seen by taking any $\beta \in n * \langle 1 \rangle$; and the second disjunct is also false, as follows by choosing $\beta = \alpha = \lambda x.0$. □

REMARK. It is worth noting that this proof explicitly uses the fact that certain lawlike sequences (i.e. $\lambda x.0$) are part of the universe of choice sequences. In the presence of FAN* (7.3) or bar-continuity (8.10) $\neg\exists$-PEM can be proved without this assumption (cf. E4.7.6, E4.8.12).

COROLLARY. *The logical principles* PEM *and* $\forall\alpha\neg\neg A \to \neg\neg\forall\alpha A$ *are not generally valid.*

PROOF. Cf. 3.4. □

6.5. PROPOSITION. WC-N$^\vee$ *proves* $\neg$SEP.

PROOF. Let us assume SEP, i.e.

$$\forall\alpha\beta(\neg(\exists x(\alpha x = 0) \wedge \exists x(\beta x = 0)) \\ \to \neg\exists x(\alpha x = 0) \vee \neg\exists x(\beta x = 0)).$$

Let γ be arbitrary; we define γ', γ'' from γ by

$$\gamma' x = \begin{cases} 0 & \text{if } \gamma(2x) = 0 \wedge \forall y < 2x(\gamma y \neq 0), \\ 1 & \text{otherwise,} \end{cases}$$

$$\gamma'' x = \begin{cases} 0 & \text{if } \gamma(2x+1) = 0 \wedge \forall y < 2x+1(\gamma y \neq 0), \\ 1 & \text{otherwise.} \end{cases}$$

Clearly $\neg(\exists x(\gamma' x = 0) \wedge \exists x(\gamma'' x = 0))$, for all γ, hence by SEP

$$\forall\gamma[\neg\exists y(\gamma' y = 0) \vee \neg\exists y(\gamma'' y = 0)]$$

and thus with WC-N$^\vee$ we find for any γ a y such that

$$\forall\delta \in \bar{\gamma}y\neg\exists y(\delta' y = 0) \vee \forall\delta \in \bar{\gamma}y\neg\exists y(\delta'' y = 0).$$

Apply this to $\gamma = \lambda x.1$. Then the first part of the disjunction becomes false for any $\delta \in (\overline{\lambda x.1})(2y) * \langle 0\rangle$ and the second half of the disjunction becomes false for $\delta \in (\overline{\lambda x.1})(2y+1) * \langle 0\rangle$, and so we have arrived at a contradiction, i.e. $\neg$SEP. □

6.6. We mention some interesting mathematical consequences of WC-N, such as

(i) If $\{X_i : i \in \mathbb{N}\}$ covers a complete separable metric space, then $\{\mathrm{Int}(X_i) : i \in \mathbb{N}\}$ is again a cover (cf. 6.3.3, 7.2.6).

(ii) As a corollary to the preceding property one obtains: each function from a complete separable metric space to a separable metric space is continuous. In particular functions from $\mathbb{R}$ to $\mathbb{R}$ are continuous (6.3.4, 7.2.7).

(iii) WC-N can also be used to prove certain classically valid results which, however, are usually proved by constructively invalid reasoning. For example, all sequentially continuous functions from a separable metric space to a metric space are continuous (cf. 7.2.8). Here "sequentially continuous" means that convergent sequences are mapped to convergent sequences. On the other hand we have

6.7. PROPOSITION. WC-N *refutes* CT.

PROOF. Apply WC-N to $\forall\alpha\exists x\forall y\exists z(Txyz \wedge \alpha y = Uz)$. □

Instead of WC-N one may also consider the more cautious principle of *weak continuity with uniqueness*

WC-N! $\quad \forall\alpha\exists!xA(\alpha, x) \rightarrow \forall\alpha\exists x\exists y\forall\beta \in \bar{\alpha}yA(\beta, x)$.

This corresponds to assuming continuity only for those $\Phi \in \mathbb{N}^{\mathbb{N}} \rightarrow \mathbb{N}$ which are already given to us as extensional, i.e. $\alpha = \beta \rightarrow \Phi\alpha = \Phi\beta$. (Here and elsewhere in this chapter, the predicates considered are assumed to be extensional in their choice parameters.) WC-N! does not obviously conflict with CT; in fact, as follows from E6.4.9 and theorem 7.2.7, WC-N! is derivable from ECT_0 + MP.

We now turn to a strengthening of weak continuity.

6.8. *Neighbourhood functions and weak continuity.* WC-N does not exhaust our intuitions about continuity. For if $\Phi \in \mathbb{N}^{\mathbb{N}} \rightarrow \mathbb{N}$ is any continuous functional on $\mathbb{N}^{\mathbb{N}}$, it seems plausible to assume that we actually *know* whether an initial segment of α suffices to compute $\Phi\alpha$. That is to say, we can actually associate with Φ a "neighbourhood function" $\varphi \in \mathbb{N} \rightarrow \mathbb{N}$ such that

$$\varphi(\bar{\alpha}n) = 0 \Leftrightarrow \bar{\alpha}n \text{ is not sufficiently long to compute } \Phi\alpha \text{ from it;}$$

$$\varphi(\bar{\alpha}n) = m + 1 \Leftrightarrow \bar{\alpha}n \text{ is sufficiently long to compute } \Phi\alpha \text{ from it, and } \Phi\alpha = m.$$

What is more, the natural way of presenting a continuous functional $\Phi \in \mathbb{N}^{\mathbb{N}} \rightarrow \mathbb{N}$ seems to be via such a neighbourhood function. Clearly, the neighbourhood function satisfies (i) $\forall\alpha\exists x(\varphi(\bar{\alpha}x) \neq 0)$, and (ii) $\forall nm(\varphi n \neq 0 \rightarrow \varphi n = \varphi(n * m))$. (i) is a consequence of Φ being everywhere defined, and (ii) is due to the fact that the value computed should not change if more information about the argument becomes available. Conversely, any φ satisfying (i) and (ii) defines a continuous $\Phi \in \mathbb{N}^{\mathbb{N}} \rightarrow \mathbb{N}$ by

$$\Phi\alpha = n := \exists m(\varphi(\bar{\alpha}m) = n + 1).$$

Thus, if we define K_0, the *class of neighbourhood functions* by

$$\gamma \in K_0 := \forall\alpha\exists n(\gamma(\bar{\alpha}n) \neq 0) \wedge \forall nm(\gamma n \neq 0 \rightarrow \gamma n = \gamma(n * m)),$$

and define

$$\gamma(\alpha) = n := \exists m(\gamma(\bar{\alpha}m) = n + 1),$$

(the continuous functional given by $\gamma \in K_0$ assigns n to α), we can now express a global form of continuity by

C-N $\quad \forall\alpha\exists x A(\alpha, x) \rightarrow \exists\gamma \in K_0 \forall\alpha A(\alpha, \gamma(\alpha))$

((*strong*) *continuity for numbers*, sometimes also denoted by CONT_0).

Remark. "$\bar{\alpha}n$ is sufficiently long for computing $\Phi\alpha$" should not be understood as merely saying that Φ is constant on $V_{\bar{\alpha}n}$, but rather that the *method* for computing Φ, by which Φ is constructively given to us, requires $\bar{\alpha}n$ for its execution.

The following example makes it clear that the first interpretation is to be excluded. Specify Φ_β depending on β by

$$\Phi_\beta(\alpha) = \begin{cases} 0 & \text{if } \forall y \leq \alpha 0(\neg\beta y = 0), \\ 1 & \text{if } \exists y \leq \alpha 0(\beta y = 0). \end{cases}$$

Clearly the method for computing $\Phi_\beta(\alpha)$ requires knowledge of $\alpha 0$ only, i.e. of $\bar{\alpha}1$. However, if $\neg\exists y(\beta y = 0)$, Φ_β is constant on $V_{\bar{\alpha}0} = V_{\langle\,\rangle}$. So if we could decide whether Φ is constant on $V_{\langle\,\rangle}$ or not, we could also decide whether $\exists y(\beta y = 0)$, for arbitrary β, i.e. $\exists$-PEM would follow.

As an illustration we give in fig. 4.1 the picture of a neighbourhood function γ for the functional $\Phi : \alpha \mapsto \alpha(\alpha 0)$.

6.9. CT_0, WC-N, C-N *and their impact on logic.* If we think of the constructive mathematical universe as consisting of lawlike mathematical objects, and if we also accept the identification of lawlike with recursive for number theoretic functions as constructively justified, we have no reason to doubt the compatibility of intuitionistic predicate logic with CT_0.

Similarly, as soon as we have decided that choice sequences are legitimate objects in intuitionistic mathematics, and constitute a domain over which one can quantify, we need not doubt the validity of intuitionistic predicate logic over this domain. The continuity axioms WC-N and C-N express properties of the quantifier combination $\forall\alpha\exists n$ not already implicit in the BHK-explanation, which are seen to hold by reflection on the way (data concerning) choice sequences are given.

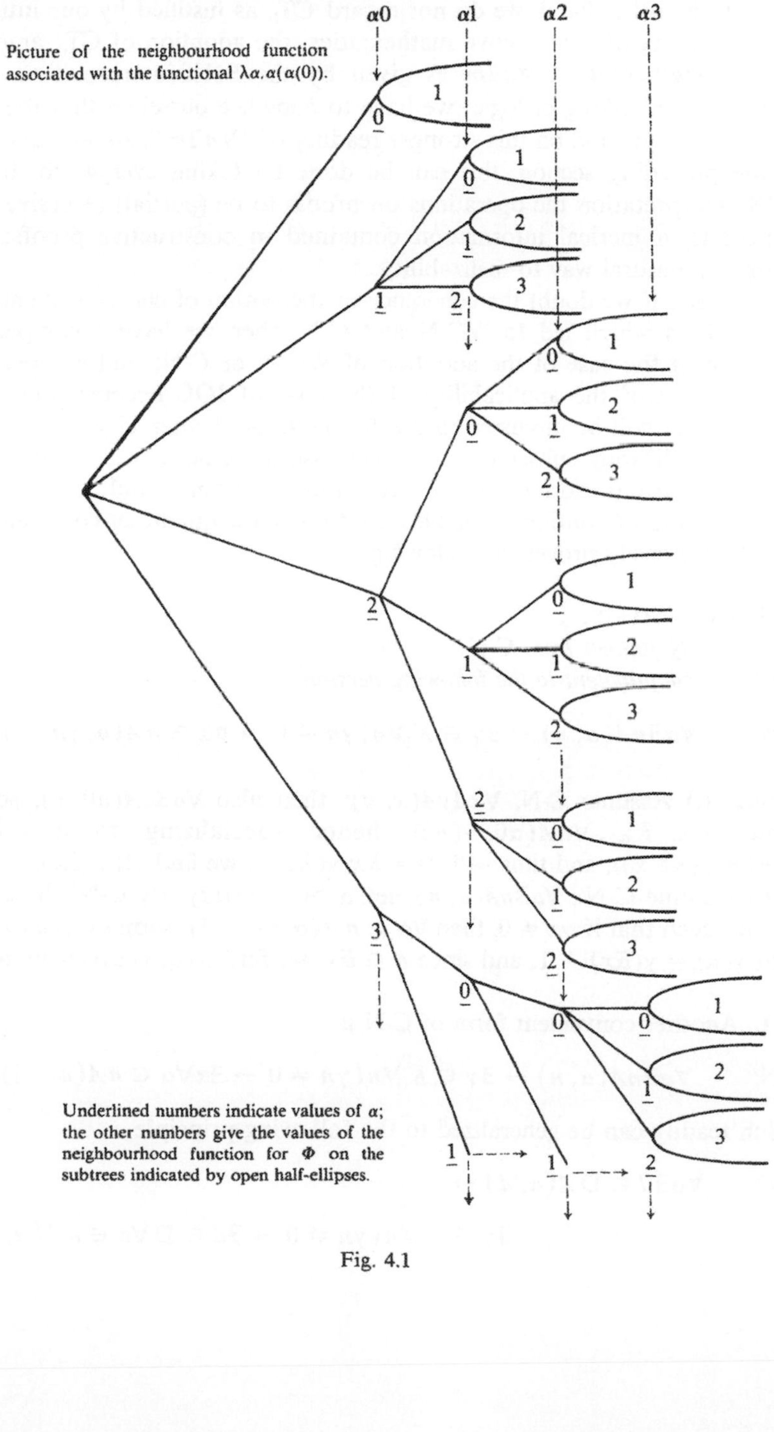
α0
α1
α2
α3
Picture of the neighbourhood function associated with the functional λα.α(α(0)).
Underlined numbers indicate values of α; the other numbers give the values of the neighbourhood function for Φ on the subtrees indicated by open half-ellipses.

Fig. 4.1

If on the other hand, we do not regard CT_0 as justified by our intuitive understanding of constructive mathematics, the adoption of CT_0 amounts to a strengthening of $\forall n \exists m$ as given by the BHK-interpretation. This requires a rethinking of logic: we have to convince ourselves that the laws of **IQC** remain valid on this stronger reading of "$\forall n \exists m$". As we have seen in the preceding section, this can be done by taking everywhere in the BHK-interpretation the operations on proofs to be (partial) recursive with respect to numerical information contained in constructive proofs; this leads in a natural way to realizability.

Similarly, if we doubt the coherence of the notion of choice sequence, or the analysis which led to WC-N and C-N, then we have a comparable situation in the case of the adoption of WC-N or C-N; and independent confirmation of the applicability of the laws of **IQC** becomes desirable. (This matter will be discussed in greater detail in chapter 12.)

In the following subsections we shall discuss some logical relationships between various forms of the continuity axioms, and one further strengthening of continuity, namely continuity for quantifier combinations $\forall \alpha \exists \beta$. One easily proves the following

6.10. PROPOSITION.
(i) AC-NN *follows from* C-N.
(ii) C-N *is equivalent to the following version*

C-N′ $\quad \forall \alpha \exists n A(\alpha, n) \to \exists \gamma \in K_0 \forall n (\gamma n \neq 0 \to \forall \alpha \in n A(\alpha, \gamma n \dot{-} 1)).$

PROOF. (i) Assume C-N, $\forall x \exists y A(x, y)$; then also $\forall \alpha \exists y A(\alpha 0, y)$, so for some $\gamma \in K_0$ $\forall \alpha A(\alpha 0, \gamma(\alpha))$, hence specializing to $\alpha = \lambda z.x$ $\forall x A(x, \gamma(\lambda z.x))$, and thus with $\beta = \lambda x.\gamma(\lambda z.x)$ we find $A(x, \beta x)$.

(ii) Assume C-N′, $\forall \alpha \exists n A(\alpha, n)$; let α be arbitrary. By C-N′ there is a $\gamma \in K_0$ such that if $\gamma n \neq 0$, then $\forall \alpha \in n A(\alpha, \gamma n \dot{-} 1)$; suppose $\gamma(\bar{\alpha} x) \neq 0$, then $\gamma(\alpha) = \gamma(\bar{\alpha} x) \dot{-} 1$; and since $\alpha \in \bar{\alpha} x$, we find $A(\alpha, \gamma(\alpha))$ etcetera. □

6.11. Another convenient form of C-N is

C-N″ $\quad \forall \alpha \exists n A(\alpha, n) \to \exists \gamma \in K_0 \forall n (\gamma n \neq 0 \to \exists x \forall \alpha \in n A(\alpha, x)),$

which readily can be generalized to the following principle

C-D $\quad \forall \alpha \exists d \in \mathrm{D} A(\alpha, d) \to$

$$\exists \gamma \in K_0 \forall n (\gamma n \neq 0 \to \exists d \in \mathrm{D} \forall \alpha \in n A(\alpha, d)),$$

where D is any domain of lawlike, completely determined objects. We have the following

PROPOSITION. C-N″ + AC-NN *is equivalent to* C-N *and to* C-N′.

PROOF. Assume AC-NN, C-N″ and let $\forall\alpha\exists nA(\alpha, n)$; then for some $\gamma \in K_0$

$$\forall n(\gamma n \neq 0 \to \exists m \forall \beta \in n A(\beta, m)), \tag{1}$$

which by the decidability of $\gamma n \neq 0$ yields

$$\forall n \exists m(\gamma n \neq 0 \to \forall \beta \in n A(\beta, m))$$

(choose m as given by (1) if $\gamma n \neq 0$, 0 otherwise). With AC-NN there is a δ such that

$$\forall n(\gamma n \neq 0 \to \forall \beta \in n A(\beta, \delta n)).$$

If we put

$$\gamma' n = \begin{cases} 0 & \text{if } \gamma n = 0, \\ \delta m + 1 & \text{if } m \leqslant n \wedge \gamma m \neq 0 \wedge \forall m' \prec m(\gamma m' = 0), \end{cases}$$

then

$$\forall n(\gamma' n \neq 0 \to \forall \alpha \in n A(\alpha, \gamma' n \dot{-} 1)). \quad \square$$

6.12. *The neighbourhood function principle.* The following principle connects C-N and WC-N

NFP $\quad \forall\alpha\exists x A(\bar{\alpha}x) \to \exists\gamma \in K_0 \forall\alpha A(\bar{\alpha}(\gamma(\alpha))),$

as becomes clear from the following

PROPOSITION. C-N *is deducible from* NFP + WC-N.

PROOF. Exercise. □

In earlier publications (e.g. Kreisel and Troelstra 1970) NFP was called SC-N ("Special Continuity for Numbers"), but we have dropped this designation, since it is not very informative and, possibly, even misleading. Formally NFP is a special case of C-N. However, it is not a typically intuitionistic continuity principle since it is classically valid. Intuitively,

NFP says that, when A is a property of initial segments of sequences, then initial segments of sequences satisfying A can be found by a neighbourhood function.

6.13. *Generalization to arbitrary spreads.* If T is any spread, we easily deduce the following generalizations of C-N and WC-N respectively:

$$\text{C-N}(T) \qquad \forall\alpha_T\exists x A(\alpha, x) \to \exists\gamma \in K_0 \forall\alpha_T A(\alpha, \gamma(\alpha)),$$

$$\text{WC-N}(T) \qquad \forall\alpha_T\exists x A(\alpha, x) \to \forall\alpha_T\exists x\exists y\forall\beta_T \in \bar{\alpha}x\, A(\beta, y).$$

To see this, let Γ be a mapping of $\mathbb{N}^{\mathbb{N}}$ onto T such that $\Gamma\alpha = \alpha$ for $\alpha \in T$ (cf. lemma 1.4), then from $\forall\alpha_T\exists x A(\alpha, x)$ we first conclude $\forall\alpha\exists x A(\Gamma\alpha, x)$, hence by C-N also $\exists\gamma \in K_0\forall\alpha A(\Gamma\alpha, \gamma(\alpha))$ and therefore $\exists\gamma \in K_0$ $\forall\alpha_T A(\alpha, \gamma(\alpha))$, etc. Of course, one can also formulate generalizations C-N$'(T)$, C-N$''(T)$ similar to C-N$'$ and C-N$''$ in 6.10, 6.11 above.

6.14. *Mathematical consequences of* C-N. From C-N (in fact, already from NFP) one obtains an intuitionistic version of Lindelöf's theorem: every open cover of a complete separable metric space has an enumerable subcover (cf. E4.6.5, E6.3.7 and theorem 7.2.10). Another example is the theorem: each $f \in \mathbb{R} \to \mathbb{R}$ is continuous and has a continuous modulus of continuity (see E6.3.8–9).

6.15. *Function continuity.* A considerably stronger form of continuity occurring in several axiomatic systems for intuitionistic analysis is function continuity

$$\forall\alpha\exists\beta A(\alpha, \beta) \to \exists\Gamma \in \text{CONT}\, \forall\alpha A(\alpha, \Gamma\alpha),$$

where CONT is the class of continuous functionals from $\mathbb{N}^{\mathbb{N}}$ to $\mathbb{N}^{\mathbb{N}}$.

We can express this in terms of neighbourhood functions as follows. Let

$$\gamma|\alpha = \beta := \forall n\big(\gamma_{\langle n\rangle}(\alpha) = \beta n\big),$$

where

$$\gamma_k := \lambda m.\gamma(k * m),$$

or more explicitly

$$\gamma|\alpha = \beta := \forall x((\gamma|\alpha)(x) = \beta x) \equiv \forall x(\lambda n.\gamma(\langle x\rangle * n)(\alpha) = \beta x).$$

Since

$$\gamma \in K_0 \leftrightarrow \exists x(\gamma = \lambda y.x + 1) \vee \big(\gamma\langle\ \rangle = 0 \wedge \forall n(\gamma_{\langle n\rangle} \in K_0)\big),$$

any $\gamma \in K_0$ may serve as a code for a sequence of continuous functionals of type $\mathbb{N}^{\mathbb{N}} \to \mathbb{N}$, precisely what is needed to code a single continuous functional of type $\mathbb{N}^{\mathbb{N}} \to \mathbb{N}^{\mathbb{N}}$. Function continuity may now be formulated as

$$\text{C-C} \qquad \forall\alpha\exists\beta A(\alpha, \beta) \to \exists\gamma \in K_0\, \forall\alpha A(\alpha, \gamma|\alpha).$$

C-C is also denoted by CONT_1. No interesting mathematical applications of C-C are known, though it is of foundational interest. We note the easy

PROPOSITION. C-C *implies* C-N. □

7. The fan theorem

7.1. By the introduction of choice sequences, Brouwer not only hoped to capture the intuition of the continuum, but also to save a core of traditional analysis. In particular, he wanted to preserve the theorem that a continuous function from a segment (= closed interval) of finite length to $\mathbb{R}$ is *uniformly* continuous.

Brouwer obtained this result as a corollary of his "bar theorem", to be discussed in the next section. We prefer to discuss first, separately, the most important corollary to the bar theorem, the so-called *fan theorem*. This corollary suffices for the theorem on continuous functions just mentioned; in fact, all the known mathematically relevant applications of the bar theorem are actually consequences of the fan theorem.

We shall first discuss some forms in which the fan theorem appears as an axiom, and show that it conflicts with CT; at the end of the section we briefly discuss the intuitive motivation for accepting the "fan theorem" as an axiom.

7.2. *The principle* $\mathrm{FAN_D}$. In its simplest form, the "fan axiom" applies to finitely branching spreads, i.e. fans; it says that if T is a fan, and for each infinite branch α in T there is an initial segment $\bar{\alpha}x$ satisfying the *decidable* property A, then there is a uniform upper bound to the x involved. For a given fan T this can be expressed as

$$\mathrm{FAN_D}(T) \qquad \forall n_T(An \vee \neg An) \wedge \forall\alpha_T\exists x A(\bar{\alpha}x) \to \exists z\forall\alpha_T\exists y \le z A(\bar{\alpha}y).$$

The general form is

$$\mathrm{FAN_D} \qquad \mathrm{fan}(T) \to \mathrm{FAN_D}(T),$$

where $\mathrm{fan}(T)$ formally expresses that the set T is a fan.

Intuitively, $\mathrm{FAN_D}(T)$ expresses a restricted form of compactness of T (with the topology generated by the basis $\{V_n : n \in T\}$, where as before $V_n := \{\alpha \in T : \alpha \in n\}$), namely that any decidable set of basis elements covering T has a finite subcover. (The decidable set of basis elements is $\{V_n : A(n) \wedge n \in T\}$, the finite subcover is found by taking $\{V_n : A(n) \wedge n \in T \wedge \mathrm{lth}(n) \leq z\}$ with z as given by $\mathrm{FAN_D}(T)$.)

We may think of $\mathrm{FAN_D}$ as an intuitionistic version of König's lemma. Classically, König's lemma and $\mathrm{FAN_D}$ are equivalent. For let T be any fan, and let us simply write α, n for α_T, n_T etc. Assume $\forall\alpha\exists xA(\bar{\alpha}x)$, then

$$T_A := \{m : \forall m' \preccurlyeq m \neg Am'\}$$

is a tree with all branches finite. Now suppose that $\neg\exists z\forall\alpha\exists y \leq zA(\bar{\alpha}y)$, i.e. $\forall z\exists\alpha\forall y \leq z\neg A(\bar{\alpha}y)$ by classical logic; this means that T_A has arbitrarily long finite branches, and hence by König's lemma there is an infinite branch in T, i.e. $\exists\alpha\forall z\neg A(\bar{\alpha}z)$, which contradicts $\forall\alpha\exists zA(\bar{\alpha}z)$. We leave the proof of the converse (i.e. $\mathrm{FAN_D}(T)$ implies König's lemma by classical logic) as an exercise.

7.3. *The principle* FAN. Full compactness of the fan T can be expressed by the following strengthening of $\mathrm{FAN_D}(T)$

$$\mathrm{FAN}(T) \qquad \forall\alpha_T\exists xA(\bar{\alpha}x) \to \exists z\forall\alpha_T\exists y \leq zA(\bar{\alpha}y).$$

To see this, let $\{W_i : i \in I\}$ be any open cover of T. Then

$$\forall\alpha_T\exists x\exists i \in I(V_{\bar{\alpha}x} \subset W_i).$$

By $\mathrm{FAN}(T)$ we find

$$\exists z\forall\alpha_T\exists y \leq z\exists i \in I(V_{\bar{\alpha}y} \subset W_i),$$

hence a fortiori

$$\exists z\forall\alpha_T\exists i \in I(V_{\bar{\alpha}z} \subset W_i).$$

There are only finitely many initial segments of length z in T, say $n_0, \ldots, n_k$; choose for each n_p a W_{i_q} such that $V_{n_p} \subset W_{i_q}$, then $\{W_{i_0}, \ldots, W_{i_k}\}$ is the required finitely indexed subcover. It is easy to see that conversely compactness of T immediately yields $\mathrm{FAN}(T)$.

The general form is

$$\mathrm{FAN} \qquad \mathrm{fan}(T) \to \mathrm{FAN}(T).$$

All the more important mathematical consequences of the fan axiom can be obtained from the version FAN. FAN is still classically valid, though this is not any longer so for the following generalization (exercise):

$$\mathrm{fan}(T) \wedge \forall \alpha_T \exists x A(\alpha, x) \rightarrow \exists z \forall \alpha_T \exists y \leq z A(\alpha, y)$$

and a fortiori not for the combination of FAN with continuity:

$$\text{FAN}^* \qquad \mathrm{fan}(T) \wedge \forall \alpha_T \exists x A(\alpha, x) \rightarrow \exists z \forall \alpha_T \exists x \forall \beta \in \bar{\alpha} z A(\beta, x).$$

We note the following

7.4. Proposition.

(i) FAN(T) *follows from* $\text{FAN}_\text{D}(T)$ + NFP (*cf.* 6.12).

(ii) *Assuming* C-N, *the principles* $\text{FAN}_\text{D}(T)$, FAN(T), FAN*(T) *are all equivalent*; FAN*(T) *is equivalent to* $\text{FAN}_\text{D}(T)$ + C-N(T).

Proof. We show (i); (ii) is proved similarly. Suppose $\forall \alpha_T \exists x A(\bar{\alpha} x)$. With NFP we find a $\gamma \in K_0$ such that

$$\forall \alpha_T A(\bar{\alpha}(\gamma(\alpha))).$$

Now consider

$$A^*(n) := \gamma n \neq 0 \wedge \mathrm{lth}(n) \geq \gamma n \dot{-} 1.$$

Obviously A^* is decidable, and $\forall \alpha_T \exists x A^*(\bar{\alpha} x)$ holds (using the two essential properties of a neighbourhood function). By $\text{FAN}_\text{D}(T)$ we find

$$\exists z \forall \alpha_T \exists y \leq z A^*(\bar{\alpha} y),$$

that is to say

$$\exists z \forall \alpha_T \exists y \leq z (\gamma(\bar{\alpha} y) \neq 0 \wedge y \geq \gamma(\bar{\alpha} y) \dot{-} 1).$$

So with (1)

$$\exists z \forall \alpha_T \exists y \leq z \exists y' \leq y A(\bar{\alpha} y'),$$

from which the required conclusion immediately follows. □

7.5. *Reduction of* FAN_D, FAN *to the case of the binary tree.* So far we stated our schemata for arbitrary finitely branching trees. But in fact the case of the binary tree T_{01} consisting of all 01-sequences is paradigmatic, as shown by the following

Proposition. FAN_D, FAN *are derivable from* $\text{FAN}_\text{D}(T_{01})$, FAN($T_{01}$) *respectively*.

PROOF. The proof is based on the following two facts (a) and (b).

(a) If T' is a subfan of the fan T, there is a continuous mapping Γ from T onto T' such that $\Gamma\alpha = \alpha$ for $\alpha \in T'$; in fact there is a $\Gamma' \in \mathbb{N} \to \mathbb{N}$ such that $\Gamma'(\bar{\alpha}x) = (\overline{\Gamma\alpha})x$ (we already indicated such a Γ in 1.4).

(b) If T is any fan, there exists a homeomorphism Φ from T onto a subfan of the binary fan. Given T and $\alpha \in T$, we can define $\Phi\alpha$ as the sequence

$$1^{\alpha 0+1}, 0, 1^{\alpha 1+1}, 0, 1^{\alpha 2+1}, 0, 1^{\alpha 3+1}, 0, \ldots$$

where 1^n stands for a sequence of n times 1. If $\beta n := \max\{\alpha n : n \in T\}$, then β is well-defined because T is a fan. $\bar{\alpha}x$ obviously determines $(\Phi\alpha)(x)$, and $(\Phi^{-1}\alpha)(x)$ can be determined from $\bar{\alpha}(\varphi x)$ where

$$\varphi(x) = \left(\sum_{i=0}^{x} \beta i\right) + 2x + 1.$$

Just as in the proof of C-N(T) from C-N (cf. 6.13), we obtain from (a) that $\mathrm{FAN_D}(T)$ implies $\mathrm{FAN_D}(T')$ etcetera.

From (b) we obtain that $\mathrm{FAN_D}(T)$ follows from $\mathrm{FAN_D}(T')$ for some subfan T' of T_{01} (and similarly for FAN, FAN*). This may be seen as follows. Let $\forall \alpha_T \exists x A(\bar{\alpha}x)$, and $\forall n_T(An \vee \neg An)$, and let us write Ψ for Φ^{-1}. Then $\forall \alpha_{T'} \exists x A((\overline{\Psi\alpha})x)$, and therefore also

$$\forall \alpha_{T'} \exists y \exists x \left[A\left((\overline{\Psi\alpha})x\right) \wedge \varphi x \leq \mathrm{lth}\,\bar{\alpha}(y)\right].$$

Since the predicate $\exists x[A((\overline{\Psi\alpha})x) \wedge \varphi x \leq \mathrm{lth}(n)]$ is a decidable predicate of n for $n \in T'$, we can apply $\mathrm{FAN_D}(T')$ and find a z such that

$$\forall \alpha_{T'} \exists y \leq z \exists x \left[A\left((\overline{\Psi\alpha})x\right) \wedge \varphi x \leq y\right].$$

Since $x < \varphi x$ for all x, also $\forall \alpha_T \exists x \leq z A(\bar{\alpha}x)$. □

REMARK. The proposition also applies, mutatis mutandis, to FAN*.

7.6. As we already pointed out, $\mathrm{ECT_0}$ + MP implies continuity of mappings from $\mathbb{N}^{\mathbb{N}}$ to $\mathbb{N}$. On the other hand, $\mathrm{FAN_D}$ is compatible with C-N but conflicts with CT. This is a corollary of the following theorem due to Kleene (1952A; see also Kleene and Vesley 1965, lemma 9.8).

THEOREM. *Let α range over recursive* 01*-sequences, n, m over finite* 01*-sequences. There is a recursive (in fact even primitive recursive) predicate R*

such that $\forall nm(R(n * m) \to Rn)$ *and*

$$\forall \alpha \exists x \neg R(\bar{\alpha}x), \tag{1}$$

$$\forall x \exists n(\text{lth}(n) = x \wedge Rn). \tag{2}$$

PROOF. Let T be Kleene's T-predicate and take for R

$$R(\bar{\alpha}x) := \forall y < x[(\exists u < x(Tyyu \wedge Uu = 0) \to \alpha y = 1)$$
$$\wedge (\exists u < x(Tyyu \wedge Uu = 1) \to \alpha y = 0)].$$

Then (2) is obviously satisfied: given x, take the recursive 01-sequence α such that

$$(\alpha y = 1) \leftrightarrow (y < x \wedge \exists u < x(Tyyu \wedge Uu = 0)).$$

As to (1), let $\alpha = \{y\}$ and suppose $Tyyu$. Now assume $R(\bar{\alpha}(u + y + 1))$. There are two possibilities, $Uu = 1$ or $Uu = 0$. If $Uu = 0$, then $\exists v < u + y + 1(Tyyv \wedge Uv = 0)$. Hence $\alpha y = 1$, so $\{y\}(y) = 1$; contradiction. Similarly $Uu = 1$ yields a contradiction. Therefore $\neg R(\bar{\alpha}(u + y + 1))$. □

COROLLARY. FAN_D *is incompatible with* CT.

PROOF. $\{n : \forall m \preccurlyeq n Rm\}$ is a recursive tree well-founded w.r.t. all recursive 01-sequences; but by (2) there is no upper bound to the length of the branches in the tree. □

REMARK. Any α such that $\forall x R(\bar{\alpha}x)$ separates $\{y : \exists y(Tyyu \wedge Uu = 0)\}$ and $\{y : \exists u(Tyyu \wedge Uu = 1)\}$, our standard example of a pair of disjoint, recursively inseparable sets. A more simple-minded definition of a recursively well-founded, not well-founded tree would be $R^*(\bar{\alpha}x) := \forall y < x (\alpha y = 1 \leftrightarrow \exists u(Tyyu \wedge Uu = 0))$, but R^* is not decidable.

7.7. *On the intuitive motivation of* FAN_D *and* FAN. We discuss $\text{FAN}_\text{D}(T_{01})$ and $\text{FAN}(T_{01})$ for the binary tree T_{01} of 01-sequences; as we have seen in 7.5, this involves no loss of generality. So let n, m range over finite 01-sequences and α, β over infinite ones. For any decidable predicate A such that $\forall \alpha \exists x A(\bar{\alpha}x)$, $A^* := \{n : \forall m \preccurlyeq n \neg Am\}$ is well-founded. That is to say, if we follow any path through the binary tree (follow an arbitrary 01-choice sequence) we always meet the *bar* $P := \{n : An \wedge \forall m \prec n \neg Am\}$.

As an illustration, we have drawn the picture of the bar associated with $A(n) := \mathrm{lth}(n) > 2^{(n)_0+(n)_1}$ (the black nodes represent the bar).

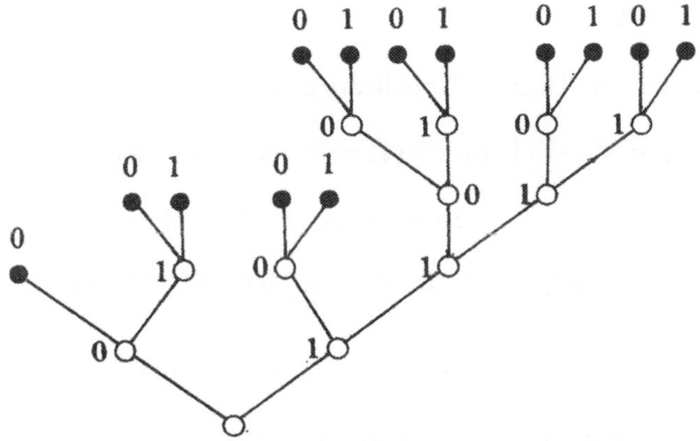

We can build a class $\mathscr{T}$ of well-founded binary trees using the following two principles

$$\{\langle\ \rangle\} \in \mathscr{T}, \tag{1}$$

if T_0, T_1 belong to $\mathscr{T}$ then so does

$$\Sigma(T_0, T_1) := \{\langle 0\rangle * n : n \in T_0\} \cup \{\langle 1\rangle * n : n \in T_1\} \cup \{\langle\ \rangle\}. \tag{2}$$

The fact that a tree is to be put in $\mathscr{T}$ on account of (1) and (2) only (in other words, $\mathscr{T}$ is the least class satisfying (i) and (ii)) can be expressed by the following principle

$$\{\langle\ \rangle\} \in \mathscr{X} \wedge \forall X_0 X_1 (X_0 \in \mathscr{X} \wedge X_1 \in \mathscr{X} \to \Sigma(X_0, X_1) \in \mathscr{X}) \\ \to (\mathscr{T} \subset \mathscr{X}) \tag{3}$$

where X_0, X_1 range over sets of finite sequences.

It is easy to see that each element of $\mathscr{T}$ has finite depth (take for $X \in \mathscr{X}$ in (3): X is a binary tree of finite depth), i.e. there is a maximum length for the branches, for any tree from $\mathscr{T}$; and conversely, a binary tree of finite depth obviously belongs to $\mathscr{T}$. The class $\mathscr{T}$ represents an inductive notion of well-foundedness; let us simply call the trees in $\mathscr{T}$ inductive binary trees. $\mathrm{FAN_D}$ identifies ordinary well-foundedness and inductive well-foundedness for binary trees.

Indeed, it is hard to see how one can establish that a tree is well-founded with respect to arbitrary paths (i.e. paths about which nothing specific is assumed) without generating the tree inductively. The conflict of FAN and CT shows, on the other hand, that it is essential that we think of the paths as "arbitrary": about recursive paths we have too much information.

In other words, $\mathrm{FAN_D}$ strengthens the interpretation of $\forall\alpha\exists x A(\bar{\alpha}x)$ for decidable A to "A^* is inductively generated"; the latter is a purely arithmetical notion, since $\mathscr{T}$ is an r.e. set (recall that finite trees can be coded by natural numbers). If we strengthen the interpretation of this particular combination of quantifiers, it is not quite obvious that all the rules of intuitionistic predicate logic still apply (see also the discussion in 6.9). A verification that $\forall\alpha\exists x$ can actually be strengthened as indicated by FAN, while retaining the laws of intuitionistic predicate logic is provided by the method of elimination of choice sequences, a special case of which is being discussed in section 12.3.

FAN states a stronger assumption: if $\forall\alpha\exists x A(\bar{\alpha}, x)$, then the tree $A^* \equiv \{n : \forall m \preccurlyeq n \neg Am\}$ is contained in an inductive binary tree. It seems most natural to motivate both FAN and FAN* by combining C-N and $\mathrm{FAN_D}$.

8. Bar induction and generalized inductive definitions

Knowledge of this section is not necessary for understanding the chapters 5–8 on intuitionistic mathematics. However, bar induction has played a very important role in Brouwer's development of intuitionistic mathematics and occupies a prominent place in the literature, and the intuitionistic motivation for accepting the fan theorem is best understood by studying the reasons for accepting bar induction.

In the preparation of our discussion of bar induction, we first introduce the inductively generated neighbourhood functions and trees, important special examples of generalized inductive definitions. After that, we turn to bar induction; and finally we return briefly to the subject of generalized inductive definitions. The latter part may be skipped by a reader who is only interested in bar induction and its connection with the fan theorem.

8.1. Let T be any tree; then we can define

$$T^{\sim} := \{n : \forall m \prec n (m \in T)\}.$$

This tree is *regular*, i.e. it has the property

$$(n \text{ is a terminal node}) \vee \forall x(n * \langle x \rangle \in T^{\sim}).$$

Conversely, each regular tree T' determines a T such that $T^{\sim} = T'$ by

$$n \in T := n * \langle 0 \rangle \in T'.$$

A tree T is said to be *well-founded* iff each sequence "leaves the tree", i.e. $\forall \alpha \exists x \neg(\bar{\alpha}x \in T)$. Clearly $T^{\sim}$ is well-founded iff T is well-founded. $T^{\sim} \backslash T$ consists of all terminal nodes of $T^{\sim}$, if T is well-founded; $T^{\sim} \backslash T$ is a *bar* of T, i.e. for each α we can find an x with $\bar{\alpha}x \in T^{\sim} \backslash T$, $\forall n \prec \bar{\alpha}x(n \in T)$.

Let K_0 be defined as in 6.8. With every $\gamma \in K_0$ there is naturally associated a well-founded tree

$$T^{\gamma} := \{n : \gamma n = 0\},$$

a bar

$$B^{\gamma} := (T^{\gamma})^{\sim} \backslash T^{\gamma},$$

and a function δ given by

$$\delta n = \gamma n \dot{-} 1.$$

Conversely, γ can be recovered from B^{γ} and δ by

$$\gamma n = 0 \text{ if } \exists m \in B^{\gamma}(n \prec m),$$

$$\gamma n = \delta m + 1 \text{ if } m \in B^{\gamma} \wedge m \preccurlyeq n.$$

Let $K_0^* \subset K_0$ consist of the neighbourhood functions uniformly bounded by 1, i.e.

$$K_0^* := \{\lambda n.\mathrm{sg}(\gamma n) : \gamma \in K_0\},$$

then K_0^* corresponds bijectively with the set of well-founded regular trees and hence with the set of well-founded trees.

8.2. *Generating the regular well-founded trees.* How can we generate all regular well-founded trees? Certainly the following two clauses generate regular well-founded trees:

1^0) the single node tree $\{\langle\ \rangle\}$ is regular and well-founded;

2^0) if $T_0, T_1, T_2, \ldots$ is an ω-sequence of such trees, we can combine them into a new regular well-founded tree T, pictorially represented by

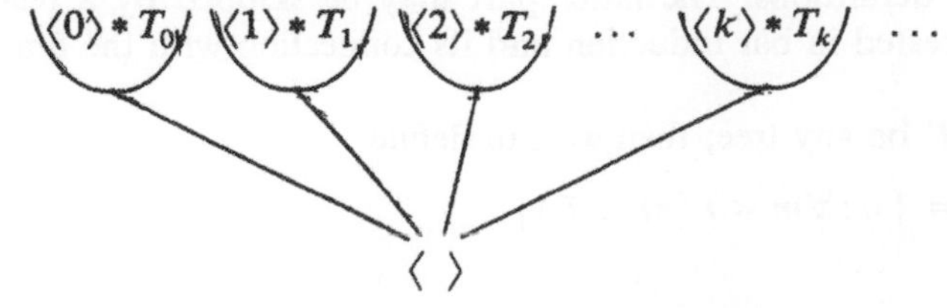

that is to say,

$$T := \Sigma_x T_x \equiv \{\langle x\rangle * n : n \in T_x\} \cup \{\langle\ \rangle\}.$$

If a regular well-founded tree is generated by repeated application of 1^0 and 2^0 only, we shall call it an *inductive tree*; let us use $\mathscr{IT}$ for the class of inductive trees.

1^0 and 2^0 above may now be reformulated as

(i) $\{\langle\ \rangle\} \in \mathscr{IT}$,

(ii) $\forall x(T_x \in \mathscr{IT}) \to \Sigma_x T_x \in \mathscr{IT}$.

$\mathscr{IT}$ is also the least class satisfying (i) and (ii), since we only put something in $\mathscr{IT}$ by virtue of its construction from (i) and (ii); so, if $\mathscr{X}$ is any class of sets of finite sequences satisfying (i) and (ii), that is to say

$$\{\langle\ \rangle\} \in \mathscr{X}, \quad \forall x(X_x \in \mathscr{X}) \to \Sigma_x X_x \in \mathscr{X},$$

then also

$$\mathscr{IT} \subset \mathscr{X}.$$

8.3. *Generating neighbourhood functions.* $\mathscr{IT}$ can equally well be described via the inductively generated elements of K_0^*; let us write K^* for the set of these elements. K^* satisfies two closure conditions parallel to (i), (ii):

(i)* $\lambda x.1 \in K^*$,

(ii)* $\forall x(\gamma_x \in K^*) \to \gamma \in K^*$,

where γ satisfies $\gamma(\langle\ \rangle) = 0$, $\gamma(\langle x\rangle * n) = \gamma_x(n)$, or equivalently

$$\begin{cases} \lambda x.1 \in K^*, \\ \gamma\langle\ \rangle = 0 \land \forall x(\lambda n.\gamma(\langle x\rangle * n) \in K^*) \to \gamma \in K^*. \end{cases} \tag{1}$$

K^* is to be the *least* class of functions satisfying these closure conditions. So if we put

$$A^*(Q, \alpha) := (\alpha = \lambda x.1) \lor (\alpha\langle\ \rangle = 0 \land \forall x(\lambda n.\alpha(\langle x\rangle * n) \in Q),$$

then a moment's reflection teaches us that $\forall\alpha(A^*(Q,\alpha) \to Q\alpha)$, being equivalent to $\forall\alpha((\alpha = \lambda x.1 \to \alpha \in Q) \land (\alpha\langle\ \rangle = 0 \land \forall x(\lambda n.\alpha(\langle x\rangle * n) \in Q) \to \alpha \in Q))$, expresses closure of Q under the closure conditions for K^*, and thus the fact that K^* is the least class satisfying these conditions is adequately expressed by

$$\forall\alpha(A^*(Q,\alpha) \to Q\alpha) \to (K^* \subset Q).$$

8.4. Only slight changes are necessary to characterize the set K of inductively generated elements of K_0, i.e. those elements $\alpha \in K_0$ for which

$\lambda n.\mathrm{sg}(\alpha n) \in K^*$. One has only to replace A^* by

$$A(Q, \alpha) :=$$

$$\exists x(\alpha = \lambda n.x + 1) \vee (\alpha 0 = 0 \wedge \forall x(\lambda n.\alpha(\langle x \rangle * n) \in Q));$$

then

K1–2 $\quad A(K, \alpha) \to K\alpha,$

K3 $\quad \forall\alpha(A(Q, \alpha) \to Q\alpha) \to (K \subset Q).$

The axiom K1–2 may be split into

K1 $\quad \lambda n.x + 1 \in K,$

K2 $\quad \alpha\langle\ \rangle = 0 \wedge \forall x(\lambda n.\alpha(\langle x \rangle * n) \in K.$

So the only difference between K and K^* is, that for K we permit arbitrary non-zero constant functions to start with, instead of only $\lambda n.1$.

The proofs of the following two propositions show how one can use K1–3 to establish facts about K.

8.5. PROPOSITION. $K \subset K_0$.

PROOF. Informally, this is rather obvious: K was, after all, introduced as an inductively generated subset of K_0. We shall give a more formal proof based on K1–3 here.

We apply K3 with $\alpha \in K_0$ for $Q(\alpha)$. We have to verify the premiss of K3. $\lambda n.x + 1 \in K_0$ is obvious; and if for all x $\lambda m.\alpha(\langle x \rangle * m) \in K_0$, and $\alpha 0 = 0$, then also $\alpha \in K_0$. For let β be any sequence, and let $\Delta\beta = \lambda x.\beta(x + 1)$. Then $\beta = \langle \beta 0 \rangle * \Delta\beta$, and $\alpha(\bar{\beta}(x + 1)) = \alpha(\langle \beta 0 \rangle * \overline{\Delta\beta}(x))$; now $\lambda m.\alpha(\langle \beta 0 \rangle * m) \in K_0$, hence for some x $\alpha(\langle \beta 0 \rangle * \overline{\Delta\beta}(x)) \neq 0$. So $\alpha(\bar{\beta}(x + 1)) \neq 0$. It is also easy to see that $\alpha n \neq 0 \to \alpha(n * m) = \alpha n$; we only have to note that $\alpha n \neq 0$ means $n = \langle x \rangle * n'$ for some x, n', so $\alpha(\langle x \rangle * n') \neq 0$, hence also $\alpha(\langle x \rangle * n') = \alpha(\langle x \rangle * n' * m) = \alpha(n * m)$. This establishes the premiss of K3 for this Q, so $K \subset K_0$. $\square$

N.B. The proof, as it stands, tacitly uses a closure condition on the range of the sequence variables: if β is a sequence, so is $\Delta\beta$. This condition is not necessary: for a proof which avoids this assumption, see E4.8.1.

8.6. *Uniform bounds on bounded subtrees.* Let $\alpha \leq \beta := \forall x(\alpha x \leq \beta x)$. The set $\{\alpha : \alpha \leq \beta\}$, for any fixed β, is a collection of functions ranging over a finitely branching tree, namely the tree T_β given by

$$n \in T_\beta := \forall x < \mathrm{lth}(n)((n)_x \leq \beta x).$$

As we shall see, for each $\alpha \in K$, the length of nodes in $\{n : \alpha n = 0 \wedge n \in T_\beta\}$ is uniformly bounded on T_β, and so is $\{\alpha n \dot{-} 1 : \alpha n \neq 0 \wedge n \in T_\beta\}$. We introduce the class K_1 by the following

DEFINITION. $K_1\alpha := \forall\beta\exists z\forall\gamma \leq \beta(\alpha(\bar{\gamma}z) \neq 0) \wedge \forall nm(\alpha n \neq 0 \rightarrow \alpha n = \alpha(n * m))$. □

PROPOSITION. $K \subset K_1$.

PROOF. Apply K3 with $K_1\alpha$ for $Q\alpha$ (exercise). □

As we just established in 8.5, $K \subset K_0$ holds; and *classically* one also easily establishes the converse:

8.7. PROPOSITION (*classical logic* + AC-NN). $K_0 \subset K$.

PROOF. Assume $\gamma \in K_0 \setminus K$, and let $\gamma n := \lambda m.\gamma(n * m)$, then $\gamma_{\langle\,\rangle} = \gamma \in K_0 \setminus K$, so $\gamma(\langle\,\rangle) = 0$ (otherwise γ would be constant and greater than zero, and therefore belong to K). For some x_0 $\gamma_{\langle x_0\rangle} \in K_0 \setminus K$, for if $\forall x(\gamma_{\langle x\rangle} \in K)$, then also $\gamma \in K$. Repeating this we find successively $x_1, x_2, \ldots$ such that

$$\gamma_{\langle x_0, x_1\rangle} \in K_0 \setminus K, \quad \gamma_{\langle x_0, x_1, x_2\rangle} \in K_0 \setminus K, \text{etc.}$$

So there is a sequence α, with $\alpha i = x_i$ such that for all y $\gamma_{\bar{\alpha}y} \in K_0 \setminus K$. But since $\gamma \in K_0$, there is a y for which $\gamma_{\bar{\alpha}y}(\langle\,\rangle) > 0$, i.e. $\gamma_{\bar{\alpha}y}$ is non-zero constant and therefore belongs to K; we have thus obtained a contradiction. □

REMARK. The application of choice above actually takes the form of DC-N, but this is equivalent to AC-NN. The proof uses PEM at each step, and is therefore of no use to us in obtaining an intuitionistic justification of $K_0 = K$.

8.8. *Intuitive motivation for* $K_0 = K$. Brouwer's so-called "bar theorem" amounts to the assumption $K_0^* = K^*$ or equivalently $K_0 = K$. The effect of equating K_0^* and K^* is to equate two notions of well-foundedness for countably branching trees. Thus FAN_D is generalized, which, as we saw, amounted to equating two notions of well-foundedness for finitely branching trees.

As in the case FAN_D, the identification of K_0 and K is plausible only if the range of the function quantifier in $\forall\beta(\alpha(\bar{\beta}x) > 0)$ in the definition of $\alpha \in K_0$ is to be taken to contain all choice sequences; if we have too much information about an arbitrary sequence in the range of the variable β, $K_0 = K$ is no longer plausible (cf. the conflict between CT and FAN_D in 7.7).

Since $K_0 = K$ sharpens the interpretation of quantifier combinations $\forall\alpha\exists n$, one would like to have metamathematical confirmation of the fact that we can safely use all the laws of intuitionistic predicate logic. As in the case of the fan theorem, such a confirmation can be obtained via the method of eliminating choice sequences (cf. 12.4.8).

Let us now derive the most important corollary of $K_0 = K$, namely the fan theorem.

8.9. THEOREM.
(i) $K_0 = K \Rightarrow FAN_D$;
(ii) $K_0 = K + \text{C-N} \Rightarrow FAN^*$.

PROOF. (i) By 7.5 it suffices to establish FAN_D in the form

$$\forall n(An \vee \neg An) \wedge \forall\alpha \leq \beta\exists x\, A(\bar{\alpha}x) \rightarrow \exists z\forall\alpha \leq \beta\exists x \leq z\, A(\bar{\alpha}x).$$

(In fact, even the special case $\beta = \gamma x.1$ suffices.) Let Γ_β be given by $\Gamma_\beta(\alpha) = \lambda x.\min(\alpha x, \beta x)$, then for $\alpha \leq \beta$ $\Gamma_\beta\alpha = \alpha$, and $\Gamma_\beta\alpha \leq \beta$ for all α. Assume $\forall n(An \vee \neg An)$, $\forall\alpha \leq \beta\exists x A(\bar{\alpha}x)$, then also $\forall\alpha\exists x A(\overline{\Gamma_\beta\alpha}(x))$. Put $A^*(\bar{\alpha}x) := A(\overline{\Gamma_\beta\alpha}(x))$. By AC-NN! we find a γ such that

$$\gamma n = 1 \leftrightarrow \exists m \preccurlyeq n(A^*m \wedge \forall m' \prec m\neg A^*m').$$

Obviously $\gamma \in K_0$ since $\forall\alpha\exists x A^*(\bar{\alpha}x)$; by proposition 8.6 $K_0 = K \subset K_1$, so we find a z such that $\forall\alpha \leq \beta(\gamma(\bar{\alpha}z) \neq 0)$, and thus also $\exists z\forall\alpha \leq \beta\exists x \leq z\, A(\bar{\alpha}x)$.

(ii) Combining (i) with the fact that $FAN_D + \text{C-N}$ yields FAN^*. □

8.10. *Bar continuity.* C-N may be combined with $K_0 = K$ into a single schema

BC-N $\quad \forall\alpha\exists nA(\alpha, n) \rightarrow \exists\gamma \in K\forall\alpha A(\alpha, \gamma(\alpha)),$

or equivalently

BC-N′ $\forall\alpha\exists n A(\alpha, n) \to \exists\gamma \in K \forall n(\gamma n > 0 \to \forall\alpha \in n\, A(\alpha, \gamma n \dot{-} 1))$.

The remainder of this section may be skipped at a first reading. The subsections 8.11–14 give some basic logical relationships between $K_0 = K$ and several forms of bar induction which appear in the literature.

8.15–17 give some supplementary comments on generalized inductive definitions, of which the introduction of K is a typical example.

Finally, in 8.18 we give a concise discussion of Brouwer's longer argument for the bar theorem.

The axiomatic basis in 8.11–14 is the system **EL** as described in 3.6.2.

8.11. *Bar induction.* (Decidable) bar induction is the following schema

$$\mathrm{BI_D} \quad \forall\alpha\exists x P(\bar{\alpha}x) \wedge \forall n(Pn \vee \neg Pn) \wedge \forall n(Pn \to Qn) \wedge \forall n(\forall y Q(n * \langle y \rangle) \to Qn) \to Q\langle\ \rangle.$$

$\mathrm{BI_D}$ may be viewed as an induction principle over trees. From $\forall n (Pn \vee \neg Pn)$ and $\forall\alpha\exists x P(\bar{\alpha}x)$ we see that $\{n : \forall m \prec n \neg Pm\}$ is a regular well-founded tree; the set $\{n : Pn \wedge \forall m \prec n \neg Pm\}$ is a bar for this tree. From $\forall n(Pn \to Qn)$ and $\forall n(\forall y Q(n * \langle y \rangle) \to Qn)$ we see that Q is a property which holds on a bar and is transmitted upwards: if Q holds for all immediate successors of a node n, it holds for n. Now $\mathrm{BI_D}$ permits us to conclude that then Q holds for the empty node.

A more general version of $\mathrm{BI_D}$ is *monotone bar induction*

$$\mathrm{BI_M} \quad \forall\alpha\exists x P(\bar{\alpha}x) \wedge \forall nm(Pn \to P(n * m)) \wedge \forall n(\forall y P(n * \langle y \rangle) \to Pn) \to P\langle\ \rangle.$$

As we shall see below, on assumption of C-N $\mathrm{BI_D}$ and $\mathrm{BI_M}$ are equivalent. It should be noted that the following apparently stronger version

$$\mathrm{BI'_M} \quad \forall\alpha\exists x P(\bar{\alpha}x) \wedge \forall nm(Pn \to P(n * m)) \wedge \forall n(Pn \to Qn) \wedge \forall n(\forall y Q(n * \langle y \rangle) \to Qn) \to Q\langle\ \rangle$$

follows from $\mathrm{BI_M}$: assume the premiss of $\mathrm{BI'_M}$, put $P'n := \forall m Q(n * m)$ and apply $\mathrm{BI_M}$ with respect to P'.

8.12. PROPOSITION. *Let* **EL**(K) *be* **EL** *with the language extended with a predicate letter* K. *In* **EL**(K) + K1–3 *we have the principle of induction over* *unsecured sequences*

IUS $\quad K\alpha \to$

$$[\forall n(\alpha n \neq 0 \to Q'n) \wedge \forall n(\forall y Q'(n * \langle y \rangle) \to Q'n) \to Q'\langle\ \rangle].$$

PROOF. Apply K3 to

$$Q\alpha \equiv \forall m[\forall n(\alpha n \neq 0 \to Q'(m * n)) \wedge \forall n(\forall y Q'(m * n * \langle y \rangle) \to Q'(m * n)) \to Q'm].$$

We find $\forall \alpha \in K Q(\alpha)$, hence for any $\alpha \in K$ $\forall m Q'm$, and so a fortiori $Q'\langle\ \rangle$. □

8.13. PROPOSITION (*logical relationships concerning* $\mathrm{BI_D}$, $\mathrm{BI_M}$). *Let* $\mathbf{EL}(K)$ *be as in* 8.12.

(i) *In* $\mathbf{EL}(K)$ + K1–3 + AC-NN! *we have* $K_0 = K \Leftrightarrow \mathrm{BI_D}$; *in fact K is definable in* $\mathbf{EL} + \mathrm{BI_D}$, *and satisfies* K1–3.

(ii) *In* $\mathbf{EL}$ + C-N *we have* $\mathrm{BI_D} \Leftrightarrow \mathrm{BI_M}$.

(iii) *In* $\mathbf{EL}(K)$ *we have* $\mathrm{BI_D}$ + C-N $\Leftrightarrow$ BC-N.

PROOF. (i) Assume $\mathbf{EL}(K)$ + K1–3 + AC-NN! + $K_0 = K$, and let the hypotheses of $\mathrm{BI_D}$ be fulfilled. By AC-NN! there is a γ such that

$$\gamma n = 0 \leftrightarrow \forall m < n \neg Pm, \quad \forall n(\gamma n \leq 1);$$

clearly $\gamma \in K_0$, hence $\gamma \in K$. For $Q'n := Qn \vee \exists m \prec n Pm$ one has

$$\forall n(\gamma n \neq 0 \to Qn'), \quad \forall n(\forall y Q'(n * \langle y \rangle) \to Q'n)$$

and applies IUS (8.12).

Conversely assume $\mathbf{EL} + \mathrm{BI_D}$; we can now prove that $\mathrm{K_0}$ satisfies the axioms K1–3 for K. It is obvious that K_0 satisfies K1–2 (this does not require $\mathrm{BI_D}$). Let Q' be any predicate such that

$$\forall n(\lambda x.n + 1 \in Q'),$$

$$\alpha 0 = 0 \wedge \forall x(\lambda n.\alpha(\langle x \rangle * n) \in Q') \to \alpha \in Q',$$

then we have to show $\gamma \in K_0 \to \gamma \in Q'$. So let $\gamma \in K_0$. Take

$$Pn := \gamma n \neq 0,$$

i.e. Pn holds iff $\gamma_n = \lambda m.\gamma(n * m)$ is constant, non-zero, and let

$$Qn := \gamma_n \in Q'.$$

Then $\forall n(Pn \vee \neg Pn)$, $\forall n(Pn \to Qn)$, $\forall \alpha \exists m(P\bar{\alpha}m)$, and obviously $\forall n(\forall y Q(n * \langle y \rangle) \to Qn)$; by $\mathbf{BI_D}$ immediately $Q\langle\ \rangle$, hence $\gamma \in Q'$.

Clearly, since K3 implies K to be the *least* predicate satisfying K1–2, we see that by the same reasoning in $\mathbf{EL}(K)$ + K1–3 + $\mathbf{BI_D}$ we must have $K = K_0$. N.B. In this direction we do not need AC-NN!.

(ii) We leave $\mathbf{BI_M} \Rightarrow \mathbf{BI_D}$ as an exercise. For the converse, assume C-N, $\mathbf{BI_D}$ and the hypotheses of $\mathbf{BI_M}$, i.e. $\forall \alpha \exists x P(\bar{\alpha}x)$, $\forall nm(Pn \to P(n * m))$, $\forall n(\forall y P(n * \langle y \rangle) \to Pn))$. By C-N we can find a $\gamma \in K_0$ such that $\forall n (\gamma n \neq 0 \to \forall \alpha \in n P(\bar{\alpha}(\gamma n \dot{-} 1)))$. Now put

$$P'n := \gamma n \neq 0 \wedge \mathrm{lth}(n) \geq \gamma n \dot{-} 1,$$

$$Q'n := Pn.$$

It is easy to see that P', Q' satisfy the hypotheses of $\mathbf{BI_D}$; thus $Q'\langle\ \rangle$, i.e. $P\langle\ \rangle$ follows.

(iii) Immediate via (i). □

8.14. Corollary. *In* $\mathbf{EL}(K)$ *we have* BC-N $\Rightarrow K_0 = K$.

Proof. Combine (iii) and (i) of the preceding proposition. □

8.15. *Generalized inductive definitions.* In $\mathcal{IT}$, K and K^* we have encountered examples of so-called generalized inductive definitions (g.i.d.'s), "generalized", since the type of closure conditions used to generate them generalizes the "finitary" closure conditions considered in section 3.6. Namely, from a countably infinite collection of elements of the inductively generated set one constructs a new element of the set; e.g. for $\mathcal{IT}$ we had the condition: if for all $x \in \mathbb{N}$ $T_x \in \mathcal{IT}$, then $\Sigma_x T_x \in \mathcal{IT}$.

An example of a g.i.d. which is familiar from recursion theory is provided by the Kleene recursive ordinals $\mathcal{O}$ with partial ordering $<_\mathcal{O}$. $<_\mathcal{O}$ is generated as the least relation satisfying

$$0 <_\mathcal{O} j(1,0) \quad (j(1,0) \neq 0 \text{ is assumed}),$$

$$x <_\mathcal{O} y \to x <_\mathcal{O} j(1, y),$$

$$x <_\mathcal{O} y \to y <_\mathcal{O} j(1, y),$$

$$x <_\mathcal{O} \{y\}(z) \wedge \forall z(\{y\}(z) <_\mathcal{O} \{y\}(z+1)) \to x <_\mathcal{O} j(2, y),$$

and $\mathcal{O} := \{x : x = 0 \vee 0 <_\mathcal{O} x\}$.

In this example $\mathcal{O}$ and $<_\mathcal{O}$ are introduced as new predicates relative to the language of arithmetic. Once $<_\mathcal{O}$ has been accepted as well-defined, we may iterate the procedure: we can introduce $<_{\mathcal{O}'}$ as the least relation satisfying the closure conditions for $<_{\mathcal{O}}$ with $<_{\mathcal{O}}$ replaced by $<_{\mathcal{O}'}$, and in addition

$$x \in \mathcal{O} \wedge x <_{\mathcal{O}} \{y\}(z) \wedge \forall z, z' \in \mathcal{O}(z <_{\mathcal{O}} z' \rightarrow \\ \{y\}(z) <_{\mathcal{O}'} \{y\}(z')) \rightarrow x <_{\mathcal{O}'} j(3, y).$$

This is an example of an *iterated* g.i.d.; that is to say, iterated relative to the language of arithmetic (it is a simple g.i.d. relative to the language of arithmetic extended with $<_{\mathcal{O}}$).

Generalized inductive definitions play an important role in definability theory, recursion theory and proof theory (cf. Aczel 1977, Buchholz et al. 1981). It is beyond the scope of this book to enter any further into the subject, except for one interesting observation. In section 3.6 we replaced (finitary) inductive definitions by explicit definitions in arithmetic via the notion of a finite (labelled) tree. The class of finite labelled trees is itself inductively definable, but in arithmetic we can straightforwardly replace induction over finite trees by ordinary induction over the *depth* of such trees. Thus the idea of the explicit definition consisted in a reduction of inductive definitions defined via $A \in \mathscr{P}$ (3.6.5) to a certain standard inductive definition, namely the class of finite labelled trees.

In the same manner we can reduce a wide class of g.i.d.'s to K or K^*, which represent countably *infinitely* branching trees; that is to say, many g.i.d.'s can be defined explicitly in terms of K or K^*. We state the precise result below, without proof.

8.16. DEFINITION. Let $\mathscr{L} = \mathscr{L}(\mathbf{EL})$, and let $\mathscr{L}(X)$ be $\mathscr{L}$ with an additional predicate letter X for a set of functions. The class $\mathscr{P}^1$ of *strongly positive formulas* $A(X, \alpha)$ in $\mathscr{L}(X)$ consists of those $A(X, \alpha)$ which can be constructed from formulas of $\mathscr{L}$ and formulas $X\varphi$ (φ a function-term of $\mathscr{L}$) by means of $\wedge, \vee, \forall x, \exists x, \exists \alpha$. □

Note that strong positivity guarantees monotonicity

$$A(P, \alpha) \wedge P \subset P' \rightarrow A(P', \alpha),$$

which can be proved purely logically.

8.17. LEMMA. *Assuming* AC_{01} *and arithmetic any strongly positive* $A(P, \alpha)$ *is equivalent to an* $A^0(P, \alpha)$ *of the form*

$$\exists \beta \forall x [R(\alpha, \beta, x) \vee (Q(\alpha, \beta, x) \wedge P\varphi[\alpha, \beta, x])],$$

where R, Q, *are in* $\mathscr{L}$, φ *a function-term of* $\mathscr{L}$. □

THEOREM. *Let* $\mathbf{IDB}_1$ *be the theory obtained by adding to* $\mathbf{EL}(K)$ *the axioms* K1–3 *and* AC_{01}, *and let* $A(P, \alpha)$ *be strongly positive in* P. *Then we can explicitly define in* $\mathbf{IDB}_1$ *a predicate* P_A *such that*

$$\forall\alpha[A(P_A, \alpha) \to P_A\alpha],$$

$$\forall\alpha[A(Q, \alpha) \to Q\alpha] \to \forall\alpha[P_A\alpha \to Q\alpha].$$

PROOF. Along the same lines as the proof of 3.6.11, using the lemma. □

8.18. *Brouwer's longer argument for the bar theorem (digression).* Let us call a set P of finite sequences a *bar* if $\forall\alpha\exists x(\bar{\alpha}x \in P)$. More generally, we shall say that *n is barred by P* (or *n is P-barred*) iff $\forall\alpha \in n\exists x(\bar{\alpha}x \in P)$. Where it is understood which set P is meant, we simply write "n is barred".

The basic assumption underlying Brouwer's proof (1927), reiterated by him for example in (1954) is that from any proof of "$\langle\ \rangle$ is barred" one can always extract an "infinitary proof-tree" which is based on three types of inference only:

(I) $n \in P$, hence n is barred (immediate inference);

(D) for all x, $n * \langle x\rangle$ is barred, hence n is barred (downwards inference);

(U) n is barred, hence $n * \langle x\rangle$ is barred (upwards inference).

Traditionally these inferences are called η-, $\digamma$- and ρ-inferences respectively ($\digamma$ is the symbol for the obsolete greek letter wau).

Under suitable assumptions on P, the U-inferences are eliminable. This is the case, for example, if P is monotone (i.e. $n \in P \wedge n \prec m \to m \in P$) or decidable (cf. E4.8.10). The following example essentially due to Kleene shows that the U-referènces are not eliminable for arbitrary P (Kleene and Vesley 1965, section 7.14; see also Dummett 1977, pp. 98–99). Take

$$P := \{\langle\ \rangle : \exists x(\beta x = 0)\} \cup \{\langle x\rangle : \beta x \neq 0\};$$

P is obviously a bar, since for any α we only have to check whether $\beta(\alpha 0) \neq 0$ or not. If $\beta(\alpha 0) \neq 0$, then $\langle \alpha 0\rangle = \bar{\alpha}1 \in P$, and if $\beta(\alpha 0) = 0$, then $\bar{\alpha}0 = \langle\ \rangle \in P$. The following "infinitary proof tree" may be extracted from this argument:

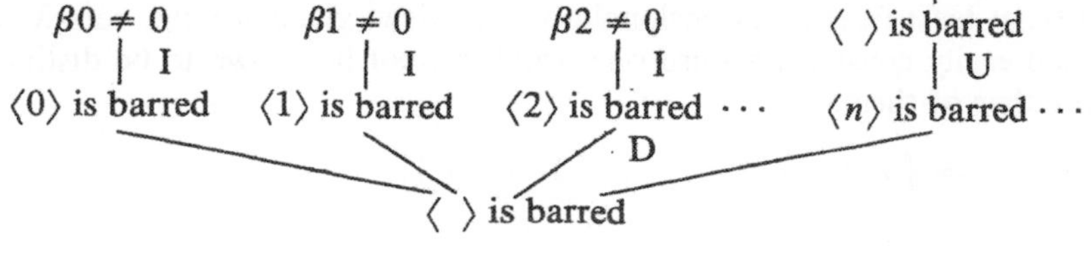

Note that P is decidable iff we can decide $\exists x(\beta x = 0)$. If we *know* that $\exists x(\beta x = 0)$, we can replace the infinitary proof tree by

$$\begin{array}{c} \beta n = 0 \\ \Big|\,\mathrm{I} \\ \langle\ \rangle \text{ is barred,} \end{array}$$

and if we know that $\neg\exists x(\beta x = 0)$, the infinitary proof-tree uses only I- and D-inferences. But as long as $\exists x(\beta x = 0)$ is unknown, we cannot indicate which tree without U-inferences is appropriate; in particular, we cannot indicate whether "$\langle 0 \rangle$ is barred" appears in the tree or not.

The same example can be used to show that the following (classically unproblematic) form of bar induction

$$\forall \alpha \exists x P(\bar{\alpha} x) \wedge \forall n(Pn \to Qn) \wedge \forall n(\forall y Q(n * \langle y \rangle) \to Qn) \to Q\langle\ \rangle$$

is not intuitionistically acceptable (cf. E4.8.11). One also readily sees that Brouwer's assumption plus the eliminability of U-inferences in, say, the monotone case leads to BI_{M}.

9. Uniformity principles and Kripke's schema

This section will be devoted to two principles which are of considerable interest because of the underlying intuitive motivation, but which will not play an important role in the mathematical developments in this book. The reader may therefore decide to skip this section until needed; the principles under discussion, the uniformity principle and Kripke's schema will turn up afterwards only in asides, digressions and exercises. We turn first to the uniformity principle.

9.1. *The uniformity principle for numbers.* This may be stated as a schema

UP $\quad \forall X \exists n A(X, n) \to \exists n \forall X A(X, n)$

(X ranging over sets of natural numbers). The idea behind this principle is as follows. Constructively, the sets of natural numbers are a very "diffuse" totality, at least from an extensional point of view: given any two sets X, Y one can easily construct a third one which cannot be shown to be distinct from either of them:

$$x \in Z := \{x \in X \wedge F) \vee (x \in Y \wedge \neg F)\},$$

where F is some (as yet) undecided statement. Thus, if one has assigned to each $X \subset \mathbb{N}$ a natural number, i.e. an element of a "very discrete" set, the only imaginable form of such an assignment seems to be a constant function. In fact, the following weaker form of the uniformity principle

$$\text{UP!} \qquad \forall X \exists ! n\, A(X, n) \rightarrow \exists n \forall X\, A(X, n)$$

(equality between subsets of $\mathbb{N}$ being interpreted extensionally) is easily proved on assumption of

$$\neg \forall P(\neg P \vee \neg\neg P),$$

where P ranges over propositions.

To see this, assume $A(X, n)$, $A(Y, m)$ and define Z_P by

$$x \in Z_P := \{x : (x \in X \wedge \neg P) \vee (x \in Y \wedge P)\},$$

then if $n \neq m$, and $A(Z_P, k)$, we find $Y \neq X$ and $k \neq m \rightarrow \neg P$, $k = m \rightarrow \neg\neg P$, and thus $\forall P(\neg P \vee \neg\neg P)$. Therefore $n = m$ and so $\forall Y\, A(Y, n)$.

9.2. *The uniformity principle for functions.* This may be stated as a schema

$$\text{UP}_1 \qquad \forall X \exists \alpha\, A(X, \alpha) \rightarrow \exists \alpha \forall X\, A(X, \alpha).$$

The motivation is similar to the one given for UP: even if functions (in particular: choice sequences) need not be completely determined, their initial segments can be determined; that is to say, $\mathbb{N}^{\mathbb{N}}$ is a separable totality which is far less "diffuse" than the collection of subsets of the natural numbers. Note also that quite trivially

$$\text{CT} \wedge \text{UP} \rightarrow \text{UP}_1$$

since by CT the function quantifier in UP_1 reduces to a number quantifier. On the other hand, $\text{UP}_1!$, with $\forall X \exists ! \alpha$ instead of $\forall X \exists \alpha$ in the premiss does not give us anything new: $\text{UP!} \rightarrow \text{UP}_1!$, because of $\forall X \exists ! \alpha\, A(X, \alpha) \leftrightarrow \forall x \forall X \exists ! y \exists \alpha[A(X, \alpha) \wedge \alpha x = y]$; assuming AC-NN for decidable predicates, this gives us $\text{UP}_1!$ from UP!.

9.3. *Parameterization principles and Kripke's schema.* We shall say that the subsets of $\mathbb{N}$ are parameterized by functions, via a (fixed) predicate A, if the following schema holds

$$\text{PF}_A \qquad \forall X \exists \alpha \forall n[n \in X \leftrightarrow A(\alpha, n)].$$

Classically, PF_A holds for a particularly simple A: take $A(\alpha, n) := (\alpha n = 0)$.

If α is a parameter for X, i.e. $\forall n[n \in X \leftrightarrow \alpha n = 0]$, α is a characteristic function for X.

Certain "creative subject" arguments by Brouwer, to be described below, lead to the assumption of "Kripke's schema"

$$\text{KS} \qquad \forall X \exists \alpha \forall n[n \in X \leftrightarrow \exists m(\alpha(n, m) = 0)],$$

so KS is $\mathrm{PF_A}$ with $A(\alpha, n) := \exists m(\alpha(n, m) = 0)$.

Brouwer's "creative subject" arguments may be viewed as an extreme expression of Brouwer's solipsistic view of mathematics. In the theory of the creative subject (= idealized mathematician; IM for short) the creative subject is viewed as carrying out his activity in ω discrete stages. Let us write "$\Box_n A$" for: "the IM has evidence for A at stage n of his mathematical activity" (elsewhere in the literature one finds $\vdash_n$ for $\Box_n$).

It is assumed that at any stage the IM knows whether he has (enough) evidence for A or not, so

$$\text{IM1} \qquad (\Box_n A) \vee \neg(\Box_n A).$$

Secondly, evidence is supposed to be cumulative:

$$\text{IM2} \qquad (\Box_m A) \wedge n > m \rightarrow (\Box_n A).$$

Finally, the solipsistic principle is assumed: *true* is, for the IM, only what he discovers to be true in the course of his own mathematical activity:

$$\text{IM3} \qquad A \leftrightarrow \exists n(\Box_n A).$$

Using IM3 as a mathematical insight by the IM is an instance of self-reflection. As an example of the use of IM3 in the construction of mathematical objects, consider α constructed as follows, for any given $X \subset \mathbb{N}$:

$$\Box_m(n \in X) \rightarrow \alpha(n, m) = 0,$$

$$\neg(\Box_m(n \in X)) \rightarrow \alpha(n, m) = 1.$$

Clearly, $n \in X \leftrightarrow \exists m(\Box_m(n \in X)) \leftrightarrow \exists m(\alpha(n, m) = 0)$, and this justifies KS. Classically of course, KS is immediately obvious – it is a weaker statement than the existence of a characteristic function for each $X \subset \mathbb{N}$. The theory of the creative subject gives rise to considerable conceptual difficulties, and has generated a fair amount of discussion in the literature, but a critical appraisal is beyond the scope of this chapter; we shall have something more to say about the "creative subject" in section 16.3.

The following proposition relates KS, continuity, CT and the uniformity principles:

9.4. PROPOSITION.
(i) KS + WC-N $\Rightarrow$ UP.
(ii) KS $\Rightarrow \neg \mathrm{UP}_1, \neg \mathrm{CT}, \neg \mathrm{CONT}_1$.

PROOF. (i) Let $\forall X \exists n A(X, n)$; take for $X := \{m : \exists n(\alpha(m, n) = 0)\}$, then

$$\forall \alpha \exists k\, A^*(\alpha, k),$$

where $A^*(\alpha, k) := A(\{n : \exists m \alpha(n, m) = 0\}, k)$. By WC-N

$$\forall \alpha \exists k n \forall \beta \in \bar{\alpha} n\, A^*(\beta, k).$$

Take $\alpha \equiv \lambda x.1$ and let k, n be such that

$$\forall \beta \in (\overline{\lambda x.1}) n\, A^*(\beta, k). \tag{1}$$

Now if $m \in X \leftrightarrow \exists n(\alpha(m, n) = 0)$, then for $n' = (\overline{\lambda x.1})n$, there is an α' such that $n' * \alpha'$ is also a parameter for X, hence from (1) $\forall X\, A(X, k)$.

(ii) apply UP_1 to KS:

$$\exists \alpha \forall X\, \forall n(n \in X \leftrightarrow \exists m(\alpha(n, m) = 0)),$$

which is obviously false. We leave the proof of the other two statements to the reader. □

9.5. *Kripke's schema and Markov's principle.* Brouwer (1948) introduced the explicit reference to the activity of the ideal mathematician in a mathematical construction in order to obtain a weak counterexample to $x \neq 0 \to x \# 0$ for a real number x. As we have seen (5.4), $\forall x \in \mathbb{R}(x \neq 0 \to x \# 0)$ is equivalent to Markov's Principle, and in an axiomatic setting Brouwer's example leads to a refutation of MP from KS:

PROPOSITION. KS + MP $\Rightarrow$ PEM.

PROOF. Let β be any function and apply KS to $A := B \vee \neg B$; by KS we find an α such that $\exists y(\alpha y = 0) \leftrightarrow A$, and since $\neg\neg A$ holds we have $\neg\neg\exists y(\alpha y = 0)$, and thus by Markov's principle $\exists y(\alpha y = 0)$, i.e. $B \vee \neg B$. □

Thus already the weakest "non-classical" assumption $\neg \forall P(P \vee \neg P)$ suffices to obtain a contradiction between KS and MP.

Kripke models for second order arithmetic, as introduced in 3.8.8 satisfy the uniformity principle at least partially, as shown by the following

9.6. Proposition. *Let $\mathscr{K}$ be a Kripke model for* **HAS** *as described in* 3.8.8, *with underlying tree* (K, $\leq$). *Then,*

(i) *If* (K, $\leq$) *is the full binary tree or the universal tree, then* UP! *holds in* $\mathscr{K}$;

(ii) *If* (K, $\leq$) *is the universal tree, then* $\mathscr{K}$ *satisfies closed instances of* UP, *i.e. those instances of* UP *where A does not contain free set variables besides X.*

Proof. We leave (i) as an exercise; the idea is to give a Kripke model version of the argument in 9.1.

(ii) Assume $k \Vdash \forall X \exists n A(X, n)$, $k \nVdash \exists n \forall X A(X, n)$, A not containing free set variables besides X. By the second assumption there are families $\mathscr{S}_n \in \mathrm{D}_2$ for each $n \in \mathbb{N}$, such that $k \nVdash A(\mathscr{S}_n, n)$. Suppose $\mathscr{S}_n \equiv \{S_{n,k} : k \in \mathrm{K}\}$, and define $\mathscr{S} \equiv \{S_k : k \in \mathrm{K}\}$ by

$$S_{k'} := \emptyset \quad \text{if} \quad \neg(k' > k),$$

$$S_{k * \langle n \rangle * k''} := S_{n, k * k''}.$$

By the first assumption $k \Vdash A(\mathscr{S}, m)$ for some m, and therefore also $k * \langle m \rangle \Vdash A(\mathscr{S}, m)$.

Now the restriction $\mathscr{K}'$ of $\mathscr{K}$ to nodes $k' \succcurlyeq k * \langle m \rangle$ is isomorphic (modulo duplication of some elements in D_2) to the model restricted to nodes $k' \succcurlyeq k$; and in $\mathscr{K}'$ $\mathscr{S}$ corresponds to $\mathscr{S}_m$. Since A does not contain set variables besides X, it follows that since $k * \langle m \rangle \Vdash A(\mathscr{S}, m)$, also $k \Vdash A(\mathscr{S}_m, m)$, contradicting our second assumption. Therefore $k \Vdash \exists n \forall X A(X, n)$. □

10. Notes

10.1. *Axioms of choice.* Probably the first place where an axiom of choice explicitly appears in formalizations of intuitionistic mathematics is Kleene (1957) (AC_{00}); Kleene and Vesley (1965) introduces AC_{01}. Kreisel's formalization of the theory of choice sequences (1963, and sections 2.5–6 of 1965) also contains choice axioms. The presentation axiom appears in the context of constructive set theory in Aczel (1978).

Further information about logical relationships between versions of the axiom of choice and dependent choice in the context of formal constructive analysis can be found in Howard and Kreisel (1966) and in Kreisel and Troelstra (1970).

10.2. *Church's thesis.* The original "thesis" was not formulated in the context of constructive mathematics, so strictly speaking, the designation "Church's thesis" for the principles discussed in section 3 is incorrect, but since the terminology is generally accepted in discussions of constructivism we adopt it here too. A thorough discussion of Church's thesis in the context of constructivism may be found in Kreisel (1970, 1972). Church's thesis, in an appropriate formulation, has been shown to be consistent with most intuitionistic formalisms studied in the literature. "Appropriate" means here that e.g. in CT the α ranges over lawlike sequences only, not over choice sequences, and that in CT_0 the predicate A should not contain choice parameters. See also Webb (1980), Shapiro (1981).

10.3. *Realizability.* Realizability by numbers was formulated by Kleene (1945); his original version was the non-formalized one. The formalized treatment of realizability for intuitionistic arithmetic was carried through by Nelson (1947).

In his retrospective survey Kleene (1973) states that his inspiration for the notion was not the BHK-interpretation, but rather certain formulations in Hilbert and Bernays (1934, p. 32).

In retrospect it seems clear enough that the ideas taken from Hilbert–Bernays and the BHK-interpretation are actually very close. More information on realizability can be found in Troelstra (1973 sections 3.2, 3.3) and Beeson (1985).

In Kleene (1960) the self-realizing character of almost negative formulas was demonstrated; in Troelstra (1971) this fact formed the basis of the axiomatic characterization of the realizable formulas by means of ECT_0. It is interesting to note that $\mathbf{HA} + \dot{C}T_0! \nvdash CT_0$ (Lifschitz 1979) and $\mathbf{HA} + CT_0 \nvdash ECT_0$ (Beeson, cf. Troelstra 1973, 3.4.14).

Dragalin (1969) uses ECT_0' (5.5) to characterize *true* realizable formulas with the help of transfinite progressions.

Kleene (1952, section 82) also considered a variant of realizability ($\vdash$-realizability) where in the clauses for $\vee, \exists, \rightarrow$ conditions involving formal deducibility were inserted, e.g.

$$n\,\mathbf{r}\,(A \vee B) \text{ iff}$$
$$(j_1 n = 0 \text{ and } j_2 n\,\mathbf{r}\,A \text{ and } \vdash A), \text{ or } (j_1 n = 1 \text{ and } j_2 n\,\mathbf{r}\,B \text{ and } \vdash B)$$

etcetera; from this notion Kleene afterwards (1962) extracted the Kleene slash (E3.5.3).

For his realizability by functions (Kleene and Vesley 1965) Kleene introduced in (1969) the formalized version, together with the **q**-variant which is similar to $\vdash$ -realizability, except that formal derivability has been replaced by truth. Troelstra (1971) studies the corresponding, but simpler, **q**-realizability by numbers. **q**-realizability as defined in E4.4.7 is slightly different; the relationships can be expressed by "Kleene's version of **q**-realizability" : "**q**-realizability of E4.4.7" = Kleene slash : Aczel slash. It was Grayson who observed that the realizability of E4.4.7 did the same job as Kleene's variant. The notion of E4.4.7 has the advantage of being closed under deducibility.

More about realizability may be found in sections 9.5 and 13.8.

10.4. *Markov's principle.* First announced in Markov (1954), according to a quotation by Shanin (1958A) (cf. Kleene and Vesley 1965, section 10). Kushner (1973) gives an even earlier published source which could not be checked by us.

Myhill (1963) gives an argument based on some general principles for an abstract theory of constructions refuting Markov's principle. The argument bears some similarity to the refutation of Markov's principle under the assumption of Kripke's schema in 9.5.

Logical relationships involving Markov's principle are discussed in Luckhardt (1977, 1977A). Our PT in 5.3 corresponds to principle (α) in Luckhardt (1977).

Information on the Dialectica interpretation may be found in Troelstra (1973, section 3.5), and in the "Introductory Note to Gödel 1958, 1972" in Gödel (1987).

10.5. *Choice sequences.* Information on the history of choice sequences can be found in Troelstra (1977, 1982A, 1983); of these, (1977) and (1983) also contain a fairly complete bibliography on the subject. The lawless sequences are discussed in much greater detail in chapter 12.

The first formulation of continuity, corresponding to our WC-N, is found in Heyting (1930B). Kleene (1957) contains a combination of the fan theorem with continuity. In Kreisel (1963) (later summarized in Kreisel 1965, sections 2.5–6) continuity appears in the form of bar-continuity (8.10). Kleene and Vesley (1965) is based on CONT_1 (Brouwer's principle for functions) which specializes to C-N (Brouwer's principle for numbers).

More information on logical relationships between various forms of continuity and choice axioms is contained in Kleene and Vesley (1965) and Kreisel and Troelstra (1970).

Brouwer used continuity for choice sequences in lectures as early as 1916–1917; the first clear-cut statements of continuity, fan theorem and bar theorem appear in Brouwer (1924). A very clear formulation of the bar induction principle is contained in footnote 7 of Brouwer (1927).

10.6. *Uniformity, parameterization and Kripke's schema.* The uniformity principle UP is explicitly formulated in Troelstra (1973A); before that date Kreisel (1971) drew attention to the fact that DP and ED for **IQC** imply remarkable "uniformity rules" for intuitionistic validity. More information on UP and UP! may be found scattered in the papers de Jongh and Smorynski (1976), van Dalen (1977, 1978) and Troelstra (1977A).

As far as we know, the name "Kripke's schema" first appears in Myhill (1967), and in Myhill's contribution to the discussion in Kreisel (1967), for the following weak version

$$\text{WKS} \quad \exists\alpha[(\exists x(\alpha x = 0) \to A) \wedge (\neg\exists x(\alpha x = 0) \to \neg A)].$$

WKS $\to \neg$CT is due to Kripke (Kreisel 1970, Note 10; cf. also Myhill 1967, final paragraph). Myhill also mentions (1967, p. 295) the strong version but notes that it is not needed for the reproduction of Brouwer's counterexamples.

Kreisel (1967) contains the earliest axiomatization of the theory of the creative subject. In Kreisel's axiomatization the possibility that there is more than one creative subject is left open: there is a parameter Σ ranging over creative subjects; the crucial principle corresponding to our IM3 there takes the form

$$(A \to \forall\Sigma\neg\neg\exists m(\Box_{\Sigma,m} A)) \wedge \forall\Sigma(\exists m(\Box_{\Sigma,m} A) \to A)$$

(the intention, not explicitly stated in the paper, is that $A \leftrightarrow \exists\Sigma m(\Box_{\Sigma,m} A)$). This corresponds to WKS. Troelstra (1969) gives a survey of the theory of the creative subject.

WKS is in general weaker than KS (Krol' 1978A), and BI + C-N + KS is consistent relative to elementary analysis (Krol' 1978), a fact which is far from evident from the intuitive justification of BI and C-N for choice sequences.

10.7. The paper Troelstra (1977A) discusses some further "non-classical" principles such as subcountability of discrete sets (SCDS); little has been published about these so far. McCarty (1986) shows that SCDS is consistent with a set theory based on intuitionistic logic (**IZF** of chapter 11).

Exercises

4.1.1. Prove lemma 1.4.

4.2.1. If we assume enough set-comprehension as admissible, we can construct for each proposition A the sets $a := \{x : x = 0 \vee (x = 1 \wedge A)\}$ and $b := \{x : (x = 0 \wedge A) \vee x = 1\}$; clearly $\forall y \in \{a, b\} \exists x (x \in y)$. Derive PEM by application of AC to this premiss (Diaconescu 1975, Goodman and Myhill 1978).

4.2.2. Show that DC-N is equivalent to AC-NN, and show that AC-ND (i.e. AC-N for a specific domain D) follows from DC-D. Show also that any instance of RDC-D be derived from DC-D′ for suitable D′, where D′ may depend on the instance of RDC-D to be derived (Kreisel and Troelstra 1970, 2.7.2).

4.3.1. Show that, assuming CT_0, the following principles of classical predicate logic are constructively false by refuting universally closed instances of them:

$$\neg\neg(P \vee Q) \to \neg\neg P \vee \neg\neg Q,$$

$$\neg\neg\exists x A \to \exists x \neg\neg A,$$

$$\neg\exists x \neg A \to \forall x A,$$

$$\neg\forall x \neg A \to \exists x A,$$

$$\forall x (A \vee B(x)) \to A \vee \forall x B(x) \quad (x \notin \mathrm{FV}(A)).$$

4.4.1. Complete the proofs of 4.5 and 4.6, and fill in the details of the remark at the end of 4.6.

4.4.2. Complete the proof of 4.10.

4.4.3. Establish the refinement mentioned in the remark at the end of 4.10: for each sentence A provable in **HA** there is a numeral $\bar{n}$ such that **HA** $\vdash \bar{n}\, \mathbf{r}\, A$. *Hint*. Make use of the fact that only true Σ_1^0-sentences can be proved in **HA** (see e.g. Troelstra 1973, 3.2.4).

4.4.4. Show that the addition of the following schema

$$\mathrm{ECT}_0!^* \qquad \forall n[An \to \exists! m B(n, m)] \to \exists k \forall n[An \to \exists m(Tknm \wedge B(n, Um))]$$

to **HA** is inconsistent. N.B. A is *not* assumed to be almost negative. This shows that, a fortiori, the generalization ECT_0^* of ECT_0 with condition "A almost negative" dropped is inconsistent. *Hint.* For the premiss of $ECT_0!^*$ take $\forall n[(\exists mTnnm \vee \neg\exists mTnnm) \to \exists k((k = 0 \wedge \exists mTnnm) \vee (k = 1 \wedge \neg\exists mTnnm))]$ (Troelstra 1973, 3.2.20, remark).

4.4.5. Let Γ_0 be the class of formulas of **HA** such that for all subformulas $A \to B$ of a $C \in \Gamma_0$ A is almost negative. Show that $\mathbf{HA} + ECT_0$ is conservative over **HA** with respect to formulas of Γ_0 (Troelstra 1973, 3.6.6).

4.4.6. Extend the result of exercise 4.4.3 to $\mathbf{HA} + ECT_0$ (Troelstra 1973, 3.2.15).

4.4.7. Define the notion of **q**-realizability as follows: the clauses (i)–(iii), (v)–(vi) in 4.2 are also adopted for **q**-realizability, with **q** everywhere replacing **r**; (iv) is replaced by

$$x\,\mathbf{q}\,(A \to B) := \forall y(y\,\mathbf{q}\,A \to \exists u(Txyu \wedge Uu\,\mathbf{q}\,B)) \wedge (A \to B).$$

Show the following
(i) $\mathbf{HA} \vdash \exists x(x\,\mathbf{q}\,A) \to A$,
(ii) $\mathbf{HA} + ECT_0 \vdash A \leftrightarrow \exists x(x\,\mathbf{q}\,A)$,
(iii) $\mathbf{HA} \vdash A \Rightarrow \mathbf{HA} \vdash \exists x(x\,\mathbf{q}\,A)$.
(Cf. remarks in 10.3.)

4.4.8. Use the result of exercise 4.4.7 to show closure of **HA** under *Church's rule*

$$CR_0 \qquad \mathbf{HA} \vdash \forall x\exists yA(x,y) \Rightarrow \mathbf{HA} \vdash \exists z\forall xA(x,\{z\}(x)),$$

(Troelstra 1973, 3.7.2). Use this result in turn to derive MR from MR_{PR} (3.5.4).

4.4.9. Show that proposition 4.5 also holds for **q**-realizability (E4.4.7) and use this to strengthen the result of E4.4.8 to closure of **HA** under

$$ECR_0 \qquad \vdash \forall x[Ax \to \exists yBxy] \Rightarrow \vdash \exists z\forall x[Ax \to E\{z\}(x) \wedge B(x,\{z\}(x))],$$

(*extended Church's rule*, Troelstra 1973, 3.7.2).

4.4.10. Show that the refinement of E4.4.3 also holds for **q**-realizability; use this to give another proof of DP and EDN for **HA** (cf. 3.5.6–10).

4.4.11. Extend **r**-realizability to **HAS** (cf. 3.8.4) as follows. To each set variable X we assign a set variable X^* such that $X \not\equiv Y \Rightarrow X^* \not\equiv Y^*$, and we add the following clauses

$$x\, \mathbf{r}\, (t \in X) := (x, t) \in X^*,$$

$$x\, \mathbf{r}\, \forall X A := \forall X^* (x\, \mathbf{r}\, A),$$

$$x\, \mathbf{r}\, \exists X A := \exists X^* (x\, \mathbf{r}\, A).$$

Prove the soundness theorem, thereby establishing consistency of **HAS** + CT_0. N.B. Note that in fact one may choose $X^* \equiv X$. (Troelstra 1973, 3.2.31.)

4.4.12. Extend **q**-realizability to **HAS** (3.8.4) as follows. We now use each set variable X to code two sets: $X_0 := \{x : 2x \in X\}$, $X_1 := \{x : 2x + 1 \in X\}$. Now $x\, \mathbf{q}\, A$ is defined as $x\, \mathbf{r}\, A$ for **HAS** except for the following clauses

$$x\, \mathbf{q}\, (t \in X) := (x, t) \in X_0,$$

$$x\, \mathbf{q}\, A \to B := \forall y (y\, \mathbf{q}\, A \to \{x\}(y)\, \mathbf{q}\, B) \wedge (A \to B)^*.$$

Here C^* is obtained from C by replacing all occurrences $t \in X$ for numerical terms t and set variables X in C by $t \in X_1$ (or $2t + 1 \in X$). Prove the soundness theorem for sentences A: **HAS** $\vdash A \Rightarrow$ **HAS** $\vdash \bar{n}\, \mathbf{q}\, A$ for some numeral $\bar{n}$. Use this to obtain Church's rule, EDN and DP for **HAS** (cf. Friedman 1977, and E4.4.7–10).

4.4.13. Let PCT be "Partial Church's Thesis"

$$\forall \pi \exists x \forall yz (Txyz \leftrightarrow \pi y = Uz),$$

where π ranges over partial functions from $\mathbb{N}$ to $\mathbb{N}$. Show that PCT conflicts with the following axiom of partial choice

$$\forall x \in A \exists ! y B(x, y) \to \exists \pi \forall x \in A (E(\pi x) \wedge B(x, \pi x))$$

(compare E4.4.4).

4.5.1. Show that
(i) $\mathrm{MP_{PR}}$ and MP are equivalent in the presence of CT_0;
(ii) MP and PT are equivalent.
Formulate a generalization PT′ of PT also permitting possibly empty sets and show
(iii) MP and PT′ are equivalent.
((i) and (ii) in Luckhardt (1977).

4.5.2. (Assuming familiarity with elementary recursion theory.) Show that $\mathrm{MP_{PR}}$ is equivalent to

$$\forall xy [\neg\neg \exists z Txyz \to \exists z Txyz],$$

and even to

$$\forall x[\neg\neg\exists zTxxz \to \exists zTxxz].$$

4.5.3. Show that ECT_0 and ECT_0' (5.5) are equivalent in the presence of MP_{PR}.

4.5.4. Show that, relative to **HA**, assuming the realizability of MP is equivalent to assuming MP itself.

4.5.5. Let IP be the schema $(\neg A \to \exists yB) \to \exists y(\neg A \to B)$ (y not free in A) and show $\mathbf{HA} + MP_{PR} + IP + CT_0 \vdash \bot$ (Troelstra 1973, 3.2.27(i)).

4.5.6. Modify realizability for **HAS** (E4.4.11) by putting as clause for $t \in X$

$$x \,\mathbf{r}\, (t \in X) := \neg\neg(x, t) \in X^*.$$

Check the soundness theorem for this variant and show moreover that "Shanin's principle"

SHP $\quad \forall X \exists Y \forall x[x \in X \leftrightarrow \exists y((y, x) \notin Y)]$

is realizable (Gordeev 1982).

4.6.1. Show that WC-N actually is equivalent to the following principle: if $\{X_n : n \in \mathbb{N}\}$ is a covering of $\mathbb{N}^{\mathbb{N}}$ by arbitrary sets, then $\{\mathrm{Int}(X_n) : n \in \mathbb{N}\}$ also covers $\mathbb{N}^{\mathbb{N}}$.

4.6.2. Give an explicit definition of the neighbourhood functions for the following functionals in $\mathbb{N}^{\mathbb{N}} \to \mathbb{N}$:

$$\Phi_1\alpha := \alpha(\alpha 0),$$

$$\Phi_2\alpha := \alpha 0 + \alpha 1 + \cdots + \alpha(\alpha 0),$$

$$\Phi_3\alpha := \alpha 1 \cdot \alpha 2,$$

$$\Phi_4\alpha := \alpha(\alpha(\alpha 0)).$$

4.6.3. Show that AC-NN follows from NFP (6.12). *Hint.* Given $A(x, y)$, consider $A^*(\bar{\alpha}x) := x > 0 \wedge A(\alpha 0, x \dot{-} 1)$.

4.6.4. Derive C-N from WC-N + NFP. *Hint.* Use the preceding exercise, and apply NFP to $\forall\alpha\exists y\exists x\forall\beta \in \bar{\alpha}yA(\beta, y)$ obtained from WC-N (6.12).

4.6.5. Prove Lindelöf's theorem (6.14) for $\mathbb{N}^{\mathbb{N}}$ from NFP. *Hint.* Let $\{W_i : i \in I\}$ be an open cover of $\mathbb{N}^{\mathbb{N}}$; apply NFP to $\forall\alpha\exists x\exists i \in I(V_{\bar{\alpha}x} \subset W_i)$.

4.6.6. Refute the schemata of E4.3.1 assuming WC-N instead of CT_0.

4.6.7. Show that C-C implies C-N.

4.6.8. Define neighbourhood functions (cf. 6.15) for the following functionals in $\mathbb{N}^\mathbb{N} \to \mathbb{N}^\mathbb{N}$:

$$\Psi_1\alpha := n * \alpha,$$

$$\Psi_2\alpha := j_1\alpha + j_2\alpha,$$

$$\Psi_3\alpha := \lambda x.\bar{\alpha}x,$$

$$\Psi_4\alpha := \lambda x.\alpha(\alpha x).$$

4.6.9. Suppose that $\Psi_\gamma \in \mathbb{N}^\mathbb{N} \to \mathbb{N}^\mathbb{N}$ is given by $\gamma \in K_0$, and $\Phi_\delta \in \mathbb{N}^\mathbb{N} \to \mathbb{N}$ by $\delta \in K_0$. Define explicitly from γ and δ a neighbourhood function for the composition $\Phi_\delta \circ \Psi_\gamma \in \mathbb{N}^\mathbb{N} \to \mathbb{N}$ (Kreisel and Troelstra 1970, 3.2.11–12).

4.6.10. Let Ψ_γ, Φ_δ be as in the preceding exercise. Show that there is no total $\Delta \in \mathbb{N}^\mathbb{N} \times \mathbb{N}^\mathbb{N} \to \mathbb{N}^\mathbb{N}$ such that for neighbourhood functions γ, δ always $\Phi_\delta \circ \Psi_\gamma = \Phi_{\Delta(\delta,\gamma)}$ (Kreisel and Troelstra 1970, 3.5).

4.7.1. Show in detail that, assuming classical logic, FAN_D implies König's lemma.

4.7.2. For the functionals Φ mentioned in exercise 4.6.2 determine (a) their maximum value on T_{01} and (b) their maximum value on the tree of sequences with values bounded by n.

Determine also a least upper bound for the length of initial segments of infinite branches needed to determine the values required under (a) and (b).

4.7.3. Show that FAN and NFP of 6.12 are classically valid.

4.7.4. Give an example showing that the principle

$$\forall\alpha \in T_{01}\exists x A(\alpha, x) \to \exists z \forall \alpha \in T_{01} \exists x \leq z A(\alpha, x)$$

is not compatible with classical logic.

4.7.5. Prove 7.4(ii).

4.7.6. Refute $\exists$-PEM by means of $FAN^*(T_{01})$ without assuming the existence of lawlike elements among the 01-choice sequences, but only the existence of a 01-sequence $\alpha \in n$ for each $n \in T_{01}$.

4.7.7. Show that CT implies that $2^{\mathbb{N}}$ and $\mathbb{N}^{\mathbb{N}}$ are homeomorphic. *Hint.* Let $T_R := \{n : \forall m \preccurlyeq n\ Rm\}$ be the Kleene tree as defined in 7.6. Let f be a function enumerating without repetition the terminal nodes of T_R. Then $\Psi \in \mathbb{N}^{\mathbb{N}} \to 2^{\mathbb{N}}$ given by

$$\Psi(\alpha) = f(\alpha 0) * f(\alpha 1) * f(\alpha 2) * f(\alpha 3) * \cdots$$

is a homeomorphism (Beeson 1985, IV, 13).

4.8.1. Prove $\forall \alpha \in K \forall \beta \forall n \exists m(\beta \in n \to (\alpha m > 0 \wedge \beta \in n * m))$ without assuming any closure conditions on the range of function variables (Kreisel 1968, p. 232).

4.8.2. Prove $K \subset K_1$ (8.6).

4.8.3. Complete the proof of 8.12.

4.8.4. Prove 8.13(ii), namely $\mathbf{EL} + \mathbf{BI}_M \vdash \mathbf{BI}_D$ in $\mathbf{EL}$ (Howard and Kreisel 1966, Theorem 4A).

4.8.5. Assuming K1–3, show $\forall \gamma \in K \forall \delta \in K_0[\forall n(\gamma n \neq 0 \to \delta n \neq 0) \to \delta \in K]$.

4.8.6. Show by induction over K that $\gamma \in K \to \lambda n.\gamma n \cdot \mathrm{sg}(\mathrm{lth}(n) \dot{-} \gamma n) \in K$ (use the preceding exercise). From this it follows that

$$\gamma \in K \to \exists \gamma' \in K(\gamma' n \neq 0 \to \gamma n = \gamma' n \wedge \gamma' n < \mathrm{lth}(n)).$$

4.8.7. Give a direct proof that BC-N $\Rightarrow K_0 = K$, using the preceding two exercises (cf. 8.14).

4.8.8. Prove the lemma and theorem in 8.17 (Kreisel and Troelstra 1970, section 4).

4.8.9. Let $\alpha \in K$; define on the corresponding tree $T_\alpha := \{n * \langle m \rangle : \alpha n = 0, m \in \mathbb{N}\}$ $k <^* k' := k \succ k'$ and show that transfinite induction holds w.r.t. $<^*$, i.e. $\forall y(\forall x <^* y F(x) \to F(y)) \to \forall x F(x)$ for x, y ranging over T_α.

4.8.10. Define P_1 and P_2 by a generalized inductive definition as follows. Put

$$A_1(n, X) := n \in P \vee \forall y(n * \langle y \rangle \in X) \vee \exists zm(n = m * \langle z \rangle \wedge m \in X),$$

$$A_2(n, X) := n \in P \vee \forall y(n * \langle y \rangle \in X);$$

then P_1, P_2 are characterized by

$$\begin{cases} A_i(n, P_i) \to P_i n, \\ \forall n(A_i(n, Q) \to n \in Q) \to (P_i \subset Q). \end{cases}$$

The assumption in Brouwer (1927) underlying the proof of the "bar theorem" amounts to

$$\langle\ \rangle \text{ is } P\text{-barred} \Rightarrow \langle\ \rangle \in P_1$$

and the eliminability of U-inferences corresponds to

$$\langle\ \rangle \in P_1 \Rightarrow \langle\ \rangle \in P_2.$$

Show that for monotone P we even have $P_1 = P_2$, and for decidable P

$$\forall n(n \in P_1 \to n \in P_2 \vee \exists m \prec n(m \in P))$$

(Martino and Giaretta 1981).

4.8.11. Show that $\forall\alpha\exists x P(\bar{\alpha}x) \wedge \forall(Pn \to Qn) \wedge \forall n(\forall y Q(n * \langle y \rangle) \to Qn) \to Q\langle\ \rangle$ implies $\exists n(\beta n = 0) \vee \neg\exists n(\beta n = 0)$, for P as in Kleene's example (8.18) and suitable Q.

4.8.12. Prove, assuming bar-continuity BC-N (8.10) plus a density principle $\forall n \exists\alpha(\alpha \in n)$ that $\neg\exists$-PEM can be proved without assuming that lawlike functions are (extensionally) equal to choice sequences.

4.8.13. Derive double bar induction from BI, where *double bar induction* is the following principle:

$$\begin{aligned} &\forall\alpha\beta\exists xy P(\bar{\alpha}x, \bar{\beta}y) \wedge \\ &\forall nn'm(P(n, m) \to P(n * n', m) \wedge \forall nmm'(P(n, m) \to P(n, m * m')) \wedge \\ &\forall n(\forall x P(n * \langle x \rangle, m) \to P(n, m)) \wedge \forall m(\forall y P(n, m * \langle y \rangle) \to P(n, m)) \\ &\qquad\qquad\qquad\qquad\qquad\qquad\qquad\qquad \to P(\langle\ \rangle, \langle\ \rangle). \end{aligned}$$

4.9.1. Show that $\mathbf{KS} \vdash \neg\mathrm{CT}$, $\mathbf{KS} \vdash \neg\mathrm{CONT}_1$ (9.4(ii), Myhill 1967).

4.9.2. Prove 9.6(i); give a counterexample to 9.6(ii) for a non-closed instance of UP (de Jongh and Smorynski 1976, p. 163).

4.9.3. Use $\mathbf{r}$-realizability (E4.4.11) for **HAS** to establish consistency of **HAS** + UP + CT_0, and $\mathbf{q}$-realizability (E.4.4.12) to derive EDN and DP (3.5.6) for **HAS** + UP + CT_0.

4.9.4. Show that C-N + $\mathrm{PF_A}$ imply UP, and that $\mathrm{PF_A}$ contradicts $\mathrm{UP_1}$. This generalizes 9.4 (Troelstra 1977A, section 5).

4.9.5. Let **H** be any extension of **EL** + KS + AC-NF in the language of **EL**, and **HS** the corresponding extension with variables for sets of natural numbers, impredicative comprehension and extensionality for sets added. Use KS to define a translation τ such that $\mathbf{HS} \vdash A \Rightarrow \mathbf{H} \vdash \tau(A)$, and $\mathbf{HA} \vdash A \Rightarrow \mathbf{EL} + \mathrm{KS} + \text{AC-NF} \vdash \tau(A)$ (van Dalen 1977).

4.9.6. Introduce functions in the models of 9.6 via their graphs, i.e. $\mathscr{S}$ is a function at node k if $k \vDash \forall n \exists ! m((n, m) \in \mathscr{S})$. Show that on this interpretation of functions KS is false in the models (de Jongh and Smorynski 1976).

4.9.4. Show that $C_2N + EL_2$ imply UP, and that EP_2 contradicts UP. This generalizes 9.4 (Troelstra 1977a, section 5).

4.9.5. Let H be any extension of EL + KS + AC-NF in the language of EL, and H^S the corresponding extension with variables for sets of natural numbers, impredicative comprehension and extensionality for sets added. Use KS to define a translation ∗ such that $H^S \vdash \varphi \leftrightarrow \varphi^*$, and H ⊢ [illegible] (van Dalen 1977).

4.9.6. Introduce functions in the models of 9.6 via their graphs, i.e. φ is a function at node k if $k \Vdash \forall n \exists ! m (\varphi(n, m) \in X)$. Show that on this interpretation of functions, KS is false in the models (de Jongh and Smorynski 1976).

CHAPTER 5

REAL NUMBERS

In this chapter we give a detailed constructive treatment of the theory of real numbers. The reader who wishes to arrive speedily at some examples of constructive mathematics may restrict himself, at a first reading, to the approach via fundamental sequences, which is developed in sections 2–4, and then continue with chapter 6.

In sections 5 and 6 we describe the introduction of reals by means of Dedekind cuts; simultaneously with the Dedekind reals ($\mathbb{R}^d$) we treat the much larger structure of the (bounded) extended reals $\mathbb{R}^{be}$, which turns out to be the uniquely determined order completion of $\mathbb{Q}$. In $\mathbb{R}^{be}$ each bounded inhabited set has a supremum. A still larger structure is $\mathbb{R}^e$, which is in many respects similar to $\mathbb{R}$ in classical mathematics.

We have chosen to introduce both the Cauchy reals and the Dedekind reals, even though on the basis of AC_{00}, a choice principle accepted by most constructivists, both structures can be shown to be isomorphic. One reason for this is that in chapter 15 we shall encounter some "natural" models for analysis in which AC_{00} fails, and where $\mathbb{R}$ and $\mathbb{R}^d$ are actually not isomorphic. In these models $\mathbb{R}^d$ is the more interesting structure of the two. Another reason is that the treatment of $\mathbb{R}^{be}$ and $\mathbb{R}^e$ is most naturally given in parallel with a treatment of $\mathbb{R}^d$.

$\mathbb{R}^d, \mathbb{R}^{be}, \mathbb{R}^e$ have unique characterizations as ordered structures extending $\mathbb{Q}$ (5.12).

Section 7 contains two metamathematical digressions: by means of the Gödel negative translation it is explained why $\mathbb{R}^e$ behaves nearly "classically", and it is shown how the distinctions between $\mathbb{R}^d$, $\mathbb{R}^e$ and $\mathbb{R}^{be}$ can be made visible in Kripke models for **HAS**.

1. Introduction

1.1. *Integers and rationals.* Given the natural numbers, the introduction of the integers $\mathbb{Z}$ and the rationals is unproblematic and can be carried out just as in traditional mathematics.

The elementary theory of arithmetical operations on $\mathbb{Z}$ and $\mathbb{Q}$ does not really carry us beyond arithmetic **HA**. To see this, note that we can enumerate $\mathbb{Z}$ and $\mathbb{Q}$ without repetitions, i.e. we can use natural numbers as codes for elements of $\mathbb{Z}$ and $\mathbb{Q}$. The operations $+, \cdot$ on $\mathbb{Z}$ and $\mathbb{Q}$ respectively correspond to certain arithmetical operations $+_{\mathbb{Z}}, \cdot_{\mathbb{Z}}$ and $+_{\mathbb{Q}}, \cdot_{\mathbb{Q}}$ which, if we have chosen our coding appropriately, are in fact primitive recursive; their basic properties such as commutativity, associativity etc. can all be proved in arithmetic.

The coding permits us to think of the basic theory of $\mathbb{Z}$ and $\mathbb{Q}$ as formulated in a definitional extension of intuitionistic first-order arithmetic **HA**. Obviously, then, we can appeal to the Gödel–Gentzen translation (cf. 3.3.4) to see that all the usual identities and inequalities are provable classically iff they are provable intuitionistically, for example the triangle inequality on $\mathbb{Q}$:

$$\forall r, s \in \mathbb{Q}(|r + s| \leq |r| + |s|).$$

The reader who worries about the introduction of equivalence classes in the traditional definitions of $\mathbb{Z}$ and $\mathbb{Q}$ should take into account that the equivalence relations under consideration are decidable and that hence the forming of equivalence classes is unproblematic. One can also avoid the introduction of equivalence classes by sticking to the original objects (pairs of natural numbers in the case of $\mathbb{Z}$) and letting the defined equivalence relation play the role of equality; this is the solution chosen by Bishop (1967) for example.

The coding of $\mathbb{Z}$ and $\mathbb{Q}$ in $\mathbb{N}$, as indicated above, avoids these problems altogether, at the expense of a more involved explicit definition of the arithmetical operations on $\mathbb{Z}$ and $\mathbb{Q}$ via the codes.

For all practical purposes we will stick to the traditional conception of integers and rationals as equivalence classes.

1.2. *Introducing real numbers.* In classical mathematics there are several methods for introducing the reals. The two principal ones are:

(a) Real numbers as equivalence classes of fundamental sequences i.e. Cauchy sequences of rationals; this method goes back to G. Cantor. The method of nested intervals is closely related; the filter completion method is more general, but in the same spirit.

(b) The method of Dedekind cuts in $\mathbb{Q}$. Here reals appear as cuts, i.e. subsets of $\mathbb{Q}$ with special properties. The axiomatic variant of this method does not define reals as cuts, but instead postulates closure of the number system under the least-upper-bound principle.

Method (a) generalizes to the well-known procedure for constructing metric completions of metric spaces, method (b) to the construction of order-completions of lattices. In the context of constructive mathematics, both (a) and (b) can be utilized. (a) can almost literally be copied from classical treatments, in the case of (b) some extra attention has to be paid to the formulation.

So far we have not mentioned the infinite decimal fractions, a very special case of method (a). The reason is that this method is too narrow. Already Brouwer (1921) observed that not all Cauchy reals have decimal expansions. Even worse, the set of reals representable by infinite decimal expansions is not closed under addition and multiplication (cf. E5.3.10).

Most of the introductory accounts of intuitionism, BCM and CRM use method (a). However, there are good reasons to study *both* methods, as we shall see later.

2. Cauchy reals and their ordering

The introduction of the set of reals $\mathbb{R}$ via Cauchy sequences of rationals (fundamental sequences) is quite similar to the classical procedure, once we have chosen our initial definitions appropriately.

2.1. CONVENTION. Throughout this chapter we shall reserve n, m, i, j, k for natural numbers, r, s, t (possibly with sub- or superscripts) for rationals, α, β, γ for (infinite) sequences of natural numbers, and x, y, z for reals.

For sequences $t_0, t_1, t_2, \ldots$ (i.e. functions on $\mathbb{N}$) we shall use the "mathematical" notation $\langle t_n \rangle_n$ instead of the more "logical" $\lambda n.t_n$. □

2.2. DEFINITION. A *fundamental sequence* is a sequence $\langle r_n \rangle_n$ of rationals, together with a sequence β, its *(Cauchy-)modulus*, such that

$$\forall k m m' \left(|r_{\beta k+m} - r_{\beta k+m'}| < 2^{-k} \right).$$

Two fundamental sequences $\langle r_n \rangle_n, \langle s_n \rangle_n$ are said to *coincide* (notation $\approx$) if

$$\forall k \exists n \forall m \left(|r_{n+m} - s_{n+m}| < 2^{-k} \right).$$

The set of *Cauchy reals* (or reals for short) $\mathbb{R}$ consists of the equivalence classes of fundamental sequences relative to $\approx$. Observe that, as in the classical theory, $\approx$ is indeed an equivalence relation. For the equivalence class of $\langle r_n \rangle_n$ we use the standard notation $\langle r_n \rangle_n / \approx$. □

REMARKS. From this definition it will be clear that the extent of $\mathbb{R}$ depends on the extent of $\mathbb{N}^{\mathbb{N}}$, the set of sequences of natural numbers. For a smooth theory, $\mathbb{N}^{\mathbb{N}}$ should obviously satisfy some closure conditions. A very weak requirement, sufficient for the developments in the first few sections (2–4) of the theory of Cauchy reals, is AC-NN! (also denoted by $\mathrm{AC}_{00}!$)

AC-NN! $\quad \forall n \exists ! m A(n, m) \to \exists \alpha \forall n A(n, \alpha n)$.

In fact, even less suffices at most places. We shall return to this issue in 4.6–7.

Again the introduction of equivalence classes is not essential, cf. Bishop (1967), but it is certainly more convenient and closer to mathematical practice.

If we can assume that our sequences satisfy the countable axiom of choice (cf. 4.2.1) AC_{00}, we can simplify our definition of fundamental sequence to the more usual

$$\forall k \exists n \forall m m' \left(|r_{n+m} - r_{n+m'}| < 2^{-k} \right). \tag{1}$$

In general, the whole theory simplifies considerably if we adopt AC_{00}. However, in this chapter we shall not assume it unless explicitly indicated. This will have the advantage of making the reader more aware of the role of AC_{00} in the development of the theory, and will permit us to apply the theory of this chapter in contexts where AC_{00} cannot be assumed (cf. chapter 15).

At first sight one might expect that $\approx$ also ought to have been defined in the explicit manner: $\langle r_n \rangle_n \approx \langle s_n \rangle_n := \exists \gamma \forall k \forall m (|r_{\gamma k+m} - s_{\gamma k+m}| < 2^{-k})$. However, we get γ for free, as may be seen from the following

2.3. PROPOSITION. *Let $\langle r_n \rangle_n, \langle s_n \rangle_n$ be fundamental sequences with Cauchy moduli α, β respectively, and assume $\langle r_n \rangle_n \approx \langle s_n \rangle_n$. Then*

$$\forall km \left(|r_{\gamma k+m} - s_{\gamma k+m}| < 2^{-k} \right)$$

for $\gamma k = \max\{\alpha(k+2), \beta(k+2)\}$.

PROOF. Let $\forall m (|r_{n+m} - s_{n+m}| < 2^{-k-1})$. Take an arbitrary $m \in \mathbb{N}$. If $\gamma k \geq n$, we are done; so let $\gamma k < n$, then $\gamma k + m' = n$ for some m', and $|r_{\gamma k+m} - s_{\gamma k+m}| \leq |r_{\gamma k+m} - r_{\gamma k+m'}| + |r_{\gamma k+m'} - s_{\gamma k+m'}| + |s_{\gamma k+m} - s_{\gamma k+m'}| < 2^{-k-2} + 2^{-k-1} + 2^{-k-2} = 2^{-k}$. □

2.4. Definition (of the ordering relation on fundamental sequences). Let $\langle r_n \rangle_n, \langle s_n \rangle_n$ be fundamental sequences, and put

$$\langle r_n \rangle_n < \langle s_n \rangle_n := \exists kn \forall m \big(s_{n+m} - r_{n+m} > 2^{-k}\big). \quad \square$$

2.5. Proposition. *$<$ between fundamental sequences is compatible with $\approx$, i.e.*

$$\langle r_n \rangle_n < \langle s_n \rangle_n \wedge \langle r_n \rangle_n \approx \langle r'_n \rangle_n \wedge \langle s_n \rangle_n \approx \langle s'_n \rangle_n \to \langle r'_n \rangle_n < \langle s'_n \rangle_n.$$

Proof. From the definition it follows that there are k and n such that

$$\forall m\big(s_{n+m} - r_{n+m} > 2^{-k}\big), \quad \forall m\big(|r_{n+m} - r'_{n+m}| < 2^{-k-2}\big),$$

$$\forall m\big(|s_{n+m} - s'_{n+m}| < 2^{-k-2}\big); \quad \text{hence } \forall m\big(s'_{n+m} - r'_{n+m} > 2^{-k-1}\big). \quad \square$$

2.6. Definition. For $r \in \mathbb{Q}$ define r^* as the fundamental sequence $\langle r_n \rangle_n$ with $r_n = r$ for all n. $\square$

Remark. $r \mapsto r^*$ is an order preserving embedding from $\mathbb{Q}$ into $\mathbb{R}$, i.e. $r < s \leftrightarrow r^* < s^*$.

2.7. Definition of $<, \leq, \#$ on $\mathbb{R}$.

$$x < y := \exists \langle r_n \rangle_n \in x\, \exists \langle s_n \rangle_n \in y (\langle r_n \rangle_n < \langle s_n \rangle_n);$$

$$x \leq y := \neg(y < x);$$

$$x \# y := (x < y) \vee (y < x).$$

As usual, $>, \geq$ are taken to be the relations converse to $<, \leq$. $\#$ is called *apartness*; $x \# y$ is read as "x is apart from y". $\square$

Some authors, such as Bishop (1967), write $\neq$ for $\#$; however, since "$\neg x = y$" is also used fairly frequently in constructive mathematics, we have reserved $\neq$ for inequality.

Warning: $x \leq y$ is not used for $x < y \vee x = y$! Usually $x \leq y$ is pronounced as "x is not greater than y" which is rather awkward. Perhaps one could use "x is *down* y".

2.8. Proposition. *Let $\langle r_n \rangle_n \in x, \langle s_n \rangle_n \in y$, $x, y \in \mathbb{R}$; then*

$$x \# y \leftrightarrow \exists kn \forall m \big(|s_{n+m} - r_{n+m}| > 2^{-k}\big).$$

Proof. Exercise. $\square$

Apartness is a positive version of inequality which plays an important role in constructive mathematics. Its basic properties are given by the following

2.9. PROPOSITION

AP1 $\neg x \# y \leftrightarrow x = y$,

AP2 $x \# y \leftrightarrow y \# x$,

AP3 $x \# y \rightarrow \forall z(x \# z \vee z \# y)$.

PROOF. AP1: $x = y \rightarrow \neg(x \# y)$ trivially follows from the definition of $\#$, so it remains to show $\neg(x \# y) \rightarrow x = y$. Let $\neg(x \# y), \langle s_n \rangle_n \in x$, $\langle t_n \rangle_n \in y$, then

$$\forall kn \neg \forall m\big(|s_{n+m} - t_{n+m}| > 2^{-k}\big). \tag{1}$$

Also, if α, β are the moduli of $\langle s_n \rangle_n, \langle t_n \rangle_n$ respectively, we have

$$\forall knm\big(|s_{\alpha k+n} - s_{\alpha k+m}| < 2^{-k}\big), \tag{2}$$

$$\forall knm\big(|t_{\beta k+n} - t_{\beta k+m}| < 2^{-k}\big). \tag{3}$$

Take $k + 2$ for k and $\max\{\alpha(k+2), \beta(k+2)\}$ for n in (1), then

$$\neg \forall m\big(|s_{n+m} - t_{n+m}| > 2^{-k-2}\big). \tag{4}$$

Now suppose that for some k and m $|s_{n+m} - t_{n+m}| > 2^{-k}$, then

$$|s_{n+m'} - t_{n+m'}| \geq |s_{n+m} - t_{n+m}| - |s_{n+m'} - s_{n+m}| - |t_{n+m'} - t_{n+m}|$$
$$\geq 2^{-k} - 2 \cdot 2^{-k-2} = 2^{-k-1}$$

for all m'; this contradicts (4), hence $\forall mk(|s_{n+m} - t_{n+m}| \leq 2^{-k})$. Hence we conclude that $x = y$.

AP2 is trivially fulfilled; we leave the proof of AP3 as an exercise. □

2.10. COROLLARY. *Equality between real numbers is stable, i.e.*

$$\forall xy(\neg\neg x = y \rightarrow x = y).$$

PROOF. By AP1 and propositional logic $x = y \leftrightarrow \neg(x \# y) \leftrightarrow \neg\neg\neg(x \# y) \leftrightarrow \neg\neg(x = y)$. □

2.11. PROPOSITION. *For all reals* x, y, z

(i) $x \leq y \wedge y \leq x \rightarrow x = y$,

(ii) $x < y \wedge y < z \rightarrow x < z$,

(iii) $x < y \to x < z \lor z < y$,
(iv) $x < y \land y \le z \to x < z$,
(v) $x \le y \land y < z \to x < z$,
(vi) $x \le y \land y \le z \to x \le z$,
(vii) $x \le y \leftrightarrow \neg\neg(x < y \lor x = y)$,
(viii) $\neg\neg(x \le y \lor y \le x)$,
(ix) $\neg\neg(x < y \lor x = y \lor y < x)$.

PROOF. (i) $(x \le y \land y \le x) \to \neg(x > y) \land \neg(y > x) \to \neg(x > y \lor y > x) \to \neg x \# y \to x = y$ (E2.1.1, AP1).

(ii) Easy from the definition.

(iii) $x < y \to x \# y$, hence $x \# z \lor z \# y$, so $(x < z \lor z < x) \lor (y < z \lor z < y)$. By (ii) this implies $(x < z \lor z < y) \lor (x < z \lor z < y)$.

(iv) If $x < y$, then by (iii) $x < z \lor z < y$; $z < y$ contradicts $y \le z$, hence $x < y \land y \le z \to x < z$.

(v) Similarly.

(vi) Let $x \le y$, $y \le z$; assume $z < x$, then $z < y$ from $x \le y$ and (iv); this contradicts $y \le z$, hence $\neg z < x$, i.e. $x \le z$.

(vii)–(ix) are left as an exercise. □

2.12. *Counterexamples.* We recall the discussion in 1.3.5–6, and in particular the definition of $x^A, \langle r_n^A \rangle_n$, or equivalently $x^\alpha, \langle r_n^\alpha \rangle_n$ for α the characteristic function of A. Clearly

$$x^\alpha \# 0 \leftrightarrow \exists k(\alpha k \ne 0), \quad x^\alpha = 0 \leftrightarrow \forall k(\alpha k = 0), \text{ so}$$

$$\exists k(\alpha k \ne 0) \lor \neg\exists k(\alpha k \ne 0) \to x^\alpha \# 0 \lor x^\alpha = 0,$$

and thus $\forall k(\alpha k = 0) \lor \neg\forall k(\alpha k = 0) \to x^\alpha = 0 \lor x^\alpha \ne 0$. Hence

$$\neg\exists\text{-PEM} \to \neg\forall x(x \# 0 \lor x = 0),$$

$$\neg\forall\text{-PEM} \to \neg\forall x(x \ne 0 \lor x = 0).$$

It follows therefore from either CT_0 or WC-N (4.3.4, 4.6.4) that

$$\neg\forall x(x \ne 0 \lor x = 0).$$

In particular, since $x \# 0 \leftrightarrow x < 0 \lor x > 0$, we also see that

$$\neg\forall x(x < 0 \lor x > 0 \lor x = 0),$$

i.e. the law of trichotomy is refutable:

$$\neg\forall xy(x < y \lor x > y \lor x = y).$$

A more refined construction leads to the refutation of $\forall xy(x \le y \vee y \le x)$. To see this, let α, β be two sequences such that $\neg(\exists k(\alpha k = 0) \wedge \exists k(\beta k = 0))$. We define $x^{\alpha,\beta}$ via a fundamental sequence $r_n^{\alpha,\beta}$ as follows

$$r_n^{\alpha,\beta} := \begin{cases} 0 & \text{if } \neg\exists k \le n(\alpha k = 0 \vee \beta k = 0), \\ 2^{-k} & \text{if } \alpha k = 0 \wedge k \le n \wedge \forall m < k(\alpha m \ne 0), \\ -2^{-k} & \text{if } \beta k = 0 \wedge k \le n \wedge \forall m < k(\beta m \ne 0). \end{cases}$$

Then
(i) $\langle r_n^{\alpha,\beta}\rangle_n$ is a fundamental sequence with modulus $\lambda n.n+1$,
(ii) $x^{\alpha,\beta} > 0 \leftrightarrow \exists m(\alpha m = 0)$,
(iii) $x^{\alpha,\beta} < 0 \leftrightarrow \exists m(\beta m = 0)$,
and therefore

$$x^{\alpha,\beta} \le 0 \vee x^{\alpha,\beta} \ge 0 \leftrightarrow \neg\exists m(\alpha m = 0) \vee \neg\exists m(\beta m = 0).$$

If we can find particular α, β such that $\neg(\exists k(\alpha k = 0) \wedge \exists k(\beta k = 0))$, while $\neg\exists m(\alpha m = 0) \vee \neg\exists m(\beta m = 0)$ is unknown, we have a weak counterexample to $x \le 0 \vee x \ge 0$.

For example, let $Bn := \exists m_1 m_2(2n + 2 = m_1 + m_2 \wedge m_1 \ne 2 \ne m_2 \wedge m_1 \text{ prime} \wedge m_2 \text{ prime})$; so $\forall n Bn$ expresses Goldbach's conjecture. Let $A(n)$ be as before in 1.3.5 and fix α, β by

$$\alpha m = 0 \leftrightarrow \neg Am \wedge \forall k \le m Bk,$$

$$\beta m = 0 \leftrightarrow \neg Bm \wedge \forall k \le m Ak.$$

Returning to the generalized form, recalling that "SEP" was the statement $\forall\alpha\beta[\neg(\exists k(\alpha k = 0) \wedge \exists k(\beta k = 0)) \to \neg\exists k(\alpha k = 0) \vee \neg\exists k(\beta k = 0)]$, we see that

$$\neg\text{SEP} \to \neg\forall\alpha\beta(x^{\alpha,\beta} \le 0 \vee x^{\alpha,\beta} \ge 0)$$

hence

$$\neg\text{SEP} \to \neg\forall xy(x \le y \vee y \le x),$$

and in particular, assuming either CT_0 or WC-N (4.3.6, 4.6.5), we find $\forall xy(x \le y \vee y \le x)$ to be false.

2.13. Definition. Let $\langle r_n\rangle_n \in x \in \mathbb{R}$, $\langle s_n\rangle_n \in y \in \mathbb{R}$; we put

$$x + y := \langle r_n + s_n\rangle_n/\approx, \quad x - y := \langle r_n - s_n\rangle_n/\approx, \quad |x| := \langle |r_n|\rangle_n/\approx.$$

□

Of course, we have to verify that this is a correct definition, that is to say we have to show that $\langle |r_n| \rangle_n$, $\langle r_n + s_n \rangle_n$ and $\langle r_n - s_n \rangle_n$ are again fundamental sequences, and that

$$\langle r_n \rangle_n \approx \langle r'_n \rangle_n \wedge \langle s_n \rangle_n \approx \langle s'_n \rangle_n \to \langle |r_n| \rangle_n \approx \langle |r'_n| \rangle_n \wedge$$
$$\langle r_n + s_n \rangle_n \approx \langle r'_n + s'_n \rangle_n \wedge \langle r_n - s_n \rangle \approx \langle r'_n - s'_n \rangle_n,$$

which is straightforward, and left to the reader.

It is virtually immediate that for all $r \in \mathbb{Q}$, $x \in \mathbb{R}$

$$r > 0 \to x < x + r.$$

The following proposition gives a convenient "positive" characterization of $\le$:

2.14. PROPOSITION. *For* $x, y \in \mathbb{R}$

$$x \le y \leftrightarrow \forall k(x < y + 2^{-k}).$$

PROOF. If $x \le y$, we have $x < y + 2^{-k}$ for all k (by $y < y + 2^{-k}$ and 2.11(v)). Conversely, assume $\forall k(x < y + 2^{-k})$, and let $x > y$, then for some m, $x > y + 2^{-m}$ (why?). Since also $x < y + 2^{-m}$, a contradiction follows. So $x \le y$. □

2.15. PROPOSITION. *Let* $\langle r_n \rangle_n \in x \in \mathbb{R}$, $y, z \in \mathbb{R}$ *and* $r \in \mathbb{Q}$.

(i) *If* $\forall m(|r_n - r_{n+m}| \le 2^{-k})$, *then* $|x - r_n| \le 2^{-k}$.

(ii) $|x - y| \le r$ *iff* $x - r \le y \le x + r$.

(iii) *If* $|x - y| \le r$ *and* $|y - z| \le r'$ *then* $|x - z| \le r + r'$.

PROOF. (i) is left to the reader.

(ii) Suppose $\langle s_n \rangle_n \in y$, then

$$|x - y| \le r \leftrightarrow \forall n(|x - y| < r + 2^{-n}) \leftrightarrow$$
$$\forall i \exists k n \forall m(r + 2^{-i} - |r_{n+m} - s_{n+m}| > 2^{-k}) \leftrightarrow$$
$$\forall i \exists k n \forall m(r_{n+m} - r - 2^{-i} + 2^{-k} < s_{n+m} < r_{n+m} + r + 2^{-i} - 2^{-k}),$$

which is equivalent to

$$\forall i(x - r - 2^{-i} < y < x + r + 2^{-i});$$

this is equivalent to $x - r \le y \le x + r$ by 2.14.

(iii) is left to the reader. □

REMARK. This proposition gives just a bare minimum of properties of $\lambda x, y.|x - y|$ which suffices for the time being; the next section presents a quite general and uniform treatment of "arithmetical" operations on $\mathbb{R}$ (such as $+, \cdot$, max, min, $\lambda x, y.|x - y|$ etc.).

2.16. *Markov's principle.* Obviously, the theory of real numbers could be considerably simplified if we would know that

$$\forall x(x \neq 0 \rightarrow x \# 0). \tag{1}$$

In traditional intuitionism (1) is rejected, and accepted in constructive recursive mathematics. Markov's principle is equivalent to (1), as was shown in 4.5.4.

3. Arithmetic on $\mathbb{R}$

3.1. DEFINITION. Let $\vec{x}, \vec{y} \in \mathbb{R}^n$ or $\vec{x}, \vec{y} \in \mathbb{Q}^n$, $r \in \mathbb{Q}$, and let $\vec{0}$ abbreviate the element $(0, 0, \ldots, 0)$ in $\mathbb{R}^n$. Then

$$|\vec{x} - \vec{y}| \leq r := \forall i(1 \leq i \leq n \rightarrow |x_i - y_i| \leq r),$$

$$|\vec{x}| \leq r := |\vec{x} - \vec{0}| \leq r := \forall i(1 \leq i \leq n \rightarrow |x_i| \leq r). \quad \Box$$

3.2. DEFINITION. Let $f \in X \rightarrow Z$, where X and Z are $\mathbb{Q}^n$ or $\mathbb{R}^n$. We put

$$M_f(m, i, k) := \forall \vec{x}, \vec{y} \in X^n(|\vec{x}| \leq m \wedge |\vec{y}| \leq m \wedge$$
$$|\vec{x} - \vec{y}| \leq 2^{-i} \rightarrow |f(\vec{x}) - f(\vec{y})| \leq 2^{-k}),$$

$$M_f(i, k) := \forall m M_f(m, i, k).$$

f is *uniformly continuous with modulus* α iff $\forall k M_f(\alpha k, k)$; f is *locally uniformly continuous with modulus* α if $\forall mk M_f(m, \alpha(m, k), k)$. $\Box$

On assumption of AC_{00}, these notions may be equivalently defined respectively by: f is *uniformly continuous* iff $\forall k \exists i M_f(i, k)$, and f is *locally uniformly continuous* iff $\forall mk \exists i M_f(m, i, k)$.

3.3. REMARK. The most obvious definition of a function from $\mathbb{R}$ to $\mathbb{R}$ is via graph predicates: a function from $\mathbb{R}$ to $\mathbb{R}$ is given by a relation $R \subset \mathbb{R} \times \mathbb{R}$ such that

$$\forall x \in \mathbb{R} \exists ! y \in \mathbb{R} R(x, y);$$

the function f described by R satisfies $\forall x \in \mathbb{R}\, R(x, fx)$. For uniformly continuous f, f is already determined by its restriction $f_Q := f \upharpoonright \mathbb{Q}$. On the assumption of

$$\mathrm{AC}_{01} \quad \forall x \exists \alpha A(x, \alpha) \rightarrow \exists \beta \forall x A(x, (\beta)_x),$$

where $(\beta)_x := \lambda y.\beta(x, y)$, f_Q is determined by some $\alpha \in \mathbb{N} \rightarrow \mathbb{N}$ such that

$$\forall n(\langle \bar{r}_{\alpha(n,m)} \rangle_m \approx f(\bar{r}_n)),$$

where $\langle \bar{r}_n \rangle_n$ is a standard enumeration of $\mathbb{Q}$ without repetitions. Presumably AC_{00} or $\mathrm{AC}_{01}!$ (with $\exists! \alpha$ in the premiss) are not sufficient for this conclusion.

3.4. Definition. Let $f \in \mathbb{Q}^n \rightarrow \mathbb{Q}$, f locally uniformly continuous. Then the canonical extension $f^* \in \mathbb{R}^n \rightarrow \mathbb{R}$ is defined by

$$f^*(\vec{x}) := (\langle f(\vec{r}_n) \rangle_n)_{\approx}$$

for any $\langle \vec{r}_n \rangle_n \in \vec{x}$. □

We leave it as an exercise to show that f^* is well-defined.

3.5. Theorem. *Let f with a modulus be as in* 3.4, *then f^* is the unique locally uniformly continuous function with modulus extending f.*

Proof. For notational simplicity we shall assume f to be unary, and uniformly continuous (instead of only locally uniformly continuous); let α be a modulus for f. We first prove that f^* is uniformly continuous with modulus α. Let $\langle r_n \rangle_n \in x$, $\langle s_n \rangle_n \in y$, $x - 2^{-\alpha k} < y < x + 2^{-\alpha k}$, and assume

$$f^*(x) > f^*(y) + 2^{-k}. \tag{1}$$

From (1) it follows that there is an n such that

$$\forall m(f(r_{n+m}) - f(s_{n+m}) - 2^{-k} > 0).$$

Since $|r_{n+m} - s_{n+m}| \leq 2^{-\alpha k} \rightarrow |f(r_{n+m}) - f(s_{n+m})| \leq 2^{-k}$, it follows that

$$\forall m(|r_{n+m} - s_{n+m}| > 2^{-\alpha k}),$$

and therefore (cf. 2.8) $x < y \vee y > x$, i.e. we can now derive that for some

$n' \geq n$

$$\forall m\left(r_{n'+m} - s_{n'+m} > 2^{-\alpha k}\right) \vee \forall m\left(s_{n'+m} - r_{n'+m} > 2^{-\alpha k}\right).$$

Thus we find $(x - 2^{-\alpha k} \geq y) \vee (y - 2^{-\alpha k} \geq x)$, which conflicts with $x - 2^{-\alpha k} < y$, $y - 2^{-\alpha k} < x$; thus (1) is false and therefore $f^*(x) \leq f^*(y) + 2^{-k}$, and similarly $f^*(y) - 2^{-k} \leq f^*(x)$.

As to the uniqueness, let f° be any other uniformly continuous extension of f, say with modulus β. If $\langle r_n \rangle_n \in x$, and for some n

$$\forall mm'\left(|r_{n+m} - r_{n+m'}| < 2^{-\gamma k}\right),$$

where $\gamma k = \max(\alpha k + 1, \beta k)$, it follows that $|x - r_n| \leq 2^{-\gamma k}$, so $|x - r_n| < 2^{-\alpha k}$, $|x - r_n| \leq 2^{-\beta k}$, and therefore

$$|f(r_n) - f^*(x)| \leq 2^{-k}, \quad |f(r_n) - f^\circ(x)| \leq 2^{-k}.$$

Thus $|f^*(x) - f^\circ(x)| \leq 2^{-k+1}$. Since this holds for all k, we find $f^*(x) = f^\circ(x)$. □

COROLLARY. *The operations* $+, \cdot, |\ |$ (*absolute value*), max, min, $\lambda x, y. |x - y|$ *can be extended from* $\mathbb{Q}$ *to* $\mathbb{R}$. □

3.6. PROPOSITION. *Let* $\varphi(x_1, \ldots, x_n) = 0$ *express any algebraic* (*polynomial*) *identity in* $+, \cdot, |\ |$, max, min, *etc. which holds for all* $(x_1, \ldots, x_n) \in \mathbb{Q}^n$; *then the same identity holds on* $\mathbb{R}^n$.

PROOF. Consider, for example, the law of distributivity, which may be expressed as

$$x(y + z) - (xy + xz) = 0. \tag{1}$$

Noting the fact that locally uniformly continuous functions are closed under composition, we see that the left hand side and the right hand side of (1) may be viewed as two distinct ternary functions coinciding over $\mathbb{Q}$; hence their extensions to $\mathbb{R}$ coincide also; thus (1) also holds over $\mathbb{R}$. □

REMARKS. (i) The practical importance of this proposition is that we can transfer identities from $\mathbb{Q}$, which are usually much easier to prove because of the decidability of $=$ on $\mathbb{Q}$, to $\mathbb{R}$.

(ii) We cannot, in general, extend 3.6 to propositional combinations of polynomial equations: it is not hard to show by means of a weak counterexample that we cannot show $xy = 0 \to x = 0 \vee y = 0$ in $\mathbb{R}$, though this is true over $\mathbb{Q}$ (exercise).

3.7. Proposition.
(i) $x > y \vee x = y \to \max(x, y) = x \wedge \min(x, y) = y$,
(ii) $x \leq y \leftrightarrow \max(x, y) = y \leftrightarrow \min(x, y) = x$,
(iii) $|x - y| = \max(x, y) - \min(x, y)$,
(iv) $x < y \leftrightarrow x + z < y + z$, $x \leq y \leftrightarrow x + z \leq y + z$,
(v) $|x - y| \leq z \leftrightarrow x - z \leq y \leq x + z$.

Proof. (i) is left to the reader.

(ii) $x \leq y \leftrightarrow \neg\neg(x < y \vee x = y)$ (2.11(vii)), hence with (i) we find $\neg\neg(\max(x, y) = y)$, and by the stability of $=$, $\max(x, y) = y$. Conversely, if $\max(x, y) = y$ and $x > y$, we have by (i) $\max(x, y) = x$, hence $x = y$ contradicting $x > y$; therefore $x \leq y$, etc.

(iii) is immediate by the corollary in 3.5.

(iv) we leave to the reader.

(v) This property generalizes 2.15(ii). Observe that

$$x \leq y \to |x - y| = y - x,$$

$$x \geq y \to |x - y| = x - y,$$

and since

$$x \leq y \to (y - x \leq z \leftrightarrow x - z \leq y \leq x + z),$$

$$y \leq x \to (x - y \leq z \leftrightarrow x - z \leq y \leq x + z),$$

we have

$$x \leq y \vee x \geq y \to (|x - y| \leq z \leftrightarrow x - z \leq y \leq x + z).$$

Now $\neg\neg(x \leq y \vee x \geq y)$ holds, therefore by propositional logic

$$\neg\neg(|x - y| \leq z) \leftrightarrow \neg\neg(x - z \leq y \wedge y \leq x + z),$$

from which (v) follows, using propositional logic and the stability of $\leq$. □

3.8. Corollary. *Let $\varphi(\vec{x})$, $\psi(\vec{x})$ be two polynomials (in $+, \cdot, |\ |$ etc.) such that $\varphi(\vec{x}) \leq \psi(\vec{x})$ holds on $\mathbb{Q}^n$; then this inequality also holds on $\mathbb{R}^n$.*

Proof. $\varphi(\vec{x}) \leq \psi(\vec{x})$ is equivalent to the polynomial identity

$$\max(\varphi(\vec{x}), \psi(\vec{x})) = \psi(\vec{x}). \quad \square$$

3.9. Proposition.
(i) $\min(x, y) \le x \le \max(x, y)$, $\min(x, y) \le y \le \max(x, y)$,
(ii) $|x + y| \le |x| + |y|$, $|x - y| \ge \big||x| - |y|\big|$.

Proof. Immediate by the preceding corollary. □

3.10. Proposition.
(i) $x \# y \to x + z \# y + z$,
(ii) $x + y \# 0 \to x \# 0 \vee y \# 0$,
(iii) $xy = 1 \to x \# 0 \wedge y \# 0$.

Proof. (i) Immediate, e.g. from 3.7(iv) and the definition of $\#$.

(ii) By (i), $x + y \# 0$ entails $x \# -y$, so by AP3 $x \# 0 \vee -y \# 0$, hence $x \# 0 \vee y \# 0$.

(iii) Let $\langle r_n \rangle_n \in x$, $\langle s_n \rangle_n \in y$, and let $xy = 1$. There is an i such that $\forall n(|r_n| < 2^i)$. Determine k such that $\forall m(|r_{k+m}s_{k+m} - 1| < 2^{-1})$. Assume that for some m' $|s_{k+m'}| \le 2^{-i-1}$, then $|r_{k+m'}s_{k+m'}| < 2^{-1}$ and hence $|r_{k+m'}s_{k+m'} - 1| > 2^{-1}$, contradiction. Therefore $\forall m(|s_{k+m}| > 2^{-i-1})$, i.e. $y \# 0$. Similarly $x \# 0$. □

3.11. Remarks. (i) Theorem 3.5 can also be applied to segments of $\mathbb{Q}$, as will be clear from the proof. In particular, for any segment $[r, r'] := \{s : r \le s \le r'\}$ such that $0 \notin [r, r']$ the inverse $\lambda x.x^{-1}$ is defined and locally uniformly continuous. Therefore an extension of $\lambda x.x^{-1}$ to $\mathbb{R}$ is uniquely defined for $\{x : x \# 0\}$ and satisfies $x \cdot x^{-1} = 1$ (by an argument as in 3.6). As follows from 3.10(iii), x 0 is a *necessary and sufficient* condition for the existence of an inverse x^{-1} such that $x \cdot x^{-1} = 1$, i.e. with an existence predicate $\boldsymbol{E}$

$$\boldsymbol{E}x^{-1} \leftrightarrow x \# 0.$$

Here we encounter our first natural example of a function which is only *partial* on $\mathbb{R}$. In classical mathematics, we can avoid the consideration of partial functions by an ad hoc device, in this case by stipulating

$$0^{-1} = 0.$$

Intuitionistically, this only makes x^{-1} defined on $\{x : x = 0 \vee x \# 0\}$, not on the whole of $\mathbb{R}$. We leave it as an exercise (E5.3.5) for the reader to show that $\lambda x.x^{-1}$ cannot be extended to a total function on $\mathbb{R}$.

(ii) For all fundamental sequences $\langle r_n \rangle_n$ we can define $(\langle r_n \rangle_n)^{-1} \equiv \langle s_n \rangle_n$ by

$$s_n := \begin{cases} 0 & \text{if } r_n = 0, \\ r_n^{-1} & \text{if } r_n \neq 0. \end{cases}$$

Then, if $\langle r_n \rangle_n \# 0$, it follows that $(\langle r_n \rangle_n)^{-1}$ is a fundamental sequence and $\langle r_n \rangle_n \cdot (\langle r_n \rangle_n)^{-1} := \langle r_n \cdot s_n \rangle_n \approx 1$.

(iii) More generally, 3.6 also applies to polynomials in functions f defined on *segments* $\{r : r_0 \leq r \leq r_1\}$ of $\mathbb{Q}$. This permits us to lift also arithmetical identities involving $^{-1}$ from $\mathbb{Q}$ to $\mathbb{R}$, whenever the expressions involved are defined, noting that $^{-1}$ is uniformly continuous in a neighbourhood of each point where it is defined.

Once we have introduced $\mathbb{R}$ with its ordering and the principal arithmetical operations, we can define many other mathematical notions in the usual way, such as metric space, limit point etcetera. We give some examples below.

3.12. DEFINITION. A *metric space* is a pair (V, ρ), V a set, ρ a mapping from $V \times V$ to $\mathbb{R}$, such that for all $x, y, z \in V$

(i) $\rho(x, y) = 0 \leftrightarrow x = y$,

(ii) $\rho(x, y) = \rho(y, x)$,

(iii) $\rho(x, y) \leq \rho(x, z) + \rho(z, y)$.

The elements of V are called the *points* of the space. □

3.13. DEFINITION. If (V, ρ) is a metric space, $x \in V$, $a \in \mathbb{R}$, then the *open ball* of *radius* a with *centre* x is

$$U(a, x) := \{y : \rho(x, y) < a\}.$$

Let $W \subset V$. The *closure* W^- of W is

$$W^- := \{y : \forall r > 0 \exists z \in W(z \in U(r, y)).$$

$x \in V$ is an *accumulation point* of W iff

$$x \in (W \setminus \{x\})^-.$$

Metric spaces can be given a topology with basis $\{U(r, x) : r > 0 \wedge x \in V\}$. If we talk in the future about the *topology of a metric space* without further explanation, we mean the topology defined above. □

In agreement with the definitions in 4.1.5 we have: W is *open* iff $\mathrm{Int}(W) = W$, and the *interior* of W may be defined for metric spaces as

$$\mathrm{Int}(W) := \{y : \exists r > 0(U(r, y) \subset W)\}.$$

A mapping f from a metric space (V, ρ) to a metric space (V', ρ') is *uniformly continuous* iff

$$\forall x > 0 \exists y > 0 \forall u, v \in V(\rho(u, v) < y \to \rho'(fu, fv) < x).$$

3.14. Definition. Let (V, ρ) be a metric space, $\langle x_n \rangle_n \subset V$, $x \in V$.

$\langle x_n \rangle_n$ is said to *converge to* x (relative to the metric ρ) with *modulus* γ (the *modulus of convergence*) iff

$$\forall kn\big(\rho(x, x_{\gamma k+n}) < 2^{-k}\big).$$

Then x is said to be the *limit* of $\langle x_n \rangle_n$, notation $x = \lim \langle x_n \rangle_n$. □

3.15. Examples. $\mathbb{R}$ itself is a metric space with metric given by

$$\rho(x, y) := |x - y|.$$

Similarly, $\mathbb{R}^n$ can be metrized by

$$\rho(\vec{x}, \vec{y}) := \max_{1 \le i \le n} |x_i - y_i|.$$

In $\mathbb{R}$, x is the limit of $\langle x_n \rangle_n$ iff for some γ

$$\forall kn\big(|x - x_{\gamma k+n}| < 2^{-k}\big).$$

If we take all subsets of X as the open sets of X, X is said to be provided with the *discrete topology*. Discrete topological spaces are usually not metrizable, but the metric induced on $\mathbb{N}$ as a subset of $\mathbb{R}$ yields the discrete topology for $\mathbb{N}$. □

4. Completeness properties and relativization

In this section we shall show that $\mathbb{R}$ is Cauchy-complete (metrically complete), but not order complete; we also discuss the relativization of the theory to suitable universes $\mathcal{U}$ of sequences.

4.1. Definition. $\langle x_n \rangle_n$ is said to be a *Cauchy sequence* of reals with *modulus* (α, β), $\alpha \in \mathbb{N}^2 \to \mathbb{N}$, $\beta \in \mathbb{N} \to \mathbb{N}$ if each x_n is given by a fundamental sequence $\langle r_{m,n}\rangle_m$ with modulus $(\alpha)_n = \lambda m.\alpha(n, m)$, and where

$$\forall kmm'\big(|x_{\beta k+m} - x_{\beta k+m'}| < 2^{-k}\big). \quad \square$$

Remark. On assumption of AC_{00} we can simply define $\langle x_n\rangle_n$ to be a Cauchy sequence of reals iff

$$\forall k\exists n\forall mm'\big(|x_{n+m} - x_{n+m'}| < 2^{-k}\big).$$

4.2. Theorem (*Cauchy completeness of* $\mathbb{R}$). *Each Cauchy sequence of reals* $\langle x_n\rangle_n$ *converges to a limit* x.

Proof. Let (α, β) be a modulus for $\langle x_n\rangle_n$, and let γ be given by

$$\gamma 0 = \beta 0, \quad \gamma(n+1) = \max(\beta(n+1), \gamma n) + 1;$$

then (α, γ) is also a modulus for $\langle x_n\rangle_n$. We put $s_n := r_{\alpha(\gamma n, n), \gamma n}$. Now $|x_{\gamma n} - s_n| < 2^{-n}$, and thus, since $\gamma(n+m)$ and $\gamma(n+m')$ are not smaller than γn,

$$\begin{aligned} &|s_{n+m} - s_{n+m'}| \\ &\quad \le |s_{n+m} - x_{\gamma(n+m)}| + |s_{n+m'} - x_{\gamma(n+m')}| + |x_{\gamma(n+m)} - x_{\gamma(n+m')}| \\ &\quad < 2^{-n} + 2^{-n} + 2^{-n} < 2^{-n+2}. \end{aligned}$$

Therefore $\langle s_n\rangle_n$ represents a real x, and $\langle x_n\rangle_n$ converges to x with a modulus $\lambda n.\gamma(n+3)$:

$$\begin{aligned} &|x - x_{\gamma(n+3)+m}| \\ &\quad \le |x - s_{n+3}| + |s_{n+3} - x_{\gamma(n+3)}| + |x_{\gamma(n+3)} - x_{\gamma(n+3)+m}| \\ &\quad < 2^{-n-1} + 2^{-n-3} + 2^{-n-3} < 2^{-n}. \quad \square \end{aligned}$$

N.B. If we adopt the simpler definition of Cauchy sequence (4.1, remark), the above argument itself requires AC_{00}.

4.3. Definition. $\langle x_n\rangle_n$ is *bounded*, if for some $m \in \mathbb{N}$ $\forall n(|x_n| < m)$. $\langle x_n\rangle_n$ is *monotone, non-decreasing* if $\forall nm(n < m \to x_n \le x_m)$, and *monotone, non-increasing* if $\forall nm(n < m \to x_n \ge x_m)$. A sequence is *monotone* iff it is monotone, non-decreasing or monotone, non-increasing. Replacing $\le, \ge$ by $<, >$ respectively yields the notions of *monotone increasing*, *monotone decreasing* and *strictly monotone* sequence. $\square$

4.4. *Bounded monotone sequences and the Bolzano–Weierstrass theorem.* The Bolzano–Weierstrass theorem states that every bounded infinite set of reals has a point of accumulation; a well-known consequence is the fact that a bounded monotone sequence of reals has a limit. Constructively this is false, as follows from the following

PROPOSITION ($\neg\exists$-PEM). *It is false that every bounded monotone sequence in* $\mathbb{N} \to \mathbb{Q}$ *has a limit in* $\mathbb{R}$.

PROOF. Define $\langle r_n^\alpha \rangle_n$ by

$$r_n^\alpha = \begin{cases} 1 - 2^{-n} & \text{if } \neg\exists k \le n(\alpha k = 0), \\ 2 - 2^{-n} & \text{if } \exists k \le n(\alpha k = 0). \end{cases}$$

Assume $\lim \langle r_n^\alpha \rangle_n = x \in \mathbb{R}$, then $x < 2$ or $x > 1$, and thus $\neg\exists k(\alpha k = 0) \vee \exists k(\alpha k = 0)$ for all α; this contradicts $\neg\exists$-PEM. □

$\neg\exists$-PEM is a consequence of either CT_0 or WC-N (4.3.4, 4.6.4). However, CT_0 permits a much stronger sort of refutation which is typical for constructive recursive mathematics; namely we can show, assuming CT_0, the existence of a bounded monotone sequence without limit.

4.5. DEFINITION. A bounded monotone sequence $\langle r_n \rangle_n$ without a limit is called a *Specker sequence*. □

THEOREM (CT_0). *A Specker sequence exists.*

PROOF. Let f be any injection of $\mathbb{N}$ into $\mathbb{N}$ with non-recursive range (e.g. let f enumerate $\{x : \exists y Txxy\}$ without repetitions) and consider $\langle r_k \rangle_k$ defined by

$$r_k = \sum_{i=0}^{k} 2^{-f(i)}. \tag{1}$$

$\langle r_k \rangle_k$ is obviously bounded and monotone; now assume $\langle r_k \rangle_k$ to satisfy a Cauchy condition, then for any given m we can find k such that

$$\forall n(|r_k - r_{k+n}| < 2^{-m}). \tag{2}$$

m is in the range of f iff the sum of one of the r_k's ends with 2^{-m}. So determine k with $\forall n(|r_k - r_{k+n}| < 2^{-m})$, then we only have to compute the first k values of f in order to test if m belongs to the range of f. Hence the range of f is decidable, and therefore, by CT_0, recursive. Contradiction. □

Later, in 6.4.8 we shall give a proof of a much stronger statement, namely the existence of a strong Specker sequence, i.e. a bounded monotone sequence staying apart from each specific $x \in \mathbb{R}$.

Note that we have also shown, in 4.4, that $\mathbb{R}$ is not order-complete: not every bounded set of reals has a supremum (least upper bound). (Take the sets $\{r_n^\alpha : n \in \mathbb{N}\}$.)

4.6. *Relativizing to a universe.* The theory we developed so far is obviously correct if we take as our basis the most general notion of sequence of natural numbers, i.e. the variables $\alpha, \beta, \ldots$ range over all "legitimate" (constructively, intuitionistically acceptable) sequences. In this case, AC_{00} and AC_{01} seem to be intuitively justified, which simplifies the theory.

The same can be said, if we take as our basis the notion of lawlike sequence or choice sequence.

More generally, we may attempt to relativize the notion of real to some universe $\mathscr{U}$ of sequences of natural numbers; for a smooth theory of reals $\mathscr{U}$ will have to satisfy certain closure conditions.

More specifically, let $\langle \bar{r}_n \rangle_n$ be a fixed enumeration of $\mathbb{Q}$. Relative to this enumeration, we may consider $\mathscr{U}$-fundamental sequences $\langle \bar{r}_{\alpha n} \rangle_n$ with modulus β, where α and β belong to $\mathscr{U}$.

Inspection shows that the positive theory of $<, \leq$ is indeed applicable to $\mathscr{U}$-reals, since it is a theory of relations, not of operations. Closure under specific operations, however, such as $+, \cdot$ requires also certain closure conditions on $\mathscr{U}$; also, in our discussion of counterexamples and Markov's principle certain closure conditions on $\mathscr{U}$ are implicit, since we appeal to $AC_{00}!$.

At first sight it might seem as if the property: $x \leq y \leftrightarrow \forall k(x < y + 2^{-k})$ implicitly appeals to closure conditions on $\mathscr{U}$, since if y is a $\mathscr{U}$-real, $y + 2^{-k}$ need not be so unless $\mathscr{U}$ satisfies certain closure conditions. However, if we let $\mathscr{U}^*$ be the universe generated by $\mathscr{U}$ by closing off under certain suitable, explicitly given closure conditions, we can guarantee that if y is a $\mathscr{U}^*$-real, then so is $y + 2^{-k}$; and for $\mathscr{U}$-reals x, y the property $x \leq y \leftrightarrow \forall k(x < y + 2^{-k})$ is nothing else but a characterization of $\leq$ on the $\mathscr{U}$-reals via $<$ for $\mathscr{U}^*$-reals.

Note that if we choose *another* fixed enumeration of the rationals, say $\langle q_n \rangle_n$, with $q_n = \bar{r}_{\varphi n}$, we may get another notion of $\mathscr{U}$-real; requiring isomorphism w.r.t. $<$, on assumption that $\varphi \in \mathcal{U}$, again imposes certain closure conditions on $\mathcal{U}$.

4.7. *Axiomatic basis.* A detailed inspection shows that the theory of $\mathbb{R}$ we developed can be axiomatized in the language of **EL** (elementary analysis, cf. 3.6.2) with the following very weak axiom of choice

$$\text{QF-AC}_{00} \quad \forall n \exists m A(n, m) \rightarrow \exists \alpha \forall n A(n, \alpha n) \ (A \text{ quantifier-free}).$$

QF-AC_{00} as an assumption on $\mathcal{U}$ expresses nothing but the fact that $\mathcal{U}$ is closed under "recursive in".

5. Dedekind reals

In this and the following section we give a separate and independent treatment of Dedekind reals because (1) they illustrate the difference between the Cauchy-completion and the order completion of $\mathbb{Q}$, and (2) Dedekind reals play an important role in sheaf models for analysis in which AC_{00} fails (cf. chapter 15).

5.1. DEFINITION. Let $S \subset \mathbb{Q}$. S is called a *weak left cut* (weak cut, or w.l.c. for short), or (*bounded*) *extended real* if

(a) $\exists r(r \in S), \exists s(s \notin S)$ (*boundedness*),
(b) $\forall rs(r < s \wedge \neg\neg s \in S \rightarrow r \in S)$ (*strong monotonicity*),
(c) $\forall r \in S \exists r' \in S(r < r')$ (*openness*).

If (a) is weakened to

(a′) $\neg\neg\exists r(r \in S), \neg\neg\exists s(s \notin S)$ (*weak boundedness*),

we obtain the *weakly bounded cuts* or *classical reals*.

S is called a *left cut* (or cut) or *Dedekind real* (in this section simply called "real") if S satisfies in addition to (a), (b), (c)

(d) $\forall rs(r < s \rightarrow r \in S \vee s \notin S)$ (*locatedness*).

The collections of Dedekind reals, extended reals and classical reals we denote by $\mathbb{R}^{\text{d}}$, $\mathbb{R}^{\text{be}}$ and $\mathbb{R}^{\text{e}}$ respectively; obviously

$$\mathbb{R}^{\text{d}} \subset \mathbb{R}^{\text{be}} \subset \mathbb{R}^{\text{e}}.$$

Equality between weak cuts is simply equality between sets of rationals. □

REMARK. (d) implies (b): let $r < r'$, $\neg\neg r' \in S$; since $r \in S \vee r' \notin S$, we see that $r \in S$. Therefore Dedekind reals may also be defined by (a), (c), (d).

One often finds cuts defined as pairs of sets of rationals (cf. exercise 5.5.3), but using left cuts we find technically less cumbersome.

Dedekind reals can be approximated by rationals – this is the content of the following proposition. Classical reals can be approximated "up to double negation".

5.2. Proposition.
(i) $\forall n \forall S \in \mathbb{R}^{\mathrm{d}} \exists r \in S \exists r' \notin S(|r - r'| < 2^{-n})$,
(ii) $\forall n \forall S \in \mathbb{R}^{\mathrm{c}} \neg\neg \exists r \in S \exists r' \in S(|r - r'| < 2^{-n})$.

Proof. (i) Let $S \in \mathbb{R}^{\mathrm{d}}$, $r \in S$, $t \notin S$ and choose $r = r_0 < r_1 < r_2 \cdots < r_k = t$ such that $r_{i+1} - r_i < 2^{-n-1}$, for $i < k$. By the locatedness of S we have

$$\forall i < k(r_i \in S \vee r_{i+1} \notin S), \tag{1}$$

and thus there exists a finite sequence $m_0, \ldots, m_{k-1} \in \{0,1\}$ such that

$$m_i = 0 \to r_i \in S,$$

$$m_i = 1 \to r_{i+1} \notin S.$$

Let j be the least i such that $m_i \neq 0$, if existing, $j = k$ otherwise. If $j = k$, we have $r_{k-1} \in S$, while it is known that $r_k \notin S$, hence (i) holds. If $j = 0$, we have $r_0 \in S$, $r_1 \notin S$, and again (i) holds; finally, if $0 < j < k - 1$, we have $r_{j-1} \in S$, $r_{j+1} \notin S$, and $|r_{j+1} - r_{j-1}| < 2^{-n}$.

(ii) Let now $S \in \mathbb{R}^{\mathrm{c}}$. Then

$$\forall rr'(r < r' \to \neg\neg(r \in S \vee r' \notin S)). \tag{2}$$

For if $r < r'$, $r \notin S$, $\neg r' \notin S$ we obtain from $r < r' \wedge \neg\neg r' \in S$ with property (b) of (5.1) that $r \in S$, contradicting $r \notin S$; hence $\neg\neg(r \in S \vee r' \notin S)$. Now assume

$$r \in S, \quad t \notin S, \tag{3}$$

and choose $r_0, \ldots, r_k$ as under (i); the proof under (i) shows that from (1) and (3) we find

$$\exists r \in S \exists r' \notin S(|r - r'| < 2^{-n});$$

therefore, since (2) implies

$$\forall i < k \neg\neg(r_i \in S \vee r_{i+1} \notin S)$$

which is equivalent to

$$\neg\neg\forall i < k(r_i \in S \vee r_{i+1} \notin S)$$

(by repeated use of $\neg\neg(A \wedge B) \leftrightarrow (\neg\neg A \wedge \neg\neg B)$), we find

$$\neg\neg\exists r \in S \exists r' \notin S(|r - r'| < 2^{-n}).$$

This follows from (3), hence from $\exists r(r \in S)$, $\exists t(t \notin S)$, and therefore also from $\neg\neg\exists r(r \in S)$ and $\neg\neg\exists t(t \notin S)$, which hold for any $S \in \mathbb{R}^{\mathrm{e}}$. □

5.3. PROPOSITION. *Let* $S \subset \mathbb{Q}$. *Then* $S \in \mathbb{R}^{\mathrm{d}}$ *iff* S *is strongly monotone, open and satisfies* 5.2(i).

PROOF. The direction from left to right has already been proved. Conversely, assume 5.1(b), (c) and 5.2(i); we have to show 5.1(d). If $r < s$, determine $r' \in S$, $r'' \notin S$, $|r' - r''| < s - r$, then either $r' \leq r$, so $r'' < s$ and hence $s \notin S$ (5.1(b)), or $r' > r$, and then $r \in S$ (5.1(b)). □

5.4. DEFINITION.

$$S \leq T := S \subset T,$$

$$S + r := \{r + r' : r' \in S\}$$

$$S < T := \exists r > 0(S + r \leq T). \quad \square$$

5.5. PROPOSITION. *For* $S, T \in \mathbb{R}^{\mathrm{d}}$

$$S < T \leftrightarrow \exists s(s \in T \wedge s \notin S). \tag{1}$$

PROOF. Exercise. □

REMARK. As the proposition shows, it is possible to use (1) to define $<$ on $\mathbb{R}^{\mathrm{d}}$. We might be inclined to think that we could also use this definition on $\mathbb{R}^{\mathrm{be}}$ or $\mathbb{R}^{\mathrm{e}}$; however, the resulting notion is too strong, at least if we insist at the same time on the usual properties of $<$ in relation to addition. For we want $\forall n(S < S + n^{-1})$; however, on definition (1) this implies $S \in \mathbb{R}^{\mathrm{d}}$, since if $s \in S + n^{-1}$, $s \notin S$, we see that $s - n^{-1} \in S$, $s \notin S$, and since we can find such an s for all n, it follows that S is located (5.3).

5.6. PROPOSITION. *For all* $S, S', S'', T \in \mathbb{R}^{\mathrm{e}}$

(i) $S < S' \wedge S' \leq S'' \rightarrow S < S''$;

(ii) $S \leq S' \wedge S' < S'' \rightarrow S < S''$;

(iii) $S \leq T \leftrightarrow \neg(T < S)$;

(iv) $S \leq T \wedge T \leq S \rightarrow S = T$;

(v) $S \leq S' \wedge S' \leq S'' \rightarrow S \leq S''$.

Proof. Exercise. □

(iii) shows that the relation between $\leq$ and $<$ is the same as for the Cauchy reals.

5.7. Theorem. $\mathbb{R}^e, \mathbb{R}^{be}$ *are order-complete, that is to say*
(i) *any inhabited, weakly bounded subset of* $\mathbb{R}^e$ *has a supremum (least upper bound) in* $\mathbb{R}^e$;
(ii) *any inhabited, bounded subset of* $\mathbb{R}^{be}$ *has a supremum (least upper bound) in* $\mathbb{R}^{be}$.
Here a set $\mathscr{X} \subset \mathbb{R}^{be}$ *is (uniformly) bounded iff* $\exists r r' \forall X \in \mathscr{X}(r \in X \wedge r' \notin X)$ *and weakly bounded iff it is not not bounded (= not unbounded).*

Proof. The proofs of (i) and (ii) are completely similar. If $\mathscr{X}$ is any inhabited subset of $\mathbb{R}^e$, put

$$S := \operatorname{Int}\left(\left(\bigcup\{X : X \in \mathscr{X}\}\right)^{cc}\right).$$

We leave it as an exercise for the reader to verify that S is in fact the desired supremum. □

Remark. In 5.12 we shall show that $\mathbb{R}^e, \mathbb{R}^{be}$ can in fact be uniquely characterized as the order completions of $\mathbb{Q}$, w.r.t. weakly bounded and bounded sets respectively.

5.8. Proposition (*definition by cases*). *Let A be an arbitrary assertion, and let* $S_0, S_1 \subset \mathbb{Q}$ *be such that if A, then* $S_0 \in \mathbb{R}^e$, *and if* $\neg A$, *then* $S_1 \in \mathbb{R}^e$, *then*
(i) *there is a unique* $S \in \mathbb{R}^e$ *such that*

$$S = \begin{cases} S_0 & \text{if } A, \\ S_1 & \text{if } \neg A; \end{cases} \tag{1}$$

(ii) *if it is known that* $S_0, S_1 \in \mathbb{R}^{be}$ *(independently of the truth of* $A, \neg A$), *then there is a unique* $S \in \mathbb{R}^{be}$ *satisfying* (1).

Proof. Take

$$S := \operatorname{Int}\{r : \neg\neg((r \in S_0 \wedge A) \vee (r \in S_1 \wedge \neg A))\}.$$

We leave the verification as an exercise. □

In the next subsections we show that $\mathbb{R}$ and $\mathbb{R}^{\mathrm{d}}$ are isomorphic; AC-NN suffices.

5.9. DEFINITION. $S \subset \mathbb{Q}$ is called a *strong Dedekind real* if S satisfies boundedness (a), openness (c), and the following strong form of locatedness

$$(\mathrm{d}') \qquad \exists \alpha \in \mathbb{Q}^2 \to \mathbb{N} \forall r r' (r < r' \to (\alpha(r, r') = 0 \wedge r \in S) \vee (\alpha(r, r') \neq 0 \wedge r' \notin S)).$$

In the presence of AC_{00}, a Dedekind real is always a strong Dedekind real. □

5.10. PROPOSITION. *The collection of strong Dedekind reals with* $<$ *inherited from* $\mathbb{R}^{\mathrm{d}}$ *is order-isomorphic to* $\mathbb{R}$, *by an isomorphism which is the identity on* $\mathbb{Q}$.

Therefore, assuming AC_{00}, $\mathbb{R}$ *and* $\mathbb{R}^{\mathrm{d}}$ *are order isomorphic via an isomorphism which is the identity on* $\mathbb{Q}$.

PROOF. Given any Cauchy real x, $\langle r_n \rangle_n \in x$, we put

$$\varphi(x) := \mathrm{Int}\{r : \exists k \forall n (r_{k+n} > r)\}.$$

Then $\varphi(x)$ is a strong Dedekind real and leaves all rationals fixed.

Conversely, let S be a strong Dedekind real, with α as required by (d′). Assume $p \in \mathbb{Z}$ to be such that $p \in S$, $p + 1 \notin S$. Choose a $k \in \mathbb{N}$, and let $r_i = p + i \cdot 2^{-k}$, $i \leq 2^k + 1$. Without loss of generality we may assume $\alpha(r_0, r_1) = 0$; obviously $\alpha(p+1, p+1+r) \neq 0$ for all $r > 0$.

Put $s_k = r_j$ for $j := \max\{i : \alpha(r_i, r_{i+1}) = 0\}$. Then $r_j \in S$, $r_{j+2} \notin S$, hence $|S - s_k| \leq 2^{-k-1}$. Therefore $\langle s_k \rangle_k$ is a fundamental sequence; let ψ be the mapping assigning $\langle s_k \rangle_k / \approx$ to S. It remains to be shown that φ, ψ are each others inverse and preserve $<$; we leave this as an exercise. □

The next proposition shows how already a quite weak non-classical principle, namely $\neg\forall P(\neg P \vee \neg\neg P)$, enforces that $\mathbb{R}^{\mathrm{d}}$ and $\mathbb{R}^{\mathrm{be}}$ do not coincide. So, if we assume AC_{00} and $\neg\forall P(\neg P \vee \neg\neg P)$, then the Cauchy completion of $\mathbb{Q}$ (i.e. $\mathbb{R}$) and the order completion of $\mathbb{Q}$ (i.e. $\mathbb{R}^{\mathrm{be}}$) are non-isomorphic. This shows that the two completions are quite different.

5.11. PROPOSITION. *$\forall P(\neg P \vee \neg\neg P)$ is equivalent to each of the following assertions*

$$\mathbb{R}^{\mathrm{be}} = \mathbb{R}^{\mathrm{d}},$$

$$\exists f \in \mathbb{R}^{\mathrm{be}} \to \mathbb{R}^{\mathrm{d}}\ (f \text{ is non-constant}),\ i.e.$$

$$\exists f \in \mathbb{R}^{\mathrm{be}} \to \mathbb{R}^{\mathrm{d}} \exists x \in \mathbb{R}^{\mathrm{be}} \exists y \in \mathbb{R}^{\mathrm{be}} (f(x) \# f(y)).$$

Hence

$$\neg\forall P(\neg P \vee \neg\neg P) \to \forall f \in \mathbb{R}^{\mathrm{be}} \to \mathbb{R}^{\mathrm{d}}\ (f \text{ is constant}).$$

PROOF. (i) Assume $\forall P(\neg P \vee \neg\neg P)$, and let $S \in \mathbb{R}^{\mathrm{be}}$, $r < s$; take r' with $r < r' < s$, and by assumption $\neg r' \in S \vee \neg\neg r' \in S$; if $\neg r' \in S$, then $s \notin S$; if $\neg\neg r' \in S$, then $r \in S$. Hence $\forall rs(r < s \to r \in S \vee s \notin S)$, so $S \in \mathbb{R}^{d}$, and thus $\mathbb{R}^{\mathrm{be}} = \mathbb{R}^{\mathrm{d}}$.

(ii) Assume $\mathbb{R}^{\mathrm{be}} = \mathbb{R}^{\mathrm{d}}$, then $\lambda x.x$ is a non-constant function from $\mathbb{R}^{\mathrm{be}}$ to $\mathbb{R}^{\mathrm{d}}$.

(iii) Suppose $x, y \in \mathbb{R}^{\mathrm{be}}$, $f(x) \# f(y)$. By 5.8 we can define, for each proposition P, a number $z_P \in \mathbb{R}^{\mathrm{be}}$ such that

$$z_P = \begin{cases} x & \text{if } \neg P, \\ y & \text{if } P. \end{cases}$$

By the properties of #

$$\forall P(f(z_P) \# f(x) \vee f(z_P) \# f(y))$$

and thus $\forall P(\neg P \vee \neg\neg P)$. □

REMARK. $\neg\forall P(\neg P \vee \neg\neg P)$ is weaker than $\neg\forall$-PEM, i.e. $\neg\forall\alpha(\neg\exists x(\alpha x = 0) \vee \neg\neg\exists x(\alpha x = 0))$, which is a corollary of either WC-N or CT_0 (4.3.4, 4.6.4). □

We now turn to the characterization of $\mathbb{R}^{\mathrm{d}}, \mathbb{R}^{\mathrm{be}}, \mathbb{R}^{\mathrm{e}}$ as ordered structures.

5.12. THEOREM (*characterization of $\mathbb{R}^{\mathrm{d}}, \mathbb{R}^{\mathrm{be}}, \mathbb{R}^{\mathrm{e}}$ as ordered structures*).

(i) $(\mathbb{R}^{\mathrm{be}}, <)$ *is, up to an isomorphism which is the identity on $\mathbb{Q}$, the unique strict partially ordered structure $(X, <)$ extending $(\mathbb{Q}, <)$ such that for all $x, y, z \in X$, with $x \le y := \neg(y < x)$:*
   (a) $x < z \wedge y \le z \to x < z$;
   (b) $x \le y \wedge y < z \to x < z$;

(c) $x \le y \wedge y \le x \rightarrow x = y$;
(d) $x < y \leftrightarrow \exists r > 0 \forall r' \le x(r + r' \le y)$;
(e) $\exists rr'(r < x < r')$;
(f) *every bounded subset of* $\mathbb{Q}$ *has a l.u.b. in* X.

(ii) $(\mathbb{R}^e, <)$ *is the unique strict partially ordered structure* $(X, <)$ *extending* $(\mathbb{Q}, <)$ *such that* (a)–(d) *as under* (i) *hold, and in addition*
(e′) $\neg\neg\exists rr'(r < x < r')$;
(f′) *every weakly bounded subset of* $\mathbb{R}$ *has a l.u.b. in* X.

(iii) $(\mathbb{R}^d, <)$ *is the unique strict partially ordered structure* $(X, <)$ *extending* $(\mathbb{Q}, <)$ *such that* (a)–(e) *of* (i) *hold, and in addition*
(f″) *each located, bounded downwards monotone subset of* $\mathbb{Q}$ *has a l.u.b. in* X;
(h) $\forall x \in X \forall rs(r < s \rightarrow r < x \vee x < s)$.

Here $Y \subset X$ *is* located *iff* $\forall rs(r < s \rightarrow r \in Y \vee s \notin Y)$.

PROOF. We shall establish (i) by proving the lemmas 5.13 and 5.14 below. The proofs of (ii) and (iii) require minor modifications which we shall leave as an exercise to the reader. □

5.13. LEMMA. *Let* $(X, <)$ *satisfy the conditions under* (i) *in theorem* 5.12. *Then each* $v \in X$ *is uniquely determined by* $\{r : r < v\}$, *in fact*

$$\{r : r < v\} \subset \{r : r < v'\} \leftrightarrow v \le v'$$

for all $v, v' \in X$.

PROOF. The implication from right to left is immediate by property (a). For the converse, we first note that

$$\forall x, y \in X(x < y \rightarrow x \le y). \tag{1}$$

Assume $v' < v$. By (d) there is an $r > 0$ such that $\forall r' \le v'(r + r' \le v)$. Now choose an r'' such that $0 < r'' < r$, then

$$\forall r' \le v'(r'' + r' < v). \tag{2}$$

Choosing some fixed $r' \le v'$, which is possible by (e), we may take $r'' + r'$ for r' in (2), since $r'' + r' \le v'$ (cf. (1)). This yields $2 \cdot r'' + r' \le v'$, and by induction $\forall n(n \cdot r'' + r' \le v')$.

By (e) there is an r^* such that $v' < r^*$, and so $\forall n(n \cdot r'' + r' \le r^*)$ which contradicts the archimedean ordering of $\mathbb{Q}$. Hence $\neg(v' < v)$, i.e. $v \le v'$. □

This lemma reduces the proof of 5.12(i) to the special case of partially ordered systems $(\mathscr{X}, \subset^+)$ such that (i) $\mathscr{X}$ is a collection of subsets of $\mathbb{Q}$, $X \subset^+ Y := \exists r > 0(\{x + r : x \in X\} \subset Y)$, (ii) the rationals are embedded into $\mathscr{X}$ by $r \mapsto r^\wedge := \{r' : r' < r\}$, and (iii) $\forall X \in \mathscr{X}(X = \{r : r < X\})$. $\subset$ corresponds to $\leq$. The proof of 5.12(i) is then completed by

5.14. Lemma. *Let $(\mathscr{X}, \subset^+)$ be a family of sets of rationals ordered by strong inclusion as indicated, and assume each bounded $S \subset \mathbb{Q}$ to have a l.u.b. in $\mathscr{X}$; then $(\mathscr{X}, \subset^+)$ is isomorphic to $\mathbb{R}^{\mathrm{be}}$.*

Proof. Exercise. □

6. Arithmetic of Dedekind reals and extended reals

6.1. The developments in this section run parallel to those for the Cauchy reals. The principal difference is that we have to do some extra work in proving the existence of a unique canonical extension of locally uniformly continuous functions.

We define provisionally, for $r \in \mathbb{Q}$, $S, S' \in \mathbb{R}^{\mathrm{e}}$

$$|S - S'| \leq r := S - r \subset S' \subset S + r.$$

One readily sees that

$$|S - S'| \leq r \wedge |S' - S''| \leq r' \rightarrow |S - S''| \leq r + r'.$$

For $|S - S'| < r$ we can take $S - r < S' < S + r$. If $\vec{S}, \vec{S'} \in (\mathbb{R}^{\mathrm{e}})^n$, we define $|\vec{S} - \vec{S'}| \leq r$ coordinate-wise as in 3.1.

The formulas $M_f(m, i, k)$, $M_f(i, k)$, and the notions of continuity, uniform continuity and locally uniform continuity are defined as in 3.2, except that X and Z now may range over $\mathbb{Q}, \mathbb{R}^{\mathrm{d}}, \mathbb{R}^{\mathrm{be}}, \mathbb{R}^{\mathrm{e}}$.

6.2. Theorem. *Let $f \in \mathbb{Q}^n \rightarrow \mathbb{R}^{\mathrm{d}}$, f locally uniformly continuous on $\mathbb{Q}^n$; then there is a uniquely determined canonical extension $f^* \in (\mathbb{R}^{\mathrm{d}})^n \rightarrow \mathbb{R}^{\mathrm{d}}$, given by*

$$f^*(\vec{S}) := \mathrm{Int}\{r : \exists \vec{q} \in \vec{S} \forall \vec{q}' \in \vec{S}(\vec{q}' > \vec{q} \rightarrow r < f(\vec{q}'))\},$$

where $(q_1', \ldots, q_n') > (q_1, \ldots, q_n) := \forall i(1 \leq i \leq n \rightarrow q_i' > q_i)$.

Proof. The proof is split into two lemmas. For notational simplicity we shall take $n = 1$ and assume f to be uniformly continuous, instead of only locally uniformly continuous. □

6.3. LEMMA. *f^* is uniformly continuous and satisfies*

$$\forall ik\big(M_f(i,k) \to M_{f^*}(i+2,k)\big).$$

PROOF. Assume $M_f(i,k)$, $|S - S'| \le 2^{-i-2}$. We have to show

$$f^*(S) - 2^{-k} \subset f^*(S') \subset f^*(S) + 2^{-k}.$$

Let $r' \in f^*(S')$, then for some $r'' > r'$ and some $q \in S'$

$$\forall q' \in S'(q' > q \to r'' < f(q')). \tag{1}$$

Suppose $q'' \in S$, $q'' > q - 2^{-i-2}$; we wish to show $r'' - 2^{-k} < f(q'')$. From $|S - S'| \le 2^{-i-2}$ we get that $q'' - 2^{-i-2} \in S'$. We have either $q'' > q + 2^{-i-2}$, or $q - 2^{-i-1} < q'' \le q + 2^{-i-2}$. In the first case, by (1), $r'' < f(q'' - 2^{-i-2})$, hence $r'' - 2^{-k} < f(q'')$. In the second case we can find a $q^* > q$, $|q^* - q''| < 2^{-i-2}$, $q^* \in S'$; then $r'' < f(q^*)$, so

$$r'' - 2^{-k} < f(q^*) - |f(q'') - f(q^*)| \le f(q'').$$

Thus we have shown

$$\forall q \in S \forall q'' \in S\big(q'' > q \to r'' - 2^{-k} < f(q'')\big), \tag{2}$$

hence $r' \in f^*(S) + 2^{-k}$. This establishes $f^*(S') \subset f^*(S) + 2^{-k}$; $f^*(S) - 2^{-k} \subset f^*(S')$ is proved similarly. □

6.4. LEMMA. *Let $g \in \mathbb{R}^{\mathrm{d}} \to \mathbb{R}^{\mathrm{d}}$ be any uniformly continuous extension of f, then $g = f^*$.*

PROOF. Let $M_g(i,k)$, then also $M_f(i,k)$ on $\mathbb{Q}$, since g extends f; by lemma 6.3 therefore $M_{f^*}(i+2,k)$. Assume

$$r_1 \in S, \quad r_2 \notin S, \quad |r_1 - r_2| < 2^{-i-2}.$$

Then $r_1 - 2^{-i-2} \le S \le r_1 + 2^{-i-2}$, so $|r_1 - S| \le 2^{-i-2}$, hence $|f(r_1) - f^*(S)| \le 2^{-k}$, $|f(r_1) - g(S)| \le 2^{-k}$, and therefore

$$|f^*(S) - g(S)| \le 2^{-k+1}. \tag{3}$$

But since $\forall k \exists i M_g(i,k)$, this holds for arbitrary k, and thus $f^*(S) = g(S)$. □

REMARK. Clearly, 6.3 and 6.4 taken together establish 6.2. In fact it can be shown that

$$\forall ik\big(M_f(i,k) \to M_{f^*}(i,k)\big)$$

(exercise). □

For $\mathbb{R}^{\text{be}}, \mathbb{R}^{\text{e}}$ we have the corresponding

6.5. Theorem. *Let $f \in \mathbb{Q}^n \to \mathbb{R}^{\text{i}}$ be locally uniformly continuous on $\mathbb{Q}^n$, then there is a uniquely determined canonical extension $f^* \in (\mathbb{R}^{\text{i}})^n \to \mathbb{R}^{\text{i}}$ (where* i *is* be *or* e), *defined by*

$$f^*(\vec{S}) := \text{Int}\{r : \neg\neg\exists\vec{q} \in \vec{S}\forall\vec{q}' \in S(\vec{q}' > \vec{q} \to r < f(\vec{q}'))\}.$$

Proof. The proof follows the same track as for theorem 6.2; we have to make slight adaptations in the proofs of lemmas 6.3, 6.4.

Adaptation of the proof of 6.3. If $M_f(i, k)$, $|S - S'| \le 2^{-i-2}$, and $r' \in f^*(S')$, we only have that for some $r'' > r'$

$$\neg\neg\exists q \in S'\forall q' \in S'(q' > q \to r'' < f(q')). \tag{4}$$

Now assume for $q \in S'$ (1) as before, and deduce (2) as before. Then $\exists q \in S'\forall q \in S'(q' > q \to r'' < f(q'))$ implies (2), hence from (4) we find $\neg\neg$(2), i.e.

$$\neg\neg\exists q \in S\forall q'' \in S(q'' > q \to r'' - 2^{-k} < f(q'')),$$

and then $r' \in f^*(S) + 2^{-k}$ etc. as before. □

Adaptation of the proof of 6.4. Assuming $M_g(i, k)$ and $\exists r_1 \in S\exists r_2 \notin S$ $(|r_1 - r_2| < 2^{-i-2})$, the proof of lemma 6.4 shows (3); now from $\neg\neg\exists r_1 \in S$ $\exists r_2 \notin S(|r_1 - r_2| < 2^{-i-2})$ we find $\neg\neg$(3), which is equivalent to (3); and since $S \in \mathbb{R}^{\text{e}}$, $\neg\neg\exists r_1 \in S\exists r_2 \notin S(|r_1 - r_2| < 2^{-i-2})$ holds, etc. □

Remark. The reader should note that the adaptation of the proof of 6.2 has a routine character: all that is needed is the assertion in appropriate places of some double negations, with an appeal to laws such as $(A \to B) \to (\neg\neg A \to \neg\neg B)$. From a logical point of view this becomes understandable if we realize that the adaptation corresponds to an application of the (Gödel–Gentzen) negative translation to the proof of 6.2; cf. section 2.3 and 7.4 below.

Note also that the definition of f^* in 6.5 coincides on $\mathbb{R}^{\text{d}}$ with the definition of f^* in 6.2. □

6.6. *Arithmetic on* $\mathbb{R}^{\text{d}}, \mathbb{R}^{\text{e}}, \mathbb{R}^{\text{be}}$. Functions such as $+, \cdot, \max, \min, |\ |$ may now be extended to $\mathbb{R}^{\text{d}}, \mathbb{R}^{\text{e}}, \mathbb{R}^{\text{be}}$ with preservation of identities valid on $\mathbb{Q}$ as indicated before in the case of Cauchy reals (cf. 3.6); details are left to the reader. □

6.7. DEFINITION (of apartness). Let $S, T \in \mathbb{R}^d$; define

$$S \# T := \exists k(|S - T| \geq 2^{-k}). \quad \square$$

6.8. PROPOSITION. *For* $S, S', S'' \in \mathbb{R}^d$

(i) $S \# S' \to S' \# S$;

(ii) $S = S' \leftrightarrow \neg(S \# S')$;

(iii) $S \# S' \leftrightarrow (S < S') \vee (S' < S)$;

(iv) $S \# S' \to (S \# S'') \vee (S'' \# S')$.

PROOF. Left to the reader. □

N.B. Of course, the definition of $\#$ also extends to $\mathbb{R}^e$, but in this case the crucial properties (iii), (iv) cannot be proved. □

6.9. *Discontinuous functions.* As noted already in chapter 1, we cannot give an example of an everywhere defined discontinuous $f \in \mathbb{R} \to \mathbb{R}$. However, if we permit the functions to take values in $\mathbb{R}^e$ or $\mathbb{R}^{be}$, we again encounter the phenomenon of functions total on $\mathbb{R}$ (or $\mathbb{R}^{be}, \mathbb{R}^e$) which are discontinuous. For example, there is a total $f \in \mathbb{R} \to \mathbb{R}^{be}$ such that $f(0) = 0$, $f(x) = 1$ for $x \neq 0$. Also we can define a total $f' \in \mathbb{R} \to \mathbb{R}^e$ such that $f'(0) = 0$, $f'(x) = x^{-1}$ for $x \# 0$; however, there is no such $f' \in \mathbb{R} \to \mathbb{R}^{be}$ (exercise). □

7. Two metamathematical digressions

7.1. *Kripke models for* $\mathbb{R}$. In 3.3.8–10 we illustrated the use of Kripke models in the study of intuitionistic second-order arithmetic by studying a particular example. Now we return to the same example and show that it can also be used to obtain some insight in the various extensions of $\mathbb{R}$.

Given some standard enumeration of the rationals, it is clear how sets of rationals ought to be interpreted in the Kripke models of 3.8.8, namely as families $\mathcal{S} \equiv \{S_k : k \in K\}$ of sets of rationals such that

$$k \preccurlyeq k' \to S_k \subset S'_{k'},$$

and of course

$$k \Vdash r \in \mathcal{S} := r \in S_k.$$

What do the elements of $\mathbb{R}^d$ look like in the model? The boundedness condition has to be realized at the roots of the trees considered, i.e. there are $r, s \in \mathbb{Q}$ such that $r \in \mathscr{S}$, $s \notin \mathscr{S}$ are forced everywhere, i.e. $\forall k \in K(r \in S_k \wedge s \notin S_k)$. The condition of openness translates into openness of the S_k at each node.

Monotonicity is guaranteed by letting the S_k be monotone at each node. Strong monotonicity corresponds to a more involved closure condition in the models: if $r < s$, and the nodes $k' \succ k$ where $s \in S_{k'}$ are *dense* above k (i.e. $\forall k'' \succcurlyeq k \exists k' \succcurlyeq k''(s \in S_{k'})$) then $r \in S_k$.

Locatedness is a very strong condition, since existential quantifiers are "instantly" realized at each node separately; i.e. the decisions $r \in S \vee s \notin S$ for all $r < s$ are already made at the root and are hence not subject to changes at higher nodes.

So a Dedekind real, i.e. an element of $\mathbb{R}^{\mathrm{d}}$, is in fact represented by a *constant* family of left cuts $\mathscr{S} \equiv \{S_k : S_k = S \text{ for all } k \in K\}$, with $S \in \mathbb{R}^{\mathrm{d}}$. Therefore $\mathbb{R}^{\mathrm{d}}$ is, in a manner of speaking, in these models interpreted by itself.

7.2. *Kripke models for* $\mathbb{R}^{\mathrm{be}}$. For elements of $\mathbb{R}^{\mathrm{be}}$, the outcome is different: dropping the locatedness permits also non-constant families $\mathscr{S}$ as elements of $\mathbb{R}^{\mathrm{be}}$ in the model. For example, in the model over the binary tree there occurs the element shown in fig. 5.1. The element $\mathscr{S}$ is specified by indicating S_k at each node; we write $r^\wedge$ for $\{s : s < r\}$. Obviously, the $\mathscr{S}$ in the picture is distinct from all $\mathscr{S}$ representing elements of $\mathbb{R}^{\mathrm{d}}$ in the model. On the other hand, though clearly $\mathscr{K} \Vdash 0 \leq \mathscr{S} \leq 1$, no Dedekind real in the model is apart from $\mathscr{S}$, that is we have

$$\mathscr{K} \Vdash \forall \mathscr{S}' \in \mathbb{R}^{\mathrm{d}} \cap [0,1] \neg \exists k \left(|\mathscr{S}' - \mathscr{S}| > 2^{-k}\right)$$

(exercise).

7.3. *Singleton reals.* It is instructive to consider also the intermediate possibility of the "bounded singleton reals" $\mathbb{R}^{\mathrm{bs}}$ which may be defined by

$$S \in \mathbb{R}^{\mathrm{bs}} := S \in \mathbb{R}^{\mathrm{be}} \wedge \neg\neg S \in \mathbb{R}^{\mathrm{d}}.$$

In our models an element of $\mathbb{R}^{\mathrm{bs}}$ is a family $\mathscr{S}$ of $\mathbb{R}^{\mathrm{be}}$ such that above each node there is a node where $\mathscr{S}$ becomes constant (i.e. becomes, locally, an element of $\mathbb{R}^{\mathrm{d}}$ in the model); in short, the nodes where $\mathscr{S} \in \mathbb{R}^{\mathrm{d}}$ is forced are *dense* in the tree. Similarly one defines

$$S \in \mathbb{R}^{\mathrm{s}} := S \in \mathbb{R}^{\mathrm{e}} \wedge \neg\neg S \in \mathbb{R}^{\mathrm{d}}.$$

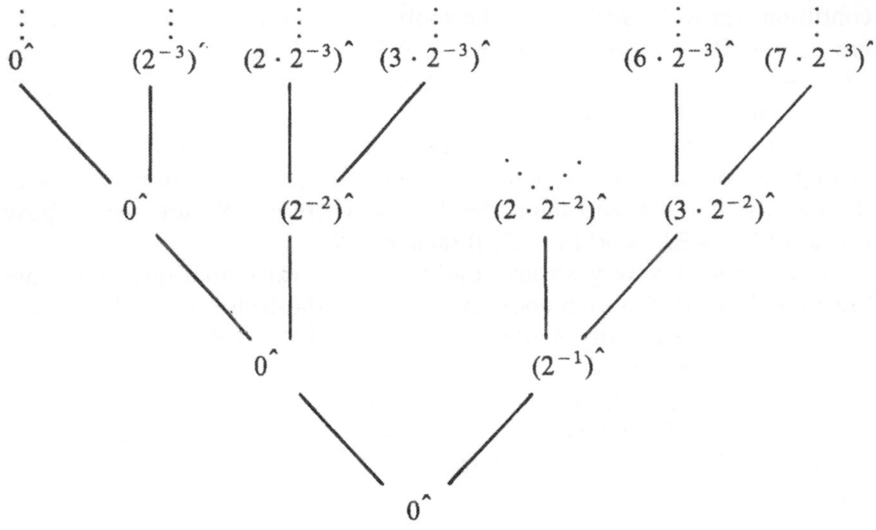

Fig. 5.1

Fig. 5.2 shows two examples of elements of $\mathbb{R}^{bs}$. More information on singleton reals is found in van Dalen (1982) and Troelstra (1982), section 4.

7.4. $\mathbb{R}^e$ *as the image of* $\mathbb{R}^d$ *under the negative translation.* The results in sections 5 and 6 show us that $\mathbb{R}^e$ behaves in many respects just as $\mathbb{R}^d$ does in classical analysis.

There is a general logical explanation of this fact: $\mathbb{R}^e$ can be obtained as the negative translation of $\mathbb{R}^d$ (cf. section 2.3). The setting is e.g. the language of intuitionistic second-order arithmetic **HAS** (cf. 3.8.4). We have

$$S \in \mathbb{R}^d := \exists s(s \in S) \wedge \exists s'(s' \notin S) \wedge$$
$$\forall sr(s < r \wedge r \in S \to s \in S) \wedge \forall s \in S \exists s' \in S(s < s').$$

The right-hand side of this definition translates (modulo the equivalences $\neg\neg\exists xA \leftrightarrow \neg\forall x\neg A$, $\neg\neg s < r \leftrightarrow s < r$ etc.) into

$$\neg\neg\exists s(s \in S^{cc}) \wedge \neg\neg\exists s'(s' \notin S^{cc})$$
$$\wedge \forall sr(s < r \wedge r \in S^{cc} \to s \in S^{cc})$$
$$\wedge \forall s \in S^{cc} \neg\neg\exists s' \in S^{cc}(s < s'). \tag{1}$$

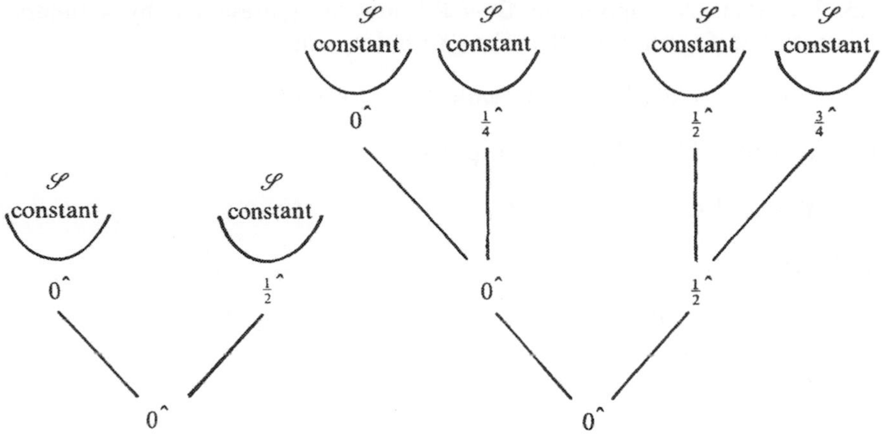

Fig. 5.2

That is to say, S^{cc} is weakly bounded, monotone and weakly open (a monotone $X \subset \mathbb{Q}$ is weakly open iff $\forall r \in X \neg\neg\exists r' \in X(r < r')$); of course, S^{cc} is also stable, i.e. $\neg\neg r \in S^{cc} \to r \in S^{cc}$. We can now prove the following

Proposition. *Let* $\mathbb{R}^e_1$ *consist of the subsets* $S \subset \mathbb{Q}$, *which are weakly bounded, monotone, stable and weakly open; then* $\mathbb{R}^e_1$ *is order-isomorphic to* $\mathbb{R}^e$ *by* $\varphi \in \mathbb{R}^e_1 \to \mathbb{R}^e$, $\varphi^{-1} \in \mathbb{R}^e \to \mathbb{R}^e_1$, *where*

$$\varphi(S) := \text{Int}(S), \quad \varphi^{-1}(S) := S^{cc};$$

φ *leaves* $\mathbb{Q}$ *(essentially) unchanged.*

Proof. Routine and left as an exercise. □

Note also that the preservation of order plus the fact that $\mathbb{Q}$ is left essentially unchanged by the negative translation has as a result that arithmetical relations are also unaffected by the negative translation. In the next subsection we shall show by an example how the negative translation may be applied to obtain facts concerning $\mathbb{R}^e$.

7.5. EXAMPLE. A mapping in $\mathbb{Q} \to \mathbb{R}^d$ may be represented by a binary relation $X \subset \mathbb{Q} \times \mathbb{Q}$ such that Fun(X) holds, where

$$\mathrm{Fun}(X) := \forall r \big(X_r \in \mathbb{R}^d \big) \quad \text{with } X_r := \{ r' : X(r, r') \}.$$

Let, as before, $|X_r - X_{r'}| \le 2^{-k}$ express

$$X_r - 2^{-k} \subset X_{r'} \subset X_r + 2^{-k}$$

and put

$$M_X(i, k) := \forall r r' \big(|r - r'| \le 2^{-i} \to |X_r - X_{r'}| \le 2^{-k} \big),$$

$$A_Y := \mathrm{Int}\{ r : \exists r' \in Y \forall r'' \in Y (r'' > r' \to r \in X_{r''}) \}$$

(A_Y is arithmetical in X and Y); also

$$M_A(i, k) := \forall Y, Y' \in \mathbb{R}^d \big(|Y - Y'| \le 2^{-i} \to |A_Y - A_{Y'}| \le 2^{-k} \big).$$

Now we can prove with classical logic

$$\forall k \exists i M_X(i, k) \wedge \mathrm{Fun}(X) \to \forall X \in \mathbb{R}^d \big(A_X \in \mathbb{R}^d \big)$$

$$\wedge \forall ik \big(M_X(i, k) \to M_A(i, k) \big) \wedge \forall r \big(X_r = A_{r^\wedge} \big); \tag{1}$$

this states that A_Y as a function of a real Y represents a uniformly continuous extension of $f \in \mathbb{Q} \to \mathbb{R}^d$ coded by X; A_Y has the same modulus of uniform continuity as f has.

Under the negative translation this is transformed into the corresponding statement for $\mathbb{R}^e$, except that the premiss $\forall k \exists i M_X(i, k)$ is replaced by the even weaker assumption $\forall k \neg\neg \exists i M_X(i, k)$. Thus one establishes theorem 6.5 for $\mathbb{R}^e$ (minus the uniqueness of the extension).

In fact, the uniqueness can also be established by this method, but that requires the extension of the negative translation to a language containing variables for functions in $\mathbb{R}^d \to \mathbb{R}^d$ (cf. Troelstra 1980, section 7).

8. Notes

8.1. *Definition of* $\mathbb{R}$ *and* $\mathbb{R}^d$. The material in section 2 is standard in the literature on constructive mathematics. Bishop (1967) and Bishop and Bridges (1985) prefer working with fundamental sequences with a fixed rate of convergence, but the difference is inessential. Note however that a proof of isomorphism of the reals as defined here, and the reals defined via fundamental sequences with a fixed rate of convergence, requires some closure conditions on $\mathbb{N} \to \mathbb{N}$ (closure under "recursive in" is amply sufficient); this becomes relevant if we wish to study definitions of $\mathbb{R}$ relative to some universe $\mathcal{U} \subset \mathbb{N} \to \mathbb{N}$ for which those conditions are not automatically satisfied (think e.g. of the case $\mathcal{U} = \mathrm{LS}$, the universe of lawless sequences, cf. 4.6.2).

The Dedekind definition of the reals, i.e. $\mathbb{R}^d$, made its appearance already in Brouwer (1926), but as long as countable choice is assumed, or built into the definition by a strong definition of locatedness ((d′) in 5.9) there is not much to choose between the two, since then $\mathbb{R}$ and $\mathbb{R}^d$ are isomorphic (5.10).

The apartness relation in the real plane was introduced by Brouwer (1919, "örtlich verschieden"); the basic properties AP1–3 were formulated by Heyting (1925). Brouwer also observed the stability of $=$ on $\mathbb{R}^2$ (1923).

8.2. *Order completeness.* The failure of the theorem that every monotone sequence of reals has a real limit, and hence the failure of the Bolzano–Weierstrass theorem, stimulated the search for extensions of $\mathbb{R}$ or $\mathbb{R}^d$ with better properties of order completeness. A survey of these attempts may be found in Troelstra (1980, 1982). $\mathbb{R}^{be}$ turned out to be the canonical answer to the question: what is the order completion of $\mathbb{Q}$? If one requires order completeness with respect to weakly bounded subsets of $\mathbb{Q}$, the answer is $\mathbb{R}^e$.

$\mathbb{R}^{be}$ was investigated by Staples (1971); Troelstra (1980) contains a rather detailed exposition of the basic properties of $\mathbb{R}^{be}$ and $\mathbb{R}^e$, which formed the basis for the presentation in the present chapter. Lorenzen (1955) uses Dedekind cuts in a predicative framework where the use of classical logic is justified via the negative translation. Thus the reals considered there are close to $\mathbb{R}^e$ from our point of view. Our definition of $\mathbb{R}^{be}$ differs from the one used by Staples in two respects: we only work with left cuts, instead of left and right cuts (cf. E5.5.3) and instead of a defined equivalence relation on cuts we simply use set-theoretic equality; for this reason we have required strong monotonicity ((b) in 5.1) instead of monotonicity only.

Certainly, $\mathbb{R}^{be}$ is far less important than $\mathbb{R}$ in constructive mathematics. Nevertheless $\mathbb{R}^{be}$ has its uses, for example in studying duals of Banach spaces and reflexivity of such spaces in a manner which follows the classical theory more closely than in the treatment of e.g. Bishop and Bridges (1985, section 7.6); see for example Troelstra (1980, section 6).

8.3. *Specker sequences.* Specker (1949) gave the first proof of theorem 4.5; the particularly simple proof presented here is due to Rice (1954). The existence of strong Specker sequences was shown as part of the proof of theorem 4.3 of Zaslavskij (1962). More recently Rice's construction has been investigated in Kushner (1981, 1983).

8.4. *Reals in CRM.* It is an oversimplification to say that the various notions of real number in the literature on CRM are obtained by taking some classical definition and interpreting the logical operations according to numerical realizability (section 4.4).

Indeed, the notion of "FR-number" or "real duplex" corresponds to our notion of Cauchy real interpreted by realizability, and the pseudo-numbers defined by sequences of rationals $\langle r_n \rangle_n$ satisfying

$$\forall k \neg\neg \exists n \forall m m' \left(|r_{n+m} - r_{n+m'}| < 2^{-k} \right), \tag{1}$$

modulo an equivalence relation $\asymp$ given by

$$\langle r_n \rangle_n \asymp \langle s_n \rangle_n := \forall k \neg\neg \exists n \forall m \left(|r_{n+m} - s_{n+m}| < 2^{-k} \right) \tag{2}$$

correspond to the non-oscillating sequences of the intuitionistic literature (e.g. Dijkman 1952) modulo $\asymp$. Realizability is in the latter case irrelevant since (1) and (2) are almost negative, i.e. truth equals realizability in this case (cf. 4.4.5).

However, in the literature on CRM one also studies "F-numbers" given by an algorithm for a fundamental sequence $\langle r_n \rangle_n$, but the algorithm for the modulus of convergence is not given, that is to say operations on F-numbers are algorithms acting on the algorithm for the sequence only, not on the algorithm for its modulus (see e.g. Shanin 1962). Thus considering F-numbers in CRM is the same as considering FR-numbers while restricting attention to those algorithmic operations depending on the code of the sequence itself only.

It hardly needs saying that such a notion is rather awkward from the logical point of view.

Exercises

5.2.1. Prove AP3 in proposition 2.9.

5.2.2. Prove 2.11(vii)–(ix).

5.2.3. Check that the operations in 2.13 are well-defined; prove 2.15(i), (iii).

5.2.4. Give a weak counterexample to the assertion "each real number is either rational or not rational".

5.2.5. Give an example of a real x such that x is not known to be rational, though $\neg\neg(x$ is rational).

5.2.6. Prove that

$$e = \sum_{n=0}^{\infty} 1/n!$$

has a positive measure of irrationality, i.e. there is a $\varphi \in \mathbb{N}^2 \to \mathbb{N}$ such that $|e - p \cdot q^{-1}| > \varphi(p, q)^{-1}$ for all $p, q \in \mathbb{N}$ (Heyting 1956, 2.3).

5.3.1. Show that f^* in 3.4 is well-defined.

5.3.2. Give a weak counterexample to $\forall xy(xy = 0 \to x = 0 \vee y = 0)$.

5.3.3. Prove 3.7(i), (iv).

5.3.4. Show that for all $x, y, z \in \mathbb{R}$: $x < y \leftrightarrow x + z < y + z$, $x < y \wedge z > 0 \to x \cdot z < y \cdot z$, $x < y \wedge z < 0 \to x \cdot z > y \cdot z$.

5.3.5. Show that we cannot expect to find a total extension of $\lambda x.x^{-1}$ (3.11).

5.3.6. Show that $x \# y := \rho(x, y) > 0$ defines an apartness relation on a metric space (V, ρ) (i.e. $\#$ satisfies AP1–3 as in 2.9).

5.3.7. Show that, under the usual definition of the metric for separable Hilbert space l^2, l^2 is also constructively a metric space. Define intuitionistically a suitable metric on Baire space yielding the usual topology (cf. 4.1.5).

5.3.8. Let (V_n, ρ_n) be a countable sequence of metric spaces; define an appropriate metric on $\Pi\{V_n : n \in \mathbb{N}\}$ (cf. 4.1.5).

5.3.9. Let $\mathcal{R}$ be the collection of fundamental sequences and let $x_0, x_1 \in \mathcal{R}$, $A := \{x_0, x_1\}$, $B := \{(x_0)_\approx, (x_1)_\approx\}$. Then $\forall y \in B \exists x \in A(x \in y)$. Show by means of a weak counterexample that we cannot expect to prove $\exists f \in B \to A \forall y \in A(f(y) \in y)$ for suitably chosen x_0, x_1. In fact, $\forall y \in B \exists x \in A(x \in y) \to \exists f \in A \to B \forall y \in B(f(y) \in y)$ for arbitrary two-element sets A, B as above implies $\exists$-PEM.

5.3.10. Show, by means of a weak counterexample, that the set of reals representable by infinite decimal fractions is not closed under addition.

5.4.1. Show that $\exists$-PEM suffices to prove "every bounded monotone sequence of reals has a limit".

5.4.2. We call a set *infinite* if it contains a subset in 1–1 correspondence with $\mathbb{N}$. Prove constructively: let A be any bounded infinite set of reals, and let $m, n \in \mathbb{N}$; then there is an interval of length 2^{-n} containing at least m elements of A (Brouwer 1952; Heyting 1956, 3.4.6).

5.4.3. We call a sequence of reals $\langle x_n \rangle_n$ *non-oscillating* iff $\forall k \neg\neg\exists n \forall m(|x_n - x_{n+m}| < 2^{-k})$. Show that a bounded monotone sequence is non-oscillating.

5.4.4. A sequence of reals $\langle x_n \rangle_n$ is said to be *negatively convergent* iff $\neg\neg\exists x \forall k \exists n \forall m$ $(|x - x_{n+m}| < 2^{-k})$ (Dijkman 1952). Give an example of a non-oscillating sequence which cannot be shown to be negatively convergent (see e.g. Heyting 1956, 1.3.2, example (ii)). What can we say on the basis of CT_0?

5.4.5. Is "$\langle x_n \rangle_n$ is negatively convergent" a weaker statement than $\exists x \neg\neg \forall k \neg_n \forall m(|x - x_{n+m}| < 2^{-k})$?

5.5.1. Prove 5.5 and 5.6.

5.5.2. Complete the proofs of 5.7, 5.8 and 5.10.

5.5.3. An equivalent definition of $\mathbb{R}^{\mathrm{d}}$ is as follows. A *Dedekind cut* is a pair (S, T) with $S \subset \mathbb{Q}$, $T \subset \mathbb{Q}$ such that
(1) $\forall r \in S \forall r' \in T(r < r'))$ (disjointness),
(2) $\exists rr'(r \in S \wedge r' \in T)$ (boundedness),
(3) $\forall r \in S \exists r' \in S(r < r'), \forall r \in T \exists r' \in T(r' < r)$ (openness),
(4) $\forall rr'(r < r' \to r \in S \vee r' \in T)$ (locatedness).
Define now $\leq$ and $<$ and show how this notion of Dedekind cuts yields a system order-isomorphic to the system of our original definition. Can you adapt this symmetric definition to $\mathbb{R}^{\mathrm{be}}, \mathbb{R}^{\mathrm{e}}$?

5.5.4. Assuming Kripke's schema KS (4.9.3), show that each element of $\mathbb{R}^{\mathrm{be}}$ is the supremum of a bounded monotone increasing sequence of rationals.

5.5.5. Assuming CT_0, show that not every element of $\mathbb{R}^{\mathrm{be}}$ is the weak limit of a non-oscillating sequence. Here $X \in \mathbb{R}^{\mathrm{be}}$ is said to be the weak limit of a sequence $\langle x_n \rangle_n$, iff

$$\forall k \neg\neg\exists n \forall m\big(|X - x_n| < 2^{-k}\big)$$

(cf. section 2 of Troelstra 1982).

5.5.6. Complete the proof of 5.12 by proving the second lemma 5.14 and (ii), (iii) of 5.12.

5.6.1. Establish the remark after 6.4: $\forall ik(M_f(i, k) \to M_{f^*}(i, k))$.

5.6.2. Give proofs of 6.5 in full and prove 6.8.

5.6.3. Show that we cannot expect to find an $f' \in \mathbb{R} \to \mathbb{R}^{be}$ extending $\lambda x.x^{-1}$. *Hint.* Show that such an f' cannot exist if we assume WC-N (Troelstra 1980, 5.6).

5.6.4. Show that $\mathbb{R}^e$ and $\mathbb{R}^{be}$ do not carry an apartness relation, i.e. there is no binary relation that satisfies AP1–3 of 2.9. *Hint.* Cf. 5.11.

5.7.1. Prove the statement at the end of 7.2, and show that $\mathcal{S}$ cannot be a singleton real in $\mathcal{X}$.

5.7.2. Define $\mathbb{R}^{s*}$ to be the collection of "near-singletons" among the subsets of $\mathbb{R}$, i.e.,

$$\mathcal{X} \in \mathbb{R}^{s*} := \forall S, T \in \mathcal{X}(S = T) \wedge \neg\neg\exists S(S \in \mathcal{X}).$$

For $\mathcal{X}, \mathcal{X}' \in \mathbb{R}^{s*}$ put

$$\mathcal{X} \leq \mathcal{X}' := \forall S \in \mathcal{X} \forall T \in \mathcal{X}'(S \leq T).$$

Show that $\mathbb{R}^{s*}$ is order-isomorphic to $\mathbb{R}^s$ by an isomorphism mapping $\{r^\wedge\}$ to $r^\wedge$ for all $r \in \mathbb{Q}$ (Troelstra 1982, 4.2).

5.7.3. Show that every $X \in \mathbb{R}^e$, respectively any $X \in \mathbb{R}^{be}$ can be obtained as the limit of a Cauchy sequence of elements in $\mathbb{R}^s$, respectively $\mathbb{R}^{bs}$ (van der Hoeven in Troelstra 1982, 4.3).

5.7.4. Prove the proposition in 7.4.

CHAPTER 6

SOME ELEMENTARY ANALYSIS

In this chapter we present some examples of constructive analysis, so that the reader can see some constructive mathematics "at work"; no systematic treatment is attempted. The first two sections contain some basics, such as the intermediate value theorem, differentiability, Riemann integration and Taylor series. These sections presuppose sections 5.1–5.4.

In section 3 we derive some consequences of the weak continuity principle WC-N and the fan theorem; this material relies on the sections 4.6–4.7.

Finally, section 4 contains some examples of analysis in constructive recursive mathematics; specifically, we give a number of "pathological" results based on the non-compactness of $[0,1]$ under the assumption of CT_0 (such as the existence of a continuous, but not uniformly continuous real-valued function on $[0,1]$) and a positive result requiring $ECT_0 + MP$, namely that all functions $f \in \mathbb{R} \to \mathbb{R}$ are continuous. This section presupposes sections 4.3 and 4.5, and for the final result, based on $ECT_0 + MP$, also section 4.4 is needed.

Our notions of "continuous function" and "uniformly continuous function" are the usual ones; cf. 5.3.13, 5.3.15, 5.3.2. As long as we assume AC_{00} we do not have to introduce the modulus of continuity explicitly. The basic properties of continuous functions from traditional mathematics, such as closure under addition and multiplication, are also constructively correct and we will apply them below without proving them. The reader can easily supply proofs, usually by (almost) copying the classical proof.

1. Intermediate-value and supremum theorems

1.1. Definition. Let $a, b \in \mathbb{R}$, $a \leq b$. The *open interval* (a, b) and the *closed interval* (or *segment*) $[a, b]$ are defined as

$$(a, b) := \{x : a < x < b\},$$

$$[a, b] := \{x : a \leq x \leq b\}.$$

We can define *half-open* intervals $(a, b], [a, b)$ in the same way. Extending $\mathbb{R}$ to $\mathbb{R}^*$ by addition of $-\infty, +\infty$ with $\forall x \in \mathbb{R}(-\infty < x < \infty)$, we can define $(-\infty, b), (-\infty, b], (b, \infty), [b, \infty)$; $\mathbb{R} \equiv (-\infty, \infty)$ as above. □

REMARK. The definitions of (a, b), $[a, b]$ can be extended to arbitrary pairs a, b, i.e. without the condition $a \leq b$: if $c = \min(a, b)$, $d = \max(a, b)$, we put

$$(a, b) := (c, d), \quad [a, b] := [c, d],$$

where (c, d) and $[c, d]$ are defined as before.

We shall now study the constructive validity of the intermediate-value theorem. We first present a

1.2. *Counterexample to the intermediate-value theorem.* The intermediate value theorem cannot be proved in the strong form, familiar from classical analysis, with $\exists x \in [a, b](f(x) = c)$ as its conclusion. A weak counterexample is easily constructed: let a be a real number for which $a \leq 0$ or $a \geq 0$ is unknown, and define a uniformly continuous f on $[0, 3]$ by

$$f(x) := \min(x - 1, 0) + \max(0, x - 2)$$

(see fig. 6.1) and let

$$f^*(x) := f(x) + a.$$

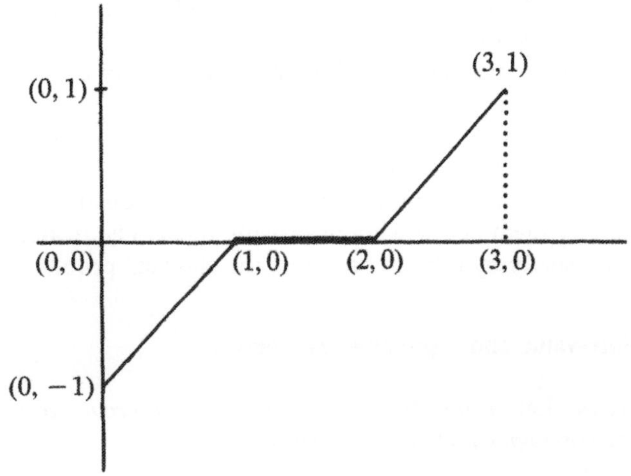

Fig. 6.1

Then, obviously, if $f^*(x) = 0$ for some $x \in [0, 3]$, we can decide whether $x < 2$ or $x > 1$; but in the first case $a \geq 0$, and in the second case $a \leq 0$. Therefore we are unable to determine x with any desired degree of accuracy. It is routine to transform this weak counterexample into a proper refutation of the universally quantified statement on the assumption of $\neg$SEP (cf. 4.3.6, 5.2.12).

Note that in terms of classical mathematics the counterexample shows that intermediate values cannot be found continuously from the data, i.e. the function itself as element of a metrizable function space. Here the metric is the familiar supremum metric:

$$\rho(f, g) := \sup\{|f(x) - g(x)| : x \in [a, b]\};$$

as we shall see below (1.8), this metric is constructively well-defined. □

1.3. PROPOSITION (*connectedness of the interval*). *Let $a < b$, and let A, B be inhabited open subsets of $[a, b]$ covering $[a, b]$, i.e. $[a, b] = A \cup B$; then there is a real $d \in A \cap B$.*

PROOF. We construct Cauchy sequences $\langle a_n \rangle_n \subset A$, $\langle b_n \rangle_n \subset B$, $\lim\langle a_n \rangle_n = \lim\langle b_n \rangle_n = c$ as follows. A and B are inhabited, so let $a_0 \in A$, $b_0 \in B$. Put $c_0 = \frac{1}{2}(a_0 + b_0)$; then $c_0 \in A$ or $c_0 \in B$. If $c_0 \in A$, we can take $a_1 := c_0$, $b_1 := b_0$, and if $c_0 \in B$, we may take $a_1 := a_0$, $b_1 := c_0$. Then $|a_1 - b_1| = \frac{1}{2}|a_0 - b_0|$ and $a_1 \in A$, $b_1 \in B$. Repeating this we find sequences $\langle a_n \rangle_n \subset A$, $\langle b_n \rangle_n \subset B$ with $|a_n - b_n| \leq 2^{-n}|a_0 - b_0|$. Obviously $\langle a_n \rangle_n$ and $\langle b_n \rangle_n$ converge, so let $c = \lim\langle a_n \rangle_n = \lim\langle b_n \rangle_n$, $c \in A$ or $c \in B$; say $c \in A$. A is open, and therefore $U(2^{-k}, c) \subset A$ for some k, hence for n sufficiently large, $b_n \in U(2^{-k}, c) \subset A$, so $b_n \in A \cap B$ and we may take $d := b_n$. □

1.4. THEOREM (*intermediate-value theorem*). *Let $f \in [a, b] \to \mathbb{R}$ be continuous, $a < b$, $f(a) \leq c \leq f(b)$, then*

$$\forall k \exists x \in [a, b](|f(x) - c| < 2^{-k}).$$

PROOF. $A \equiv \{x : f(x) < c + 2^{-k}\}$ and $B \equiv \{x : f(x) > c - 2^{-k}\}$ are open because of the continuity of f, and $a \in \{x : f(x) < c + 2^{-k}\}$, $b \in \{x ; f(x) > c - 2^{-k}\}$. $\{A, B\}$ covers $[a, b]$, since $c - 2^{-k} < c + 2^{-k}$, and so $f(x) < c + 2^{-k}$ or $c - 2^{-k} < f(x)$. By the preceding lemma we can find an x in the intersection, which therefore satisfies $c - 2^{-k} < f(x) < c + 2^{-k}$. □

The following version of the intermediate value theorem retains the strong conclusion of the classical version, but strengthens the hypothesis:

1.5. THEOREM. *Let f be continuous on $[a, b]$, $a < b$, $f(a) < 0 < f(b)$, and such that $\forall y, z \in [a, b](y < z \to \exists x \in [y, z](f(x) \# 0))$. Then $\exists x \in [a, b](f(x) = 0)$.*

PROOF. The proof is a slight variation on the well-known bisection argument: x is obtained as the common limit of $\langle x_n \rangle_n, \langle y_n \rangle_n$ with $\forall n(x_n \le x_{n+1} < y_{n+1} \le y_n)$, $f(x_n) < 0$, $f(y_n) > 0$ for all n. For x_0, y_0 we take a and b respectively. Suppose we have already constructed x_n, y_n. Find a z in a segment I with midpoint $\frac{1}{2}(x_n + y_n)$ and length $\frac{1}{2}(y_n - x_n)$ such that $f(z) \# 0$; then $f(z) < 0$ or $f(z) > 0$. If $f(z) < 0$, take $x_{n+1} := z$, $y_{n+1} := y_n$; if $f(z) > 0$, let $x_{n+1} := x_n$, $y_{n+1} := z$. In both cases $|x_{n+1} - y_{n+1}| \le \frac{3}{4}|x_n - y_n|$. Hence $\langle x_n \rangle_n$ and $\langle y_n \rangle_n$ have the same limit. Because of the continuity of f, $f(\lim\langle x_n \rangle_n) \le 0$, $f(\lim\langle y_n \rangle_n) \ge 0$, so $f(x) = 0$. □

REMARK. The condition of the theorem is satisfied, for example, by all polynomials (see 2.7).

Next we shall show that a uniformly continuous function on a segment of finite length always does have a supremum, though the supremum is not necessarily assumed.

1.6. DEFINITION. We recall that x is a *supremum* (sup) or *least upper bound* (l.u.b.) of a set $X \subset \mathbb{R}$ iff

$$\forall y \in X(y \le x) \wedge \forall k \exists y \in X(y > x - 2^{-k}).$$

Similarly for an *infimum* (inf) or *greatest lower bound* (g.l.b), we require $\forall k \exists y \in X(x + 2^{-k} > y)$. The *supremum* (*infimum*) of a function f on $[a, b]$ is the sup (inf) of $\{fx : x \in [a, b]\}$ (notation: $\sup(f), \inf(f)$).

$X \subset \mathbb{R}$ is *totally bounded* if for each $k \in \mathbb{N}$ there are $x_0, \ldots, x_{n-1} \in X$ such that $\forall x \in X \exists i < n(|x - x_i| < 2^{-k})$. □

1.7. LEMMA. *A totally bounded $X \subset \mathbb{R}$ has an infimum and a supremum.*

PROOF. For each k let $\{x_{k,0}, \ldots, x_{k,n(k)}\}$ be such that $\forall x \in X \exists i \le n(k)$ $(|x - x_{k,i}| < 2^{-k})$. Let $x_k := \max\{x_{k,i} : i \le n(k)\}$. Now it is easy to see that $\forall x \in X \forall k(x < x_k + 2^{-k})$.

From $\forall x \in X \forall k(x < x_k + 2^{-k})$ we conclude $x_{k+1} < x_k + 2^{-k}$ and $x_k < x_{k+1} + 2^{-k-1}$. So $|x_k - x_{k+1}| < 2^{-k}$. Hence we get $|x_k - x_{k+n}| <$

$2^{-k} + 2^{-k-1} + 2^{-k-n-1} < 2^{-k+1}$. $\langle x_n \rangle_n$ is therefore a Cauchy sequence with limit x. We leave it as an exercise to show that $x = \sup(X)$. □

1.8. THEOREM. *Let* $f \in [a, b] \to \mathbb{R}$ *be uniformly continuous on* $a < b$, *then* $\sup(f)$ *and* $\inf(f)$ *exist.*

PROOF. Let α be a modulus for f such that $\forall x, y \in [a, b] \forall k(|x - y| < 2^{-\alpha k} \to |f(x) - f(y)| < 2^{-k})$. $\{fx : x \in [a, b]\}$ is totally bounded: let n be such that $(b - a)n^{-1} < 2^{-\alpha k}$, and let $x_i := a + i(b - a)n^{-1}$ for $i < n$. Then by uniform continuity $\forall x \in [a, b] \exists i < n(|f(x) - f(x_i)| < 2^{-n})$. Now apply the preceding lemma. □

1.9. COROLLARY. *For* $f \in [a, b]$ *uniformly continuous,* $a < b$, *we have* $\forall k \exists x \in [a, b](\sup(f) - f(x) < 2^{-k})$ *and* $\forall k \exists x \in [a, b](f(x) - \inf(f) < 2^{-k})$. □

1.10. *Counterexample to the existence of a maximum.* It is again easy to show that we cannot strengthen the conclusion of the preceding theorem to $\exists x \in [a, b](f(x) = \sup(f))$. Let a be any real such that $a \le 0$ or $a \ge 0$ is unknown, and define f on $[0, 1]$ by

$$f(x) := ax.$$

It is easy to see that $\sup(f) = \max(0, a)$: if $a < 0$, then $\sup(f) = 0$; if $a = 0$ or $a > 0$, $\sup(f) = a$, and since, by 5.2.11(ix), $\neg\neg(a = 0 \vee a < 0 \vee a > 0)$, also $\neg\neg\sup(f) = \max(0, a)$. Hence, by the stability of equality for reals (5.2.10), $\sup(f) = \max(0, a)$. Now assume that for some $x \in [0, 1]$ $f(x) = \sup(f)$; then $x > 0 \vee x < 1$, which implies $a \ge 0$ or $a \le 0$.

The transformation into a refutation, on the basis of $\neg$SEP (cf. 4.3.6, 5.2.12), is now routine. There is also a classical counterpart: we cannot determine the real x for which $f(x) = \sup(f)$ continuously in f.

2. Differentiation and integration

2.1. DEFINITION. $f \in \mathbb{R} \to \mathbb{R}$ is *differentiable at* x_0 (with *modulus* α) if for some $a \in \mathbb{R}$

$$\forall k \forall x\big(|x_0 - x| < 2^{-\alpha k} \to |a(x - x_0) - (fx - fx_0)| \le 2^{-k}|x - x_0|\big).$$

The uniquely determined a is called the *derivative of* f *at* x_0. More important is the following notion.

Let $a < b$; $f \in [a, b] \to \mathbb{R}$ is *uniformly differentiable* on $[a, b]$, with *derivative* f', iff for some α

$$\forall k \forall x, y \in [a, b]\big(|x - y| < 2^{-\alpha k} \to$$
$$|f'(x)(y - x) - (fy - fx)| < 2^{-k}|y - x|\big).$$

For iterated derivatives we shall use the familiar notation: $f^{(0)} := f$, $f^{(1)} := f'$, $f^{(n+1)} := (f^{(n)})'$. □

REMARK. The notion "f is differentiable on $[a, b]$", defined by $\forall x \in [a, b]$ (f is differentiable at x) does not lead to a smooth theory; we need the (classically equivalent) notion of uniform differentiability.

We met with a similar phenomenon in the theory of continuous $f \in \mathbb{R} \to \mathbb{R}$, where the requirement of uniform continuity on all bounded closed intervals leads to a smoother theory.

2.2. PROPOSITION. *A uniformly differentiable f has a uniformly continuous derivative f', and is itself uniformly continuous.*

PROOF. Exercise. □

2.3. PROPOSITION. *Let f, g be uniformly differentiable on bounded intervals with derivatives f' and g' respectively. Then*

$$(f + g)'(x) = f'(x) + g'(x),$$
$$(f \cdot g)(x) = f'(x) \cdot g(x) + f(x) \cdot g'(x),$$

and if f has a positive l.u.b. on the interval I, then on I

$$(f^{-1})'(x) = -f'(x)/f(x)^2$$

PROOF. Exercise. □

The elementary theory of differentiability is rather similar to the classical theory. We shall first establish a version of Rolle's theorem. Though weaker than the classical version, it suffices for important applications such as Taylor's series in 2.5 below.

2.4. LEMMA (*Rolle's theorem*). *Let f be uniformly differentiable on $[a, b]$, $a < b$, $f(a) = f(b)$. Then* $\inf(|f'|) = 0$, *i.e.* $\forall n \exists x \in [a, b](|f'(x)| < 2^{-n})$.

PROOF. Let $m := \inf\{|f'(x)| : x \in [a, b]\}$, this exists by 2.2 and 1.8. Suppose $m > 0$. Then in particular $f'(a) \leq -m$ or $f'(a) \geq m$; let us assume $f'(a) \geq m$. Then for all $x \in [a, b]$ $f'(x) \geq m$. For if for some x in $[a, b]$ $f'(x) < m$, then $f'(x) < m - 2^{-k}$ for some k. Now apply the intermediate value theorem to f' and $[a, x]$: there is a y such that $|f'(y) - (m - 2^{-k})| < 2^{-k-1}$, i.e. $|f'(y)| < m$, contradicting the definition of m.

Let k be such that

$$|x - y| < 2^{-k} \to |f'(x)(y - x) - (fy - fx)| \leq \tfrac{1}{2}m,$$

and choose $p \in \mathbb{N}$ such that $(b - a)p^{-1} < 2^{-k}$, and let $x_i = a + i(b - a)p^{-1}$ for $i \leq p$. Then $0 = fb - fa =$

$$\sum_{k=0}^{n-1} (fx_{k+1} - fx_k) =$$

$$\sum_{k=0}^{n-1} f'(x_k)(x_{k+1} - x_k) +$$

$$\sum_{k=0}^{n-1} (f(x_{k+1}) - f(x_k) - f'(x_k)(x_{k+1} - x_k)) \geq$$

$$\sum_{k=0}^{n-1} m(x_{k+1} - x_k) - \sum_{k=0}^{n-1} \tfrac{1}{2}m(x_{k+1} - x_k) = \tfrac{1}{2}m(b - a) > 0.$$

This is contradictory, hence $m = 0$ and therefore by 1.9 the lemma follows. □

2.5. THEOREM (*Taylor series*). *Let f be $n + 1$ times uniformly differentiable on a closed interval $[a, b]$ and let the remainder function R be defined by*

$$f(a + x) := \sum_{k=0}^{n} (f^{(k)}(a)/k!)x^k + R(a + x).$$

For all k there is a $c \in [a, b]$ such that

$$(\Delta) \qquad |R(b) - (b - c)^n(b - a)f^{(n+1)}(c)/n!| \leq 2^{-k}.$$

PROOF. (i) If $|b - a| = 0$, we may take $c = b = a$.

(ii) Let now $|b - a| > 0$. We put

$$g(x) = f(b) - \big[f(x) + (f^{(1)}(x)/1!)(b - x) + \cdots + (f^{(n)}(x)/n!)(b - x)^n)\big] - R(b)(b - x)/(b - a).$$

g is continuous and uniformly differentiable (2.2, 2.3), $g(a) = g(b) = 0$.

$$\begin{aligned} g'(x) = \{ & -f^{(1)}(x) + (f^{(1)}(x)/0!) - (f^{(2)}(x)/1!)(b-x) \\ & + (f^{(2)}(x)/1!)(b-x) - (f^{(3)}(x)/2!)(b-x)^2 \\ & + (f^{(3)}(x)/2!)(b-x)^2 - (f^{(4)}(x)/3!)(b-x)^3 + \cdots \\ & + (f^{(n)}(x)/(n-1)!)(b-x)^{n-1} \\ & - (f^{(n+1)}(x)/n!)(b-x)^n \} + R(b)/(b-a) \\ = & -(f^{(n+1)}(x)/n!)(b-x)^n + R(b)/(b-a). \end{aligned}$$

With Rolle' theorem we find therefore a $c \in [a, b]$ such that

$$|g'(c)| = \left| -(f^{(n+1)}(c)/n!)(b-c)^n + R(b)/(b-a) \right| < 2^{-k}/(b-a).$$

The preceding argument shows that

$$|b-a| = 0 \vee |b-a| \# 0 \to \sup_{c \in [a, b]} \left(|R(b) - (f^{(n+1)}(c)/n!)(b-c)^n(b-a)| \right) < 2^{-k}$$

hence, since $\neg\neg(|b-a| = 0 \vee |b-a| \# 0)$, the assertion of the theorem follows. □

2.6. LEMMA. *Let f be $n+1$ times uniformly differentiable on $[a, b]$, such that*

$$\inf_{x \in [a, b]} \left(\sum_{k=0}^{n} |f^{(k)}(x)| \right) > 0.$$

Then in each interval $[c, d] \subset [a, b]$ with $c > d$ there is an x such that $f(x) \# 0$.

PROOF.

$$C := \inf_{x \in [a, b]} \left(\sum_{i=0}^{n} |f^{(i)}(x)| \right).$$

If $c + x \in [c, d]$, we have by Taylor's development

$$f(c+x) = f(c) + \left(f^{(1)}(c)/1!\right)x + \left(f^{(2)}(c)/2!\right)x^2 + \cdots + \left(f^{(n)}(c)/n!\right)x^n + R(c+x),$$

where

$$\left|R(c+x) - \left(f^{(n+1)}(e)/(n+1)!\right)(c+x-e)^n x\right| \le 2^{-m}$$

for suitable $e \in [c, c+x]$. The functions $f, f^{(1)}, \ldots, f^{(n+1)}$ are uniformly continuous on $[a, b]$, hence there is a $D \in \mathbb{R}$ such that

$$|f^{(i)}(x)| \le D \text{ for } x \in [a, b], \quad i \le n+1.$$

On the other hand there is an $i \le n$, such that

$$|f^{(i)}(c)| > C/(n+2)$$

so $f^{(i)}(c) \# 0$. If $i = 0$, we are done. So assume $i > 0$. Let $c + x \in [c, d]$ and put

$$Q := f(c) + f^{(1)}(c)x/1! + \cdots + f^{(i)}(c)x^i/i!,$$
$$Q' := f^{(i+1)}(c)x^{i+1}/(i+1)! + \cdots + f^{(n)}(c)x^n/n! + R(c+x),$$

then

$$f(c+x) = Q + Q'.$$

Assume $|f^{(i)}(c)/i!| > 2^{-k}$ and choose $x = 2^{-p}$; then

$$|f^{(i)}(c)/i!| \cdot |x|^i > 2^{-k-pi}.$$

Also

$$Q' < |x| \cdot n \cdot D \cdot 2^{-pi} = n \cdot D \cdot 2^{-p(i+1)},$$

since we can approximate $R(c+x)$ arbitrarily close by a term of the form

$$\left(f^{(n+1)}(e)/n!\right)(c+x-e)^n x$$

with $e \in [c, c+x]$. By choosing p sufficiently large, we can achieve

$$D \cdot 2^{-p(i+1)} < 2^{-pi-k-2}.$$

Also $|Q| > 2^{-k-pi-1}$ or $|Q| < 2^{-k-pi}$. If $|Q| > 2^{-k-pi-1}$, $|f(c+x)| \ge ||Q| - |Q'|| > 2^{-k-pi-1} - 2^{-k-pi-2} = 2^{-k-pi-2}$. If $|Q| < 2^{-k-pi}$, then by

$|f^{(i)}(c)x^i/i!| > 2^{-k-pi}$, there is $j < i$ such that $|f^{(j)}(c)x^j/j!| \# 0$; now repeat the argument with j instead of i. Thus in a finite number of steps we obtain $|f(c+x)| \# 0$ for suitable x. □

2.7. Theorem. *Let f be a polynomial in x with $f(0) < 0$, $f(1) > 0$; then there is an $x \in [0,1]$ with $f(x) = 0$.*

Proof. Let $f(x) \equiv a_n x^n + a_{n-1}x^{n-1} + \cdots + a_0$. Since $f(0) < 0$ and $f(1) > 0$ we have $a_0 < 0$ and $a_n + a_{n-1} + \cdots + a_0 > 0$, hence there is an $i > 0$ such that $a_i \# 0$. If $a_n \# 0$, $f^n(x) = a_n \cdot n! \# 0$ and the hypotheses of 2.6 are fulfilled. If $a_i \# 0$, $i < n$, then

$$f^{(i)}(x) = (n!/(n-i)!)a_n x^{n-i}$$
$$+((n-1)!/(n-1-i)!)a_{n-i}x^{n-i-1} + \cdots + a_i i!,$$

and

$$\left(\inf|f^{(i)}| > 2^{-2}|a_i| \cdot i!\right) \vee \left(\inf|f^{(i)}| < 2^{-i}|a_i| \cdot i!\right).$$

In the first case we are finished; in the second case we find a $j > i$ with $a_j \# 0$ and can repeat the argument. Ultimately we find that f satisfies the hypothesis of 2.6, and so with 1.5 we find x with $f(x) = 0$. □

Corollary. $\mathbb{R}$ *is real-closed field* (*cf. chapter* 8). □

Riemann integration (2.8–2.14). We shall show that for any closed interval $[a, b]$ with $a < b$ and every uniformly continuous $f \in [a, b] \to \mathbb{R}$, the Riemann integral $\int_a^b f$ exists. The treatment is rather similar to treatments in classical mathematics, so we give an outline only, leaving a lot of details to the reader as exercises. For the time being, $[a, b]$ will be fixed.

2.8. Definition. The sequence $P \equiv \langle a_0, \ldots, a_n \rangle$ with $a_0 = a$, $a_n = b$, $a_0 \le a_1 \le \cdots \le a_n$ is called a *partition* of $[a, b]$; if $a_i = a + r_i(b-a)$, $r_i \in \mathbb{Q}$ for $i \le n$, we have a *rational* partition. n is called the *length* of the partition $\langle a_0, \ldots, a_n \rangle$, and

$$\text{mesh}(P) := \sup\{|a_{i+1} - a_i| : i \le n-1\}.$$

A partition Q is a *refinement* of P if P is a subsequence of Q. □

Notation. Let f be any uniformly continuous f on $[a, b]$; we use $S(f, P)$ to denote any sum of the form

$$\sum_{i=0}^{n-1} f(x_i)(a_{i+1} - a_i) \quad \text{with } x_i \in [a_i, a_{i+1}].$$

If P is the rational partition $\langle a_0, \ldots, a_n \rangle$ with $a_i = a + i(b - a)n^{-1}$, then

$$S(f, a, b, n) := (b - a)n^{-1}\left(\sum_{i=0}^{n-1} f(a_i) \right)$$

is an $S(f, P)$.

2.9. Theorem. *Let f be uniformly continuous on $[a, b]$ with modulus α. Then $\langle S(f, a, b, n) \rangle_n$ converges to a limit $\int_a^b f$. Moreover, for any rational partition P of $[a, b]$, with* mesh$(P) \leq 2^{-\alpha k}$ *we have for all $S(f, P)$*

$$\left| S(f, P) - \int_a^b f \right| \leq 2^{-k}(b - a).$$

Proof. The proof heavily relies on the fact that all partitions considered are rational. Let $P \equiv \langle a_0, \ldots, a_n \rangle$ be a rational partition of $[a, b]$ with mesh$(P) \leq 2^{-\alpha k}$, and let $Q \equiv \langle b_0, \ldots, b_m \rangle$ be a rational refinement of P. Let us write Σ_i for summation over the j with $a_i \leq b_j < a_{i+1}$. Then

$$\begin{aligned}
&|S(f, P) - S(f, Q)| \\
&\quad = \left| \sum_{i=0}^{n-1} f(x_i)(a_{i+1} - a_i) - \sum_{j=0}^{m-1} f(y_j)(b_{j+1} - b_j) \right| \\
&\quad = \left| \sum_{i=0}^{n-1} f(x_i)\Sigma_i(b_{j+1} - b_j) - \sum_{i=0}^{n-1} \Sigma_i f(y_j)(b_{j+1} - b_j) \right| \\
&\quad \leq \sum_{i=0}^{n-1} \Sigma_i |f(x_i) - f(y_j)|(b_{j+1} - b_j) \\
&\quad \leq \sum_{i=0}^{n-1} \Sigma_i 2^{-k}(b_{j+1} - b_j) = 2^{-k}(b - a).
\end{aligned}$$

Now let R be any other rational partition of $[a, b]$ with mesh $(R) \leq 2^{-\alpha k}$, and let Q be a common rational refinement of P and R; then

$|S(f, P) - S(f, R)| \le 2^{-k+1}(b-a)$. In particular

$$|S(f, a, b, m) - S(f, a, b, m')| \le 2^{-k+1}(b-a)$$

for all $m, m' \ge (b-a)2^{\alpha k}$, hence $\langle S(f, a, b, n)\rangle_n$ is a Cauchy sequence, etcetera. □

2.10. LEMMA. *Let* $P \equiv \langle a_0, \ldots, a_n\rangle$ *be any partition with* $\mathrm{mesh}(P) < \varepsilon$, *then for any* $\delta > 0$ *there is a rational partition* P' *of* $[a_0, a_n]$ *with* $\mathrm{mesh}(P') < \varepsilon$, $P' \equiv \langle a'_0, \ldots, a'_n\rangle$, *such that for all*

$$S(f, P) := \sum_{i=0}^{n-1} f(x_i)(a_{i+1} - a_i),\ S(f, P') := \sum_{i=0}^{n-1} f(x'_i)(a'_{i+1} - a'_i)$$

we have

$$|S(f, P') - S(f, P)| < \delta.$$

PROOF. Exercise. □

As a corollary we have

2.11. THEOREM. *For any* $S(f, P)$ *with* $\mathrm{mesh}(P) < 2^{-\alpha k}$ *we have*

$$\left|S(f, P) - \int_a^b f\right| \le 2^{-k}(b-a).\quad □$$

The following propositions and theorem are also easily proved; their proofs are left as exercises.

2.12. PROPOSITION. *Let* $a \le b \le c$, *and let* f, g *be uniformly continuous on* $[a, c]$. *Then*

(i) $\int_a^c(x \cdot f + y \cdot g) = x \cdot (\int_a^c f) + y \cdot (\int_a^c g)$;
(ii) $\forall x \in [a, c](f(x) \le g(x)) \to \int_a^c f \le \int_a^c g$;
(iii) $\int_a^c f = \int_a^b f + \int_b^c f$. □

2.13. PROPOSITION. *Let* p *be a non-negative polygonal function on* $[a, b]$, *then* $\int_a^b p$ *is the area of* $\{(x, y) : x \in [a, b],\ y \in [0, p(x)]\}$. □

2.14. THEOREM. *Let* f *be uniformly continuous on* $[a, b]$. *Then* g *defined by* $g(x) = \int_a^x f$ *is uniformly differentiable on* $[a, b]$ *with* $g'(x) = f(x)$. □

3. Consequences of WC-N and FAN

3.1. *A spread-representation of* [0, 1]. We start with the construction of a spread-representation of the interval [0, 1]. Every $x \in [0,1]$ can be approximated by a fundamental sequence $\langle \alpha(n) \cdot 2^{-n-1} \rangle_n$ such that

$$\forall n(\alpha(n) \le 2^{n+1}) \wedge \forall n(|2\alpha(n) - \alpha(n+1)| \le 1). \tag{1}$$

To see this, choose α such that $|x - \alpha(n) \cdot 2^{-n-1}| < (5/8) \cdot 2^{-n-1}$, $\alpha(n) \le 2^{n+1}$. Then $|\alpha(n) \cdot 2^{-n-1} - \alpha(n+1) \cdot 2^{-n-2}| \le |x - \alpha(n) \cdot 2^{-n-1}| + |x - \alpha(n+1) \cdot 2^{-n-2}| < (5/8)(2^{-n-1} + 2^{-n-2})$, hence $|2\alpha(n) - \alpha(n+1)| < 15 \cdot 2^{-3} < 2$. Let T be the spread such that $\alpha \in T$ iff (1) holds, and put for $\alpha \in T$

$$x_\alpha := \lim \langle \alpha(n) \cdot 2^{-n-1} \rangle_n, \quad x_\alpha^n := \alpha(n) \cdot 2^{-n-1},$$

$$V'_n := \{x_\alpha : \alpha \in n \wedge \alpha \in T\} \text{ for } n \in T.$$

T, provided with the tree topology of 4.1.5 is (homeomorphic to) the Cantor discontinuum, and this representation of [0, 1] shows in fact that [0, 1] is the image of the Cantor discontinuum under a continuous mapping φ; note that $V'_n = \varphi[V_n]$ (notation of 4.1.5).

In fact, we can show a little bit more: [0, 1] is a quotient of T (cf. Engelking 1968, p. 83). More precisely, we have the following

3.2. **Proposition.**
(i) $\forall \alpha (x_\alpha \in [0,1])$;
(ii) $\forall n \forall \alpha \in T(|x_\alpha - x_{\alpha}^{n}| \le 2^{-n-1})$, *hence* $\forall n (V'_{\bar{\alpha}(n+1)} \subset U(2^{-n-1}, x_\alpha))$ *for* $\alpha \in T$;
(iii) $\forall x \in [0,1] \exists \alpha \in T(x = x_\alpha \wedge \forall k(U(2^{-k-4}, x) \subset V'_{\bar{\alpha}k}))$.

Proof. (i) is obvious; for (ii) observe that

$$|x_\alpha - x_\alpha^n| \le \sum_{k=0}^{\infty} |x_\alpha^{n+k+1} - x_\alpha^{n+k}| \le \sum_{k=0}^{\infty} 2^{-n-k-2} \le 2^{-n-1}.$$

(iii) Suppose $|x - y| < 2^{-k-5}$, $x, y \in [0,1]$ and determine $\alpha, \beta \in T$ such that for all n $|x - \alpha(n) \cdot 2^{-n-1}| < (5/8)2^{-n-1}$, $|y - \beta(n) \cdot 2^{-n-1}| < (5/8)2^{-n-1}$, then $|2^{-k-1}\alpha(k) - 2^{-k-2}\beta(k+1)| \le |x - 2^{-k-1}\alpha(k)| + |y - x| + |y - 2^{-k-2}\beta(k+1)| < (5/8) \cdot 2^{-k-1} + (5/8) \cdot 2^{-k-2} + 2^{-k-5} = 2^{-k-1}$, so $|2\alpha(k) - \beta(k+1)| \le 1$.
Therefore $\gamma = \bar{\alpha}(k+1) * \lambda n.\beta(k+1+n) \in T$, and $x_\gamma = y \in V'_{\bar{\alpha}(k+1)}$. □

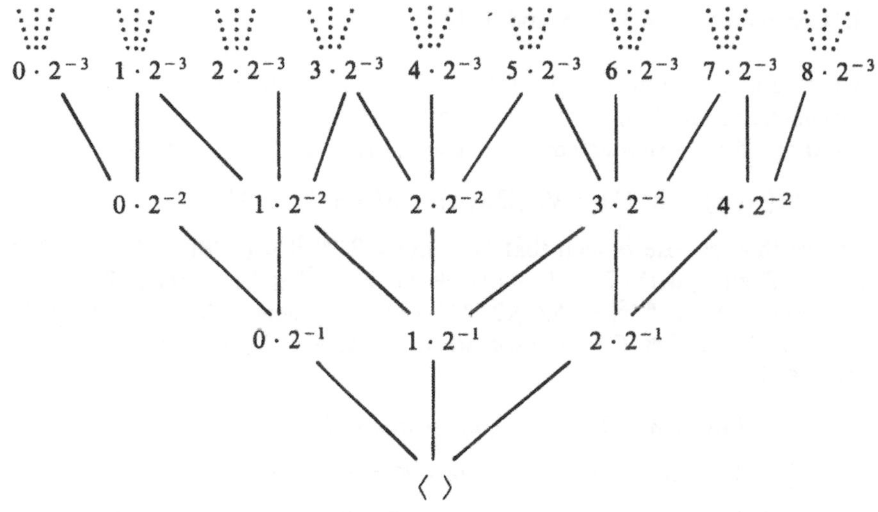

Fig. 6.2

N.B. Property (iii) is the quotient-property.

It is easy to draw a picture of T and the associated sequences $\langle \alpha(n)2^{-n-1}\rangle_n$. The infinite branches in fig. 6.2 correspond to the associated sequences. The branches passing through $k \cdot 2^{-n}$ represent the interval $[(k-1)2^{-n},(k+1)2^{-n}] \cap [0,1]$.

3.3. Proposition (WC-N). *Let* $\{X_n : n \in I\}$ *with* $I \subset \mathbb{N}$ *be a cover of* $\mathbb{R}$, *then* $\{\text{Int}(X_n) : n \in I\}$ *is again a cover (and similarly for any segment instead of* $\mathbb{R}$).

Proof. Without loss of generality let $I = \mathbb{N}$ (if necessary we can replace $\{X_n : n \in I\}$ by $\{X'_n : n \in \mathbb{N}\}$ with $X'_n := \{x : x \in X_n \wedge n \in I\}$). It suffices to show that for any cover $\langle X_n\rangle_n$ of $[-m, m]$, $\langle \text{Int}(X_n)\rangle_n$ also covers $[-m, m]$, and for this in turn it is enough to consider covers $\langle X_n\rangle_n$ of $[0,1]$. Therefore, let $[0,1] \subset \bigcup\{X_n : n \in \mathbb{N}\}$, i.e.

$$\forall \alpha \in T \exists n (x_\alpha \in X_n).$$

Let α_T, β_T range over T (notation as in 4.1.2), then with WC-N

$$\forall \alpha_T \exists mn \forall \beta_T \in \bar{\alpha} m \big(x_\beta \in X_n \big),$$

so

$$\forall \alpha_T \exists mn \forall x \in V'_{\bar{\alpha} m} \big(x \in X_n \big).$$

Now choose, by 3.2(iii), for any $x \in [0,1]$ an α such that $x = x_\alpha$, $\forall n (U(2^{-n-4}, x_\alpha) \subset V'_{\bar{\alpha} n}$; then $\forall x \exists \alpha \exists mn (U(2^{-m-4}, x_\alpha) \subset X_n \wedge x_\alpha = x)$. Hence each x belongs to the interior of some X_n. □

3.4. Theorem (WC-N). *All mappings $f \in X \to \mathbb{R}$, X a segment, are continuous.*

Proof. Fix $k \in \mathbb{N}$, let $\langle r_n \rangle_n$ be a fixed enumeration of $\mathbb{Q}$, and let

$$X_n := \big\{ x : |f(x) - r_n| < 2^{-k} \big\}.$$

Then $\langle X_n \rangle_n$ covers X, hence $\langle \mathrm{Int}(X_n) \rangle_n$ covers X, i.e.

$$\forall x \in X \exists m \forall y \in U(2^{-m}, x) \cap X \big(|f(y) - r_n| < 2^{-k} \big),$$

hence

$$\forall x \in X \exists m \forall y \in X \big(|x - y| < 2^{-m} \to |f(x) - f(y)| < 2^{-k+1} \big). \quad \square$$

The next few applications require FAN.

3.5. Theorem (FAN; *compactness of the interval*). *Let $[a, b]$ be a segment with $a < b$, and $\{W_i : i \in I\}$ an open cover of $[a, b]$; then there is a finitely indexed subcover $W_{i_0}, \ldots, W_{i_n}$.*

Proof. It suffices to prove the theorem for $[0,1]$. As noted in 4.7.3, FAN expresses compactness of T, and since we just constructed in 3.1–2 a continuous mapping φ from T onto $[0,1]$ and compactness is clearly preserved under continuous mappings, $[0,1]$ is compact.

The argument may be spelt out more formally as follows. We obviously have

$$\forall \alpha_T \exists m \exists i \in I \big(U(2^{-m}, x_\alpha) \subset W_i \big).$$

If $U(2^{-m}, x_\alpha) \subset W_i$, then because of $x_\alpha \in V'_{\bar{\alpha}m} \subset U(2^{-m}, x_\alpha)$ we find

$$\forall \alpha_T \exists m \exists i \in I(V'_{\bar{\alpha}m} \subset W_i).$$

By FAN we find a k such that

$$\forall \alpha_T \exists i \in I(V'_{\bar{\alpha}k} \subset W_i).$$

There are finitely many possible values for $\bar{\alpha}k$; selecting a W_i for each $\bar{\alpha}k$ we find $\{W_{i_0}, \ldots, W_{i_n}\}$ covering $[0, 1]$. □

3.6. THEOREM (FAN). *Each continuous* $f \in [a, b] \to \mathbb{R}$ *is uniformly continuous.*

PROOF. By the continuity of f, for fixed $n \in \mathbb{N}$

$$\forall \alpha_T \exists m \exists i (f[U(2^{-m}, x_\alpha)] \subset U(2^{-n}, r_i)),$$

where $\langle r_n \rangle_n$ is a standard enumeration of $\mathbb{Q}$ as before. Hence, since $x_\alpha \in V'_{\bar{\alpha}m} \subset U(2^{-m}, x_\alpha)$,

$$\forall \alpha_T \exists m \exists i (f[V'_{\bar{\alpha}m}] \subset U(2^{-n}, r_i)).$$

Thus we find a k such that

$$\forall \alpha_T \exists i (f[V'_{\bar{\alpha}k}] \subset U(2^{-n}, r_i)).$$

Suppose $|x - y| < 2^{-k-4}$; choose $x = x_\alpha$ such that $\forall m(U(2^{-m-4}, x_\alpha) \subset V'_{\bar{\alpha}m})$, then in particular $y \in V'_{\bar{\alpha}k}$, $f(x) = f(x_\alpha) \in U(2^{-n}, r_i)$, $f(y) \in U(2^{-n}, r_i)$ for some i, and thus $|f(x) - f(y)| < 2^{-n+1}$. Therefore

$$|x - y| < 2^{-k-4} \to |f(x) - f(y)| < 2^{-n+1}. \quad \square$$

3.7. THEOREM (FAN). *Let* $f \in [a, b] \to \mathbb{R}$ *be continuous. Then*
(i) *f has a supremum,*
(ii) $\forall x \in [a, b](f(x) > 0) \to \inf(f) > 0$.

PROOF. (i) Immediate by 3.6 and 1.8.
(ii) The cover $\langle W_n \rangle_n$ with $W_n := \{x : fx > 2^{-n}\}$ is an open cover since f is continuous; by 3.5 we find a subcover $W_{n_0}, \ldots, W_{n_p}$. Take $n := \max\{n_0, \ldots, n_p\}$, then $\forall x \in [a, b](f(x) > 2^{-n})$. □

3.8. FAN* combines FAN with WC-N, i.e. FAN* is

$$\forall \alpha_T \exists n A(\alpha, n) \to \exists n \forall \alpha_T \exists m \forall \beta \in \bar{\alpha}_T(n) A(\beta, m).$$

Assuming FAN* we can strengthen 3.6, 3.7 by dropping the condition that f has to be continuous:

THEOREM (FAN*). *Let* $f \in [a, b] \to \mathbb{R}$, *then*
(i) *f is uniformly continuous and has a supremum.*
(ii) $\forall x \in [a, b](f(x) > 0) \to \inf(f) > 0$. □

The first half of (i) is often called "Brouwer's continuity theorem".

4. Analysis in CRM: consequences of CT_0, ECT_0, MP

In the present section we shall demonstrate some of the striking divergences from traditional analysis in CRM, such as the existence of continuous, unbounded (hence not uniformly continuous) functions on [0, 1], uniformly continuous functions not assuming their supremum on [0, 1], etc.

The principal tool in many of these constructions is the notion of a singular cover.

4.1. DEFINITION. Let us denote the length of an open interval or segment I (which by definition contains its endpoints) by $|I|$. A sequence $\langle I_n \rangle_n$ is a *cover* of $X \subset \mathbb{R}$ iff $\forall x \in X \exists n(x \in I_n)$. $\langle I_n \rangle$ is an *interval cover* if each I_n is an open interval with rational endpoints, of finite length, and a *segmental cover* if each I_n is a segment of finite length with rational endpoints.

A sequence $\langle I_n \rangle_n$ of intervals or segments is called *ε-bounded* iff

$$\forall m \left(\sum_{n=0}^{m} |I_n| < \varepsilon \right).$$

The cover of an interval (a, b), a segment $[a, b]$ (resp. $\mathbb{R}$) is *singular* iff it is ε-bounded for some $\varepsilon < (b - a)$ (resp. some ε).

An interval cover $\langle I_n \rangle_n$ is *universal* iff for each sequence $\langle J_n \rangle_n$ of intervals of finite length with rational endpoints such that $\lim \langle |J_n| \rangle_n = 0$ we have $\exists nm(J_n = I_m)$.

A cover is said to have *finite overlap* if each point belongs to finitely many elements of the cover only. □

4.2. THEOREM (CT_0). *For each $\varepsilon > 0$ a universal ε-bounded interval cover of $\mathbb{R}$ exists (hence such a cover also exists for intervals and segments).*

PROOF. Let $\lambda n.\psi(m, n)$ enumerate all rational intervals of length $< 2^{-m}$, and let φ be the partial function given by $\varphi(m) \simeq \{m\}(m)$. Let γ enumerate the domain of φ without repetition (such a γ exists, why?). Put

$$I_m = \psi(n_0 + \gamma m + 1, \varphi(\gamma m)),$$

where $2^{-n_0} < \varepsilon$. Then

$$\sum_{m=0}^{k} |I_m| < \sum_{m=0}^{k} 2^{-(n_0+\gamma m+1)} \leq \sum_{m=0}^{k} 2^{-(n_0+m+1)} < 2^{-n_0}.$$

For the second inequality, arrange the numbers $\{\gamma 0, \ldots, \gamma k\}$ according to magnitude, say as $m_0 < m_1 < \cdots m_k$; and observe $p \leq m_p$ for $p \leq k$, hence $2^{-(n_0+m_p+1)} \leq 2^{-(n_0+p+1)}$ for $p \leq k$).

Now we prove universality. Let ξ enumerate rational intervals, with $\lim\langle|\xi n|\rangle_n = 0$; suppose δ to be such that $|\xi(\delta n)| < 2^{-n}$, and let $\xi' = \lambda n.\xi(\delta n)$. Then $\forall n \exists m(\xi' n = \psi(n, m))$, hence for some δ' $\forall n(\xi' n = \psi(n, \delta' n))$, therefore

$$\xi'(n_0 + n + 1) = \psi(n_0 + n + 1, \delta''(n_0 + n + 1))$$

where $\delta'' = \lambda n.\min_m(\psi(n, m) = \xi' n)$. By CT_0 we have for some k $\{k\} = \lambda m.\delta''(n_0 + m + 1)$. δ'' is total, hence φ is defined for k, and $k = \gamma m$ for suitable m. Thus

$$\xi'(n_0 + \gamma m + 1) = \psi(n_0 + \gamma m + 1, \delta''(n_0 + \gamma m + 1)).$$

$\varphi(k) = \{k\}(k) = \delta''(n_0 + \gamma m + 1)$, hence $\xi'(n_0 + \gamma m + 1) = \psi(n_0 + \gamma m + 1, \varphi(\gamma m))$. Finally, we have to show that $\langle I_n\rangle_n$ covers. Take any x; we can find $\langle I'_n\rangle_n$ with $|I'_n| < 2^{-n}$ such that $\forall n(x \in I'_n)$. By the universality $\exists nm(I'_n = I_m)$. □

4.3. REMARK. If $\langle I_n\rangle_n$ is a singular cover of $[0,1]$, say, no finite part $\{I_0, \ldots, I_m\}$ covers $[0,1]$; in fact, we can easily find a rational $r \in [0,1] \setminus (I_0 \cup \cdots \cup I_m)^-$. To see this, observe that a rational segment of length $\leq k \cdot n^{-1}$ can contain at most $k + 1$ points $i \cdot n^{-1}$ with $i \leq n$ ($k - 1$ points in the interior and 0, 1 or 2 points at the endpoints of the interval); and if $|I_0| + \cdots + |I_m| \leq k \cdot n^{-1}$, at most $k + n + 1$ points $i \cdot n^{-1}$ can be covered by $(I_0 \cup \cdots \cup I_m)^-$. For n sufficiently large we can therefore find an r of the form $i \cdot n^{-1} \in [0,1] \setminus (I_0 \cup \cdots \cup I_m)^-$. Thus the singular covers strikingly demonstrate the non-compactness of the interval in CRM. A stronger result is established in 4.9. □

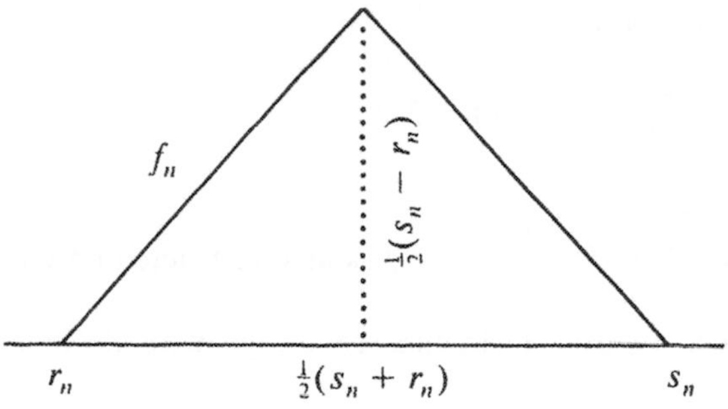

Fig. 6.3

4.4. THEOREM. (CT_0).

(i) *There exists an f on $[0,1]$ such that f is uniformly continuous, $\forall x(f(x) > 0)$, $\inf(f) = 0$; i.e. f does not assume its infimum. Similarly we can construct a uniformly continuous f which does not assume its supremum.*

(ii) *There exists a continuous function on $[0,1]$ which is unbounded (i.e. $\forall n \exists x(f(x) > n)$) and hence not uniformly continuous.*

PROOF. (i) Let $\langle I_n \rangle_n$ be a singular interval covering of $[0,1]$, and suppose $I_n = (r_n, s_n)$ (see fig. 6.3). Let f_n be the uniformly continuous function such that

$$f_n(x) = \max\{0, \tfrac{1}{2}(r_n - s_n) - |x - \tfrac{1}{2}(r_n + s_n)|\}.$$

Then

$$f(x) = \sum_{n=0}^{\infty} 2^{-n} f_n(x)$$

is everywhere defined on $[0,1]$ (observe that $|f_n(x)| \leq 1$ everywhere), and is in fact uniformly continuous. For if $|x - y| < \delta$, then $|f_n(x) - f_n(y)| < \delta$ for all n, hence

$$|f(x) - f(y)| \leq \sum_{n=0}^{\infty} |f_n(x) - f_n(y)| \cdot 2^{-n} < \delta \sum_{n=0}^{\infty} 2^{-n} = 2\delta.$$

(Why does the first inequality hold?) $\langle I_n \rangle_n$ covers, hence for each x in $[0,1]$ $x \in I_n$ for some n, i.e. $f_n(x) > 0$, hence $f(x) > 0$. On the other hand, since no subset $\{I_0, \ldots, I_n\}$ covers, we can pick $x \in [0,1] \setminus (I_0 \cup \cdots \cup I_n)$, so

$f_i(x) = 0$ for $i \leq n$, and

$$f(x) = \sum_{k=n+1}^{\infty} 2^{-k} f_k(x) \leq 2^{-n}.$$

Therefore $\inf(f) = 0$.

(ii) Take $g(x) = f(x)^{-1}$ where f is as constructed under (i). □

We can also construct singular coverings with finite overlap, via the following

4.5. Lemma. *Let $\langle I_n \rangle_n$ be an interval-covering of X, where $X = \mathbb{R}$ or a rational segment of finite length. Then there is a sequence $\langle J_n \rangle_{n \in Z}$, $Z \subset \mathbb{N}$, Z an inhabited index set such that $n \in Z \wedge m < n \to m \in Z$, consisting of rational segments such that*

(i) $\forall n \in Z \exists m (J_n \subset I_m^-)$ ($\langle J_n \rangle_n$ *refines* $\langle I_n^- \rangle_n$);
(ii) $\forall n, m \in Z (J_n \subset J_m \to n = m)$;
(iii) $\forall n, m \in Z (n \neq m \to \mathrm{Int}(J_n) \cap \mathrm{Int}(J_m) = \emptyset)$;
(iv) $\forall x \in X \exists k \exists n, m \in Z (U(2^{-k}, x) \cap X \subset (J_n \cup J_m)^-)$;
(v) *If no finite part of* $\langle I_n \rangle_n$ *covers* X, *then* $Z = \mathbb{N}$;
(vi) *If* $\langle I_n \rangle_n$ *is* ε-*bounded, then so is* $\langle J_n \rangle_n$;
(vii) *If* $Z = \mathbb{N}$ *and* $\lim \langle |I_n| \rangle_n = 0$, *then* $\lim \langle |J_n| \rangle_n = 0$.

Proof. Let $\langle I_n \rangle_n$ be as indicated. Suppose $\mathscr{J}' \equiv \{J_0, \ldots, J_k\}$ to have been constructed such that (i)–(iii) hold with $\{0, \ldots, k\}$ for Z.

Then $Y \equiv X \setminus \mathrm{Int}((J_0 \cup \cdots \cup J_k)^-)$ consists of finitely many disjoint segments, say $Y_0, \ldots, Y_m$. Now let I be any rational segment from $\langle I_n \rangle_n$, and put $\mathscr{J}'(I) \equiv \mathscr{J}' \cup \{Y_i \cap I : Y_i \cap I \text{ inhabited}, i \leq m\}$; clearly, $\mathscr{J}'(I)$ again satisfies (i)–(iii). We put

$$\mathscr{J}_0 := \{I_0^-\}, \mathscr{J}_{n+1} := \mathscr{J}_n'[I_{n+1}^-], \mathscr{J} := \bigcup \{\mathscr{J}_n : n \in \mathbb{N}\}.$$

We can enumerate $\mathscr{J}$ by successively enumerating $\mathscr{J}_0$, $\mathscr{J}_1 \setminus \mathscr{J}_0$, $\mathscr{J}_2 \setminus \mathscr{J}_1$ .... Since we cannot exclude the possibility that, for some k, for all $n \geq k$ $\mathscr{J}_{n+1} \setminus \mathscr{J}_n$ is empty, we can only state that the enumeration runs over an inhabited downwards monotone index set Z. We leave (v) as an exercise.

(vi) and (vii) are also obvious from the construction. It remains to prove (iv). Consider any $x \in X$; $x \in I_n$ say. Let $\mathscr{J}_n \equiv \{J_0, \ldots, J_k\}$; by the construction $(\bigcup \mathscr{J}_n)^- \equiv (I_0 \cup \cdots \cup I_n)^-$. Enumerating $\mathscr{J}_n$ from left to right gives a rearrangement $J_{\sigma(0)}, J_{\sigma(1)}, \ldots, J_{\sigma(k)}$. Clearly, $x \in I_n$ implies

$x \in \mathrm{Int}((J_{\sigma(i)} \cup J_{\sigma(i+1)})^-)$ for some $i < k$ (cf. fig. 6.4).

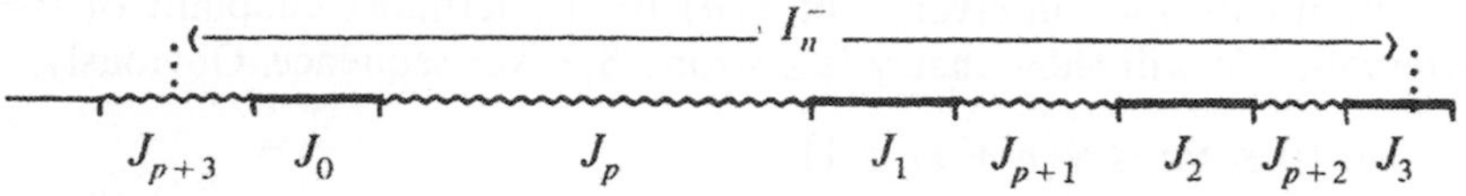

Fig. 6.4

$J_0, J_1, J_2, J_3 \in \mathscr{I}_{n-1}$; $\{J_p, J_{p+1}, J_{p+2}, J_{p+3}\} = \mathscr{I}_n \backslash \mathscr{I}_{n-1}$. □

Note that in this case the "overlap" of the J_i is at most one point.

REMARK. A sequence of rational segments $\langle J_n \rangle_n$ with property (iv) is sometimes called a *quasi-cover*. For a rather similar lemma, see E6.3.6.

4.6. PROPOSITION (CT_0). *For* $X = \mathbb{R}$, *or a rational segment and for any* $m \in \mathbb{N}$:

(i) *there exists a* 2^{-m}*-bounded quasi-cover* $\langle I_n \rangle_n$ *of* X *with* $\lim\langle |I_n| \rangle_n = 0$;

(ii) *there exists a* 2^{-m}*-bounded interval cover* $\langle J_n \rangle_n$ *with* $\lim\langle |J_n| \rangle_n = 0$ *and finite overlap, in fact*:

$$\forall x \in X \exists k n_1 n_2 n_3 \forall n y \left(y \in U(2^{-k}, x) \cap J_n \to n \in \{n_1, n_2, n_3\} \right).$$

PROOF. (i) Apply the preceding lemma to any singular cover as given by 4.2.

(ii) Construct a 2^{-m-1} quasi-cover as under (i), and let $\langle J_n \rangle_n$ enumerate without repetitions the open intervals $\mathrm{Int}((I_k \cup I_m)^-)$ for those I_k and I_m which have an endpoint in common. Then $\langle J_n \rangle_n$ is a 2^{-m}-bounded interval covering satisfying (ii) (exercise). □

4.7. DEFINITION. $\langle x_n \rangle_n$ is said to be a *strong Specker sequence* iff $\langle x_n \rangle_n$ is monotone, bounded and satisfies

$$\forall x \exists m k \forall n \geq k (x_n < x - 2^{-m} \vee x_n > x + 2^{-m})$$

or equivalently $\forall x \exists m k \forall n \geq k (x_n \notin (x - 2^{-m}, x + 2^{-m}))$. □

Of course, the limit of a strong Specker sequence exists as an element in $\mathbb{R}^{\mathrm{be}}$, which lies apart from every real.

4.8. PROPOSITION (CT_0). *Strong Specker sequences exist.*

First proof. Let $\langle I_n \rangle_n$ be a quasi-order of [0, 1] as indicated in the preceding lemma. Consider $\mathrm{Int}((0,1) \setminus (I_0 \cup I_2 \cup \cdots \cup I_n)^-)$; this will consist of finitely many open intervals. Let $\gamma(n)$ be the leftmost endpoint of these intervals. We will show that γ is a strong Specker sequence. Obviously,

$$\forall n (0 \leq \gamma n \leq \gamma(n+1) \leq 1).$$

Now let $I_n \equiv [r_n, s_n]$, and suppose

$$0 \in I_{n_0}, \quad 1 \in I_{n_1}.$$

Find k such that

$$|I_{n_0}| = s_{n_0} > 3 \cdot 2^{-k},$$

$$|I_{n_1}| = 1 - r_{n_1} > 3 \cdot 2^{-k}.$$

Noting that $1 - \frac{1}{3}(1 - r_{n_1}) = \frac{2}{3} + \frac{1}{3}r_{n_1}$, $1 - \frac{2}{3}(1 - r_{n_1}) = \frac{1}{3} + \frac{2}{3}r_{n_1}$ we see that

$$x < \tfrac{2}{3}s_{n_0} \vee \left(\tfrac{1}{3}s_{n_0} < x < \tfrac{2}{3} + \tfrac{1}{3}r_{n_1}\right) \vee \left(\tfrac{1}{3} + \tfrac{2}{3}r_{n_1} < x\right).$$

In the first case for $n > n_0$, and in the third case for $n > n_1$

$$\gamma(n) \notin (x - 2^{-k}, x + 2^{-k}).$$

In the second case, let $U(2^{-k}, x) \cap [0,1] \subset (I_{n_2} \cup I_{n_3})^-$. So for that case $\gamma(n) \notin (x - 2^{-k}, x + 2^{-k})$. Hence, for $n > \max(n_0, n_1, n_2, n_3)$, $\gamma(n) \notin (x - 2^{-k}, x + 2^{-k})$. □

SECOND PROOF (sketch). Let $\langle \delta n \rangle_n$ be the Specker sequence of E6.4.4. Take any $x \in [0, 1]$, and let $\rho(x, C)$ be the distance of x from the Cantor set C, which exists by E6.1.9.

Making use of the fact that for all $n \in \mathbb{N}$ $\rho(x, C) < n^{-1}$ or $\rho(x, C) > (n+1)^{-1}$ we construct a sequence $\langle b_n \rangle_n$ in C such that for all n

$$\rho(x, C) > (n+1)^{-1} \quad \text{and} \quad b_{n+1} = b_n, \quad \text{or} \quad \rho(x, b_{n+1}) < n^{-1}.$$

The b_n converge to a limit in C, say b. For this b $\exists mk \forall n \geq m(|b - \delta n| > 2^{-k})$. If $|x - b| < 2^{-k}$, we are done; if not, we find $|x - b| > 2^{-k-1}$, hence at some stage of our construction of $\langle b_n \rangle$ $\rho(x, C) > 2^{-k'}$, and then also $\forall n(|x - \delta n| > 2^{-k'})$. □

Strong Specker sequences can be used to give another proof of theorem 4.4.

At first sight it might appear as if the existence of singular covers prevents the development of a good measure theory in CRM, since the interval can be covered by a collection of intervals with arbitrarily small measure! However, the difficulty disappears if we realize that for singular covers $\langle I_n \rangle_n$ the sum

$$\sum_{n=0}^{\infty} |I_n|$$

does not converge to a real; if we require this sum to have a limit in $\mathbb{R}$, the cover cannot be singular. We shall prove this in 4.10, but we first need a lemma.

4.9. LEMMA. *Let $\langle I_n \rangle_n$ be a sequence of intervals contained in* $[0,1]$ *such that for some $\delta > 0$*

$$\sum_{n=0}^{\infty} |I_n| \text{ converges and is less than } 1 - \delta,\ \delta > 0.$$

Then there is an x not covered by $\langle I_n \rangle_n$, in fact

$$\exists x \in [0,1] \forall n \forall y \in I_n \big(|x - y| \geq 2^{-n-2}\delta \big). \tag{1}$$

PROOF. Assume the hypotheses of the theorem; without loss of generality we may assume the I_n to be rational (why?). Let I be a rational segment, then it is easy to see that also

$$\sum_{n=0}^{\infty} |I \cap I_n| \text{ converges;}$$

we write $\Phi(I)$ for the limit. If $I \equiv [a,b]$, let $I' := [a, \frac{1}{2}(a+b)]$, $I'' := [\frac{1}{2}(a+b), b]$. Then $\Phi(I) = \Phi(I') + \Phi(I'')$ (by a straightforward argument about convergent series), and if $\Phi(I) < |I|$, then either $\Phi(I') < |I'|$ or $\Phi(I'') < |I''|$. Therefore, starting from $J_0 := [0,1]$, which satisfies $\Phi(J_0) < |J_0| = 1$, we can construct a sequence $\langle J_n \rangle_n$ such that for all n

$$|J_n| = 2^{-n}, \quad \Phi(J_n) < |J_n|,\ J_{n+1} \subset J_n.$$

Clearly there is a unique $x \in [0,1]$ such that $\forall n (x \in J_n)$. x is not covered by $\langle I_n \rangle_n$; for if $x \in I_n$, then, since I_n is open, for some m $J_m \subset I_n$. But then $\Phi(J_m) \geq |I_n|$, which is impossible.

We can now easily refine the argument to prove (1). For suitable ε we have

$$\sum_{n=0}^{\infty} |I_n| < 1 - \delta - \varepsilon.$$

Let, for $I_n \equiv [a, b]$, $I'_n := [a - \delta 2^{-n-2}, b + \delta\ 2^{-n-2}]$, then

$$\sum_{n=0}^{\infty} |I'_n| < 1 - \varepsilon.$$

The preceding argument now applies, with ε for δ and $\langle I'_n\rangle_n$ for $\langle I_n\rangle_n$; hence $\langle I_n\rangle_n$ does not cover x for some x, and (1) holds for this x. □

4.10. COROLLARY.
(i) *If $\langle I_n\rangle_n$ is a singular cover of* $[0,1]$ *then*

$$\langle \sum_{k=0}^{n} |I_k|\rangle_n$$

is a Specker sequence.
(ii) *Let $\langle x_n\rangle_n$ be a sequence of points in* $[0,1]$, *then*

$$\exists x \in [0,1] \forall n \left(|x - x_n| > 2^{-n-3}\right).$$

(iii) *For any finite sequence $\langle x_0, \dots, x_k\rangle$*

$$\exists r \in \mathbb{Q} \cap [0,1] \forall i < k \left(|r - x_i| > 2^{-k-3}\right).$$

PROOF. (i) is immediate.
(ii) Choose a sequence $\langle I_n\rangle_n$ such that $|I_n| < 2^{-n-2}$, $x_n \in I_n$ and apply 4.9.
(iii) is immediate from (ii). □

As the following theorem, due to Kushner, shows, for Riemann integrability on an interval it is not sufficient that a function is continuous. This is an application of the existence of strong Specker sequences.

4.11. THEOREM (CT_0). *There is a function f, continuous on* $[0,1]$ *which is not integrable for any notion of integral $\mathcal{I}$ such that*
(i) *$\mathcal{I}$ has the usual value for polygonal functions;*
(ii) *$\mathcal{I}$ is monotone, i.e. if $f \geq 0$ on* $[0,1]$ *then $\mathcal{I}(f) \geq 0$;*
(iii) *$\mathcal{I}(f) - \mathcal{I}(g) = \mathcal{I}(f - g)$.*

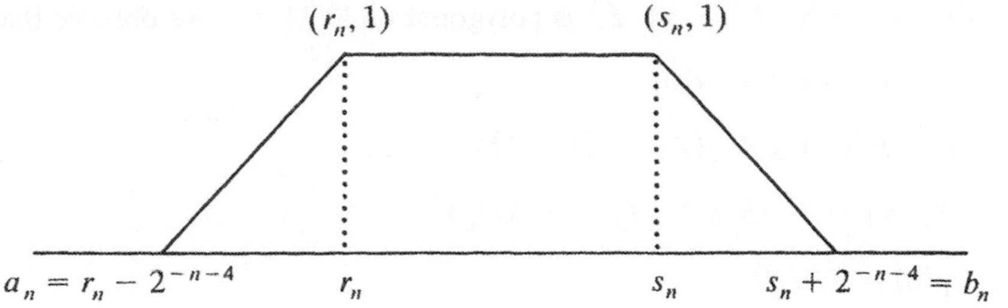

Fig. 6.5

PROOF. Let γ be a strong Specker sequence such that

$$0 \leq \gamma n \leq \gamma(n+1) \leq 1,$$

and let $\langle I_n \rangle_n$ be a quasi-cover of $[0,1]$ with bound $\frac{1}{2}$; let $I_n \equiv [r_n, s_n]$. We put

$$[a_n, b_n] := [r_n - 2^{-n-4}, s_n + 2^{-n-4}]$$

and let F_n^1 be the function on $[0,1]$ as indicated in fig. 6.5.

We write $F^1(n, x) \equiv F_n^1(x)$; F_n^1 is a polygonal function for all n.

$$F^2(n, x) := 1 - \min\left(1, \sum_{k=0}^{n} F^1(k, x)\right);$$

we write F_n^2 for $\lambda x. F^2(n, x)$. F_n^2 is again polygonal on $[0,1]$, and the slopes are given by rational coefficients. We have

$$0 \leq F_n^2(x) \leq 1,$$

$$F_n^2(x) = 0 \quad \text{for } x \in (I_0 \cup \cdots \cup I_n)^-,$$

$$F_n^2(x) = 1 \quad \text{for } x \notin \bigcup\{[a_i, b_i] : i \leq n\}.$$

Let $G(n)$ be the integral of F_n^2 on $[0,1]$; $G(n) \geq \frac{1}{4}$ since

$$\sum_{k=0}^{n} (b_k - a_k) \leq \sum_{k=0}^{\infty} (s_k - r_k) + \sum_{k=0}^{\infty} 2^{-k-3} < \tfrac{1}{2} + \tfrac{1}{4} = \tfrac{3}{4}.$$

Now put

$$F^3(0, x) \quad := (\gamma(0)/G(0)) \cdot F^2(0, x),$$

$$F^3(n+1, x) := ((\gamma(n+1) - \gamma(n))/G(n+1)) \cdot F^2(n+1, x),$$

and let $F_n^3 := \lambda x.F^3(n, x)$. F_n^3 is polygonal on $[0,1]$, and we observe that

$$0 \le F_0^3(x) \le 4 \cdot \gamma(0),$$

$$0 \le F_n^3(x) \le 4(\gamma(n) - \gamma(n-1)),$$

$$F_n^3(x) = 0 \quad \text{for } x \in (I_0 \cup \cdots \cup I_n)^-,$$

$$\gamma(0) = \mathscr{I}(F_0^3),$$

$$\gamma(n+1) - \gamma(n) = \mathscr{I}(F_{n+1}^3).$$

Finally, let

$$F(n, x) := \sum_{i=0}^{n} F^3(i, x), \quad F_n := \lambda x.F(n, x).$$

Then

$$0 \le F(n, x) \le 4 \cdot \gamma(0) + \sum_{i=1}^{n} 4(\gamma(i) - \gamma(i-1)) = 4 \cdot \gamma(n)$$

and therefore

$$\forall x \in [0,1] \forall n (0 \le F(n, x) \le 4).$$

For each $x \in [0,1]$ there are n_1, n_2 such that

$$x \in \left([r_{n_1}, s_{n_1}] \cup [r_{n_2}, s_{n_2}] \right)^-$$

with $s_{n_1} = r_{n_2}$, $r_{n_1} < x < s_{n_2}$. Then $F^3(n, x) = 0$ for $n \ge \max(n_1, n_2)$. The function

$$F(x) := \sum_{i=0}^{\max(n_1, n_2)} F^3(i, x) = \sum_{i=0}^{n} F^3(i, x) = F(n, x)$$

is therefore everywhere well-defined, and

$$F(x) = \lim \langle F_n(x) \rangle_n.$$

F is not integrable, for suppose $\mathscr{I}(F) = z$. On the one hand, $F_n(x) \le F(x)$ for all x and for all $n \in \mathbb{N}$, so, since $\mathscr{I}(F_n) = \gamma(n)$ (additivity of $\mathscr{I}$), by monotonicity $\forall n(\gamma(n) \le z)$. γ is a strong Specker sequence, hence

$$\exists \nu \in \mathbb{N} \exists n' \forall n > n' (\gamma(n) \le z - 2^{-\nu}).$$

Assume now that there exists an n' such that

$$\forall n \geq n' \forall m (\gamma(n+m) - \gamma(n)) < 2^{-\nu-3},$$

then for all k

$$\left|\sum_{i=n'+1}^{n'+k+1} F^3(i,x)\right| \leq 4 \cdot \left|\sum_{i=n'+1}^{n'+k+1} (\gamma(i) - \gamma(i-1))\right|$$
$$= 4(\gamma(n'+k+1) - \gamma(n')) < 2^{-\nu-1}.$$

Therefore, for all k

$$F(n'+k+1, x) \leq F(n', x) + 2^{-\nu-1}$$

and thus

$$F(x) \leq F(n', x) + 2^{-\nu-1}.$$

However, the integral $\mathscr{I}$ is additive and continuous, hence

$$z \leq \gamma(n') + 2^{-\nu-1},$$

contradicting our assumption. Thus, if F is integrable

$$\neg\exists n'mn(n \geq n' \to \gamma(m+n) - \gamma(n) < 2^{-\nu-3})$$

and this conflicts with the boundedness of γ. □

An interesting application of strong Specker sequences is the construction by Demuth (1969) of a uniformly continuous function on $[0,1]$ which is everywhere differentiable classically, but in no recursive real has a recursive derivative (cf. also Myhill 1971).

4.12. *Positive consequences of* ECT_0 + MP. So far, our results concerned "pathologies" occurring in CRM; these could be deduced on the basis of CT_0 alone. On the basis of ECT_0 + MP we can also prove positive results such as "all functions in $\mathbb{R} \to \mathbb{R}$ are continuous". We need some preliminaries before we can present a proof of this result.

Let $\langle r_n \rangle_n$ be a fixed standard enumeration of $\mathbb{Q}$, and let us write q_x for the real determined by a fundamental sequence $\langle r_{\{x\}(n)} \rangle_n$ satisfying

$$\forall n \left(|r_{\{x\}(n)} - r_{\{x\}(n+1)}| < 2^{-n-1}. \right. \tag{1}$$

Then

$$\forall n \left(|q_x - r_{\{x\}(n)}| < 2^{-n} \right). \tag{2}$$

Let $\mathbb{R}^*$ denote the set of code numbers x satisfying $\forall y \exists u Txyu$ and (1); $x \in \mathbb{R}^*$ can be expressed by an almost negative formula.

If $t \in \mathbb{R}$, $t' \in \mathbb{R}$, $\forall n(|t - r_{\{x\}(n)}| < 2^{-n-2}$, $\forall n(|t' - r_{\{y\}(n)}| < 2^{-n-2})$, $|t - t'| < 2^{-n-2}$, and z is code for $\overline{\{x\}}(n) * \lambda m.\{y\}(n+m)$, then $z \in \mathbb{R}^*$ and $t' = q_y = q_z$, since $|r_{\{x\}(n-1)} - r_{\{y\}(n)}| \le |q_x - r_{\{x\}(n-1)}| + |q_x - q_y| + |q_y - r_{\{y\}(n)}| < 2^{-n-1} + 2^{-n-2} + 2^{-n-2} = 2^{-n}$. Hence

$$\forall t \in \mathbb{R} \exists x \in \mathbb{R}^* \forall t' \in \mathbb{R} \forall n \in \mathbb{N}\big(|t - t'| < 2^{-n-2} \to \\ \exists z \in \mathbb{R}^*\big(t = q_x \wedge t' = q_z \wedge \overline{\{z\}}(n) = \overline{\{x\}}(n)\big)\big). \tag{3}$$

We are now ready to prove the

THEOREM (ECT_0 + MP). *All* $f \in \mathbb{R} \to \mathbb{R}$ *are continuous.*

PROOF. The argument is quite different from the proof of the corresponding result on the basis of WC-N. Let $f \in \mathbb{R} \to \mathbb{R}$; then $\forall x, y \in \mathbb{R}^*(q_x = q_y \to f(q_x) = f(q_y))$, and $\forall x \in \mathbb{R}^* \exists y \in \mathbb{R}^*(f(q_x) = q_y)$. Since, as noted above, $x \in \mathbb{R}^*$ is almost negative, there exists by ECT_0 a $z \in \mathbb{N}$ such that

$$\forall x \in \mathbb{R}^*\big(\{z\}(x) \in \mathbb{R}^* \wedge q_{\{z\}(x)} = f(q_x)\big), \tag{4}$$

$$\forall x, y \in \mathbb{R}^*\big(q_x = q_y \to q_{\{z\}(x)} = q_{\{z\}(y)}\big). \tag{5}$$

Let us write $z_Q(x, n)$ for $r_{\{\{z\}(x)\}(n)}$. Let $T_{\mathbb{R}}$ be the spread of sequences α satisfying $\forall n(|r_{\alpha n} - r_{\alpha(n+1)}| < 2^{-n-1})$, then there is a recursive θ enumerating a set of codes for functions dense in $T_{\mathbb{R}}$, i.e. the collection $\{\{\theta n\} : n \in \mathbb{N}\}$ is dense in $T_{\mathbb{R}}$. Fix $\nu \in \mathbb{N}$ and let φ be partial recursive such that

$$\{\varphi(z, y, k)\}(x) \simeq \begin{cases} \{y\}(x) & \text{if } \forall n \le x \neg(Tkkn \wedge Un = 0); \\ \{\theta m\}(x) & \text{if } u = \min_n(Tkkn \wedge Un = 0) \le x \\ & \qquad \wedge\, m = \min_n\big[\{\theta n\} \in \overline{\{y\}}(u) \\ & \qquad \wedge \big|z_Q(y, \nu) - z_Q(\theta n, \nu)\big| \ge 2^{-\nu+1}\big]; \\ \text{undefined otherwise.} \end{cases}$$

We write $\varphi(y, k)$ for $\varphi(z, y, k)$. Let ψ be partial recursive such that

$$\{\psi(z,y)\}(k) \simeq \begin{cases} 0 & \text{if } \boldsymbol{E}(\{z\}(\varphi(y,k))) \wedge \\ & \quad \left|z_Q(\varphi(y,k),\nu) - z_Q(y,\nu))\right| < 2^{-\nu+1} \\ 1 & \text{if } \boldsymbol{E}(\{z\}(\varphi(y,k))) \wedge \\ & \quad \left|z_Q(\varphi(y,k),\nu) - z_Q(y,\nu))\right| \geq 2^{-\nu+1} \\ \text{undefined otherwise.} \end{cases}$$

We write ψy for $\psi(z, y)$. Then

$$\{\psi y\}(\psi y) = 0. \tag{6}$$

For if not, then $\forall n \neg(T(\psi y, \psi y, n) \wedge Un = 0)$, hence $\{\varphi(y, \psi y)\} = \{y\}$, so $q_{\varphi(y,\psi y)} = q_y$ and $|z_Q(\varphi(y, \psi y), \nu) - z_Q(y, \nu)| \leq |z_Q(\varphi(y, \psi y), \nu) - q_{\varphi(y,\psi y)}| + |z_Q(y, \nu) - q_y| < 2^{-\nu+1}$; but then $\{\psi y\}(\psi y) = 0$ by definition; hence $\neg\neg\{\psi y\}(\psi y) = 0$ and thus by Markov's principle (6) holds.

Let m be a Gödel-number for $\lambda y.\min_v T(\psi y, \psi y, v)$ and put M_y for $\{m\}(y)$. Note that

$$y \in \mathbb{R}^* \to U(M_y) = \{\psi y\}(\psi y) = 0.$$

We now show that for any n

$$\overline{\{\theta n\}}(M_y) = \overline{\{y\}}(M_y) \to \left|z_Q(\theta n, \nu) - z_Q(y,\nu)\right| < 2^{-\nu+1}. \tag{7}$$

For suppose $\overline{\{\theta n\}}(M_y) = \overline{\{y\}}(M_y)$ and $|z_Q(\theta n, \nu) - z_Q(y, \nu)| \geq 2^{-\nu+1}$. Then there is also a *least* n satisfying these conditions, say n', and for this n' we have $q_{\varphi(y,\psi y)} \doteq q_{\theta n'}$, and therefore $|z_Q(\varphi(y, \psi y), \nu) - z_Q(y, \nu)| \geq 2^{-\nu+1}$. This means that $\{\psi y\}(\psi y) = 1$, by the definition of ψ; contradiction, hence (7).

Similarly for any w, n

$$\overline{\{\theta n\}}(M_w) = \overline{\{w\}}(M_w) \to |z_Q(\theta n, \nu) - z_Q(w, \nu)| < 2^{-\nu+1}, \tag{8}$$

and therefore, if $\overline{\{y\}}(M_y) = \overline{\{w\}}(M_y)$, let $m' = \max(M_y, M_w)$ and choose n such that $\{\theta n\} \in \overline{\{w\}}(m')$; then (7) and (8) both hold for n, and thus

$$\overline{\{y\}}(M_y) = \overline{\{w\}}(M_y) \to \left|z_Q(y,\nu) - z_Q(w,\nu)\right| < 2^{-\nu+2},$$

and so

$$\overline{\{y\}}\,(M_y) = \overline{\{w\}}\,(M_y) \to |q_{\{z\}(y)} - q_{\{z\}(w)}| < 2^{-\nu+1} + 2^{-\nu+2}. \tag{9}$$

From (7) and the observation (3) above we see that

$$\forall k \forall t \in \mathbb{R} \exists n \forall t' \in \mathbb{R}\big(|t - t'| < 2^{-n} \to |ft - ft'| < 2^{-k}\big). \quad \square$$

REMARK. The theorem may also be seen as a classical result about recursively given real functions on [0, 1], where "recursively given" means that a z can be found such that (4) and (5) hold. The only non-classical element is in the use of ECT_0 to show that all real functions on [0, 1] are recursively given.

4.13. *Recursive analysis.* This approach has a somewhat hybrid character; problems with parameters are required to be solved recursively in the parameters, but in other respects classical logic is applied. Specifically, a *sentence* $\exists x A x$ may be proved by classical means, but for a formula with parameters, e.g. $\exists x A(x, y)$ one requires a proof giving x recursively in y.

Thus a bisection argument can be used to show that for any specific continuous f with $f(0) < 0, f(1) > 0$ there is indeed a (recursive) real x with $f(x) = 0$; but there is no solution recursively in data for f. The second assertion will be obvious from the discussion of the counterexample in 1.2; as to the first, we argue as follows.

Let f map recursive reals to recursive reals, and suppose $f(0) < 0, f(1) > 0$. If $f(\frac{1}{2}) = 0$, we are done. If $f(\frac{1}{2}) \neq 0$, then, by Markov's principle, either $f(\frac{1}{2}) > 0$ or $f(\frac{1}{2}) < 0$. Suppose $f(\frac{1}{2}) > 0$; we test $f(\frac{1}{4})$. If this is zero, we are done; if not, we continue as before. Thus we either meet a bisection point of the form $k2^{-n}$ with $f(k2^{-n}) = 0$, or for all bisection points considered $f(k2^{-n}) \neq 0$; in the latter case the *recursive* sequence of bisection points has a recursive real x as a limit for which $f(x) = 0$. $\square$

5. Notes

5.1. The contents of section 1 are on the whole standard. The proof of 1.3 is found in Grayson (1981, 7.2) as well as in Mandelkern (1976). The proof of 1.4 is due to J. Mines and F. Richman (in Mandelkern 1982); earlier proofs assumed uniform continuity. Theorem 1.5 is exercise 13 in chapter 2 of Bishop (1967).

The proof of 2.4 follows Bishop and Bridges (1985). 2.6 and 2.7 correspond to exercises 14 and 15 in chapter 2 of Bishop (1967). The treatment of Riemann integration is standard.

5.2. Already Brouwer (1919, 1924) gave standard representations as continuous images of Cantor spaces ("finite Mengen", "fans") similar to the one presented in 3.1 for [0, 1]; the particular presentation given here is the same as in Heyting (1956, 3.3.3–4).

3.6 and the first half of 3.8(i) are already stated by Brouwer (1924); since Brouwer did not isolate the role of the principle WC-N from FAN and FAN*, we do not find in his writings analogues to 3.3, 3.4.

3.3 and 3.4 for complete separable metric spaces appear as 2.21, 2.23 in Troelstra (1966), and also in Troelstra (1967).

3.5 is a special case of the intuitionistic "Heine–Borel" theorem proved by Brouwer (1926A; Heyting 1956, 5.2.2).

The theorem of E6.3.8–9 is due to Veldman (1981, in the leaflet "Stellingen").

5.3. Singular interval coverings were discovered by Kreisel and Lacombe (1957), and independently by Tsejtin and Zaslavskij (1962).

Theorem 4.2 is found in the latter paper (cf. also our remarks in 5.8.3). Singular covers show that in Constructive Recursive Mathematics [0, 1] is not compact; for Cantor space (the binary tree) this is an immediate consequence of Kleene's recursively well-founded, not well-founded recursive tree (4.7.6, theorem).

Julian and Richman (1984) give information on proofs of 4.4(i) in the literature; in addition they show that in BCM the statement "each uniformly continuous everywhere positive function on [0, 1] has a positive infimum" is in fact equivalent to the fan theorem.

5.4. A proof of 4.8 is contained in the proof of theorem 4.3 of Zaslavskij (1962). In our presentation we follow Kushner (1973, section 8.1, theorem 6).

4.9 and 4.10(i) are taken from Tsejtin and Zaslavskij (1962); our exposition follows Kushner (1973, section 8.1, lemma 1).

5.5. *The KLST-theorem.* Tsejtin (1959, 1962) proved that in CRM every mapping from a complete separable metric space into a separable metric space is continuous; the special case of mappings from $\mathbb{N}^{\mathbb{N}} \to \mathbb{N}$ (E6.4.9) was discovered independently by Kreisel et al. (1957, 1959). Thus the frequently used name "Kreisel–Lacombe–Shoenfield theorem", abbreviated KLS, is perhaps more appropriately named the "Kreisel–Lacombe–Shoenfield–Tsejtin theorem" (KLST).

The proofs use the creativity of the diagonal set $\{x : \exists y Txxy\}$, avoiding a direct appeal to the recursion theorem; our proof in 4.12 for the special, but representative case of mappings from $\mathbb{R}$ into $\mathbb{R}$ is based on the same idea.

Gandy (1962, for mappings in $\mathbb{N}^{\mathbb{N}} \to \mathbb{N}$) and Y. Moschovakis (1964, the general case) give proofs based on the recursion theorem. A proof for the general case of mappings from complete separable metric spaces to separable metric spaces based on Moschovakis' idea is found in the exercises E7.2.2–5. Kushner (1973, chapter 9) contains some illuminating comments on the method of proof.

Finally, it is to be noted that as early as 1954 Markov had obtained a weaker theorem to the effect that mappings from $\mathbb{R}$ to $\mathbb{R}$ cannot have positive discontinuities (for an exposition see Kushner 1973, section 9.2, theorem 6).

Exercises

6.1.1. Let us call a function $f \in X \to \mathbb{R}$ $(X \subset \mathbb{R})$ *strictly monotone* if $\forall x, y \in X(x < y \to fx < fy)$. Show that a strictly monotone and continuous $f \in [0,1] \to \mathbb{R}$ satisfies the intermediate value theorem in the strong form, i.e. for all $c \in \mathbb{R}$ $(f(0) \le c \le f(1) \to \exists x(f(x) = c)$.

6.1.2. Let us call $f \in X \to \mathbb{R}$, $X \subset \mathbb{R}$, *strongly monotone* if $\forall x, y \in X(fx < fy \to x < y)$. Show that a strongly monotone and continuous $f \in [0,1] \to \mathbb{R}$ is uniformly continuous (Mandelkern 1982).

6.1.3. Let $[0,1]^{\text{be}} := \{x \in \mathbb{R}^{\text{be}} : 0 \le x \le 1\}$. Suppose $f \in [0,1] \to \mathbb{R}$ to be uniformly continuous, $f(0) < 0$, $f(1) > 0$; show that for the canonical extension $f^* \in [0,1]^{\text{be}} \to \mathbb{R}^{\text{be}}$ of f we have $\exists x \in [0,1]^{\text{be}}(f^*(x) = 0)$.

6.1.4. Give an example of an $f \in [0,1] \to \mathbb{R}$ which assumes its local suprema (i.e. suprema of f on sufficiently small segments) but which cannot be shown to assume its supremum on $[0,1]$.

6.1.5. In the notation of exercise 6.1.3 above, show that for a uniformly continuous $f \in [0,1] \to \mathbb{R}$ one has $\exists x \in [0,1]^{\text{be}}(f^*(x) = \sup(f))$.

6.1.6. Show by means of a weak counterexample that $[0,1] \cup [1,2]$ is not closed (in particular, we cannot show $[0,1] \cup [1,2] = [0,2]$). Strengthen this under the assumption of $\neg$SEP to a refutation of "for all segments I_1, I_2, if $I_1 \cap I_2$ is inhabited, then $I_1 \cup I_2$ is a segment".

6.1.7. Show that $[0,1] \cup [1,2]$ is a continuous image of $[0,1]$ (therefore we cannot expect to prove constructively that the continuous image of a closed set is always closed cf. exercise 6.1.6). Is compactness (in the sense of "every open cover has a finite subcover") preserved under continuous mappings? Indicate a property distinguishing $[0,1) \cup [1,2]$ from $[0,1] \cup [1,2]$.

6.1.8. Let $X \subset \mathbb{R}$ be totally bounded. Show that X is located, i.e. a distance function $\rho(x, X)$ for $x \in \mathbb{R}$ is defined such that $\forall y \in X(\rho(x, X) \leq |y - x|)$ and $\forall k \exists y \in X(\rho(x, X) < |x - y| + 2^{-k})$.

6.1.9. The reals of the form

$$\sum_{n=0}^{\infty} 2\alpha \cdot 3^{-n-1},$$

with $\alpha \in \mathbb{N} \to \{0,1\}$, constitute the Cantor discontinuum C as a subset of $[0,1]$ (obtained by the well-known procedure of repeated deletion of the open middle third of segments). Show that C is located (cf. the preceding exercise).

6.2.1. Prove 2.2. *Hint:* for the second part use the first part to conclude that $\sup|f'|$ exists on $[a, b]$.

6.2.2. Prove 2.3.

6.2.3. Show that "f is uniformly differentiable with derivative f'" is equivalent to

$$\exists\alpha\forall k\forall x,y \in [a,b]\big(|x-y| < 2^{-\alpha k} \wedge x \# y \to$$
$$|f'(x) - (fy - fx)/(y - x)| < 2^{-k}\big).$$

6.2.4. Complete the proof of 2.9 and give proofs of 2.10, 2.12–14.

6.3.1. Let us call $f \in \mathbb{Q} \to \mathbb{R}$ sequentially continuous, if whenever $\lim\langle x_n\rangle_n = x$, then $\lim\langle f(x_n)\rangle_n = f(x)$. Show on the basis of WC-N that a sequentially continuous f is continuous. *Hint:* consider any $r \in \mathbb{Q}$ and construct a suitable spread T' such that under a mapping φ given by $\varphi\alpha = \langle r_{\alpha n}\rangle_n$ (relative to a standard enumeration $\langle r_n\rangle_n$ of $\mathbb{Q}$) all fundamental sequences converging to r within a certain fixed prescribed rate of convergence are represented; apply WC-N with respect to this spread (Troelstra 1967).

6.3.2. $X \subset \mathbb{R}$ is separable iff there is a $\langle x_n\rangle_n \subset X$ such that $\forall x \in X\forall k\exists x_n(|x - x_n| < 2^{-k})$. Generalize the result of the preceding exercise to sequentially continuous $f \in X \to \mathbb{R}$ (Troelstra 1967).

6.3.3. Let $f \in \mathbb{R} \to \mathbb{R}$ be continuous and sequentially differentiable at x, that is, whenever $\lim\langle x_n\rangle_n = x$ and $\forall n(x_n \# x)$ then $\lim\langle(f(x_n) - f(x))/(x_n - x)\rangle_n$ exists. Show that f is differentiable at x (van Rootselaar 1952).

6.3.4. Assume $\neg\exists$-PEM and show that every $f \in \mathbb{R} \to \mathbb{N}$ must be constant.
Hint: prove first that there is no function defined on a closed subset $X \subset \mathbb{R}$ such that for some $\langle a_n\rangle_n \subset X$, $f(a_n) = 0$ for all n and $f(\lim\langle a_n\rangle_n) = 1$.

6.3.5. Assume $\neg\exists$-PEM and show that any $f \in \mathbb{R} \to \mathbb{R}$ is weakly continuous in the following sense: there is no sequence $\langle x_n \rangle_n, \lim\langle x_n \rangle_n = x$ such that $\forall n(|fx_n - fx| \geq 2^{-k})$ for some k (cf. the preceding exercise).

6.3.6. Let $\langle I_n \rangle_n$ be any sequence of rational intervals of finite length covering $\mathbb{R}$. Show how to construct a covering $\langle J_n \rangle_n$ of rational open intervals such that
(i) $\forall n \exists m(J_n \subset I_m)$ ($\langle J_n \rangle_n$ refines $\langle I_n \rangle_n$),
(ii) $\forall nm(J_n \subset J_m \to n = m)$,
(iii) $\forall x \in \mathbb{R} \exists mm' \forall n(x \in J_n \to n = m \vee n = m')$ ($\langle J_n \rangle_n$ has finite overlap).

6.3.7. Assuming C-N, prove the following version of "Lindelöf's theorem" for $\mathbb{R}$: if $\{W_i : i \in X\}$ is any open cover of $\mathbb{R}$, then there is an enumerable subcover $\{W_{\varphi(n)} : n \in \mathbb{N}\}$ for some $\varphi \in \mathbb{N} \to X$ (Troelstra 1966, 2.2.6; Troelstra 1968, Theorem 8).

6.3.8. Assume C-N. Show that any $f \in \mathbb{R} \to \mathbb{R}$ is continuous with a continuous modulus of continuity, i.e. for all k there is a continuous $g_k \in \mathbb{R} \to \mathbb{R}^+$ (where $\mathbb{R}^+ := \{x : x > 0\}$) such that $\forall xy(|x - y| < g_k(x) \to |f(x) - f(y)| < 2^{-k})$. *Hint:* use the two preceding exercises to construct a cover $\langle (s_n, t_n) \rangle_n$ with finite overlap for $\mathbb{R}$, $s_n, t_n \in \mathbb{Q}$, such that $\forall x, y \in (s_n, t_n)(|f(x) - f(y)| < 2^{-k})$. Define

$$g_k(x) := \sup\{[\min(\max(x - s_n, 0), \max(t_n - x, 0))]) : n \in \mathbb{N}\}.$$

6.3.9. Refine the result of the preceding exercise to obtain: there exists a continuous $g(y, x)$ with $y > 0 \to g(y, x) > 0$, $y \leq 0 \to g(y, x) = 0$, and

$$\forall xy \forall z > 0(|x - y| < g(z, x) \to |fx - fy| < z).$$

6.3.10. Let

$$S := \sum_{n=0}^{\infty} x_n, \quad S_\sigma := \sum_{n=0}^{\infty} x_{\sigma(n)}, \quad S^* := \sum_{n=0}^{\infty} |x_n|,$$

where σ is a permutation of $\mathbb{N}$, i.e. a bijection from $\mathbb{N}$ to $\mathbb{N}$. S converges *absolutely* iff S^* converges. Show that (i) If S converges absolutely, then so does S_σ. We say that S *diverges* iff

$$\forall m \exists k \forall k' \geq k\left(\left|\sum_{n=0}^{k'} x_n\right| \geq m\right).$$

Show that (ii) If S converges and S^* diverges, then S_σ can take any value a in $\mathbb{R}^* = \mathbb{R} \cup \{-\infty, +\infty\}$ for a suitably chosen permutation σ depending only on a.

6.3.11. Let S, S_σ be as in the preceding exercise. Assuming WC-N, show that if S_σ converges for all permutations σ, then S is absolutely convergent. *Hint:* first prove the theorem under the assumption $\forall n(x_n \in \mathbb{Q})$; construct a spread representing all permutations of $\mathbb{N} \to \mathbb{N}$ (W. Gielen; a proof is in Troelstra 1977, 6.11.)

6.4.1. Verify in detail the inequality

$$|f(x) - f(y)| \leq \sum_{n=0}^{\infty} 2^{-n}|f_n(x) - f_n(y)|$$

in the proof of 4.4.

6.4.2. Show in 4.5: if no finite part of $\langle I_n \rangle_n$ covers, then $Z = \mathbb{N}$.

6.4.3. Prove 4.6(ii).

6.4.4. Define

$$\delta(n) := \sum_{k=0}^{n} 2\alpha_n(k) \cdot 3^{-k-1},$$

where α_n is the characteristic function of $\exists u \leq n(Txxu \wedge Uu = 0)$, i.e. $\alpha_n x = 1 \leftrightarrow \exists u \leq n(Txxu \wedge Uu = 0)$. Show that $\langle \delta(n) \rangle_n$ is a Specker sequence, and that for the Cantor discontinuum C we have

$$\forall y \in C \exists kn \forall k' \geq k(|\delta(k') - y| \geq 2^{-n})$$

(compare the proof of 4.7.6, theorem).

6.4.5. Construct a strictly monotone strong Specker sequence.

6.4.6. Construct a Specker sequence assuming the existence of a lawless 01-sequence (cf. 4.6.2).

6.4.7. Show, using strong Specker sequences instead of singular coverings, how to construct
(i) an $f \in [0,1] \to \mathbb{R}$ which is continuous and unbounded;
(ii) an $f \in [0,1] \to [0,1]$ which is continuous but not uniformly continuous;
(iii) an $f \in [0,1] \to [0,1]$ with $\inf(f) = 0, \forall x \in [0,1](f(x) > 0)$.

6.4.8. Prove the assertion of exercise 6.3.11 on the assumption of ECT_0 + MP (Beeson 1976).

6.4.9. Give a simplified version of theorem 4.12 establishing the special case: all functions from $\mathbb{N}^{\mathbb{N}}$ to $\mathbb{N}$ are continuous, where $\mathbb{N}^{\mathbb{N}}$ has the usual product topology and $\mathbb{N}$ the discrete topology (cf. 4.1.5, 5.3.15).

BIBLIOGRAPHY

Below we have only listed those items which have been referred to in this book. A more complete bibliography of constructivism (in particular the items classified under F50–65) is in Müller (1987).

Full titles of journals can easily be recovered from Müller (1987) or the Mathematical Reviews. The exceptions concern certain frequently cited journals for which we use the abbreviations below:

JSL the Journal of Symbolic Logic,
AML Annals of Mathematical Logic, continued as
APAL Annals of Pure and Applied Logic,
ZMLGM Zeitschrift für mathematische Logik und Grundlagen der Mathematik,
Arch.ML. Archiv für mathematische Logik und Grundlagenforschung.

In addition we use in the title of certain proceedings the abbreviation
LMPS Logic, Methodology and Philosophy of Science.

Congress proceedings and the papers in such proceedings appear under the year of publication, not the year of the congress. If a volume of a journal appeared over more than one calendar year, papers in the volume are listed under the year appearing on the title page of the completed volume, which is as a rule also the year in which the last issue of the volume appeared (occasionally there are awkward exceptions!).

Items with more than one author are listed in full under the author which is named first in the title of the official publication; for the coauthors there is an indirect reference. E.g. under "Bernays" we find "Bernays, P. see *Hilbert and Bernays (1934)*". We have refrained from giving indirect references in the case of multiple editors.

A few items we only know from reviews or references by others. In such cases the reference is followed by "(n.v.)".

Aberth, O.
(1980) *Computable analysis* (McGraw-Hill, New York).

Aczel, P.H.G.
(1968) Saturated intuitionistic theories, in: *H.A. Schmidt et al. (1968)* pp. 1–11.
(1977) An introduction to inductive definitions, in: *Barwise (1977)*, pp. 739–782.
(1978) The type-theoretic interpretation of constructive set theory, in: A. Macintyre, L. Pacholski and J. Paris, eds., *Logic Colloquium '77* (North-Holland, Amsterdam) pp. 55–66.

Allen, S.F. see *Constable et al. (1986)*.

Barwise, K.J.
(1977) ed., *Handbook of Mathematical Logic* (North-Holland, Amsterdam).

Barwise, K.J., H.J. Keisler, and K. Kunen
(1980) eds., *The Kleene Symposium* (North-Holland, Amsterdam).

Beeson, M.J.
(1976) Half a note on Riemann's theorem about absolute convergence. Unpublished note.
(1985) *Foundations of Constructive Mathematics* (Springer, Berlin).

Benthem Jutting, L.S. van
(1977) Checking Landau's "Grundlagen" in the AUTOMATH system, PhD thesis, Technische Hogeschool, Eindhoven.

Bernays, P. see *Hilbert and Bernays (1934)*.

Bishop, E.
(1967) *Foundations of Constructive Analysis* (McGraw-Hill, New York).

Bishop, E. and D. Bridges
(1985) *Constructive Analysis* (Springer, Berlin). Second, thoroughly rewritten edition of *Bishop (1967)*.

Boffa, M., D. van Dalen, and K. McAloon
(1979) eds., *Logic colloquium '78* (North-Holland, Amsterdam).

Boileau, A. and A. Joyal
(1981) La logique des topos. *JSL* 46, 6–16.

Borel, E.
(1914) *Leçons sur la Théorie des Fonctions* (Gauthier-Villars, Paris) (second edition). Unchanged with an additional note VII in the third edition of 1928; the fourth edition of 1950 contains a further note VIII.
(1947) Sur l'illusion des définitions numériques, *C.R. Acad. Sci.* 224, 765–767.

Bridges, D.S.
(1979) *Constructive Functional Analysis* (Pitman, London).

Bridges, D.S. see also *Bishop and Bridges (1985)*.

Bromley, H.M. see *Constable et al. (1986)*.

Brouwer, L.E.J.
(1907) *Over de Grondslagen der Wiskunde* (Dutch), PhD thesis, University of Amsterdam, Maas and van Suchtelen, Amsterdam; reprinted with additional material, ed. D. van Dalen (Mathematisch Centrum, Amsterdam, 1981).
(1908) Over de onbetrouwbaarheid der logische principes (Dutch), *Tijdschrift voor Wijsbegeerte* 2, 152–158.
(1912) *Intuïtionisme en formalisme* (Dutch), Inaugural address (Noordhoff, Groningen). Also *Wiskundig Tijdschrift* 9, 180–211. Translation: Intuitionism and formalism, *Bull. Amer. Math. Soc.* 20, 81–96.
(1914) Review of Schoenflies's "Die Entwicklung der Mengenlehre und ihre Anwendungen, erste Hälfte", *Jahresber. Dtsch. Math.-Ver.* 23(2), 78–83 italics.

(1918) Begründung der Mengenlehre unabhängig vom logischen Satz vom ausgeschlossenen Dritten I: Allgemeine Mengenlehre, *Nederl. Akad. Wetensch. Verh. Tweede Afd. Nat.* 12/5.

(1919) Begründung der Mengenlehre unabhängig vom logischen Satz vom ausgeschlossenen Dritten II: Theorie der Punktmengen, *Nederl. Akad. Wetensch. Verh. Tweede Afd. Nat.* 12/7.

(1921) Besitzt jede reelle Zahl eine Dezimalbruchentwickelung? *Nederl. Akad. Wetensch. Verslagen* 29, 803–812. Also in *Math. Ann.* 83, 201–210.

(1923) Intuïtionistische splitsing van mathematische grondbegrippen (Dutch), *Nederl. Akad. Wetensch. Verslagen* 32, 877–880. German translation *Jahresber. Dtsch. Math.-Ver.* 33, 251–256.

(1923A) Begründung der Funktionenlehre unabhängig vom logischen Satz vom ausgeschlossenen Dritten, *Nederl. Akad. Wetensch. Verh. Tweede Afd. Nat.* 13/2.

(1924) Beweis dass jede volle Funktion gleichmässigstetig ist, *Nederl. Akad. Wetensch. Proc.* 27, 189–193.

(1926) Zur Begründung der intuitionistischen Mathematik II, *Math. Ann.* 95, 453–472.

(1926A) Die intuitionistische Form des Heine-Borelschen Theorems, *Nederl. Akad. Wetensch. Proc.* 29, 866–867.

(1927) Über Definitionsbereiche von Funktionen, *Math. Ann.* 97, 60–75.

(1929) Mathematik, Wissenschaft und Sprache, *Monatshefte für Mathematik und Physik* 36, 153–164. A slightly different and expanded version appeared in Dutch: Willen, weten, spreken, *Euclides* 9 (1933) 177–193. In *Brouwer (1975)* there is a translation of section III of the 1933 version; in Heyting's notes to *Brouwer (1929)* the principal differences with section I and II of the 1933 version are indicated.

(1948) Essentieel-negatieve eigenschappen (Dutch), *Indagationes Math.* 10, 322–323.

(1949) Consciousness, philosophy and mathematics in: E.W. Beth, H.J. Pos, H.J.A. Hollak, eds., *Library of the Tenth International Congress of Philosophy, August 1948, Amsterdam* (North-Holland, Amsterdam) vol. 1, pp. 1235–1249.

(1952) Over accumulatiekernen van oneindige kernsoorten (Dutch), *Indagationes Math.* 14, 439–441.

(1954) Points and spaces, *Can. J. Math.* 6, 1–17.

(1975) *Collected Works I*, ed. A. Heyting (North-Holland, Amsterdam).

(1981) *Brouwer's Cambridge Lectures on Intuitionism*, ed. D. van Dalen (Cambridge University Press, Cambridge).

Buchholz, W., S. Feferman, W. Pohlers and W. Sieg

(1981) *Iterated Inductive Definitions and Subsystems of Analysis: Recent Proof-Theoretical Studies* (Springer, Berlin).

Cellucci, C.

(1969) Un' osservazione sul theorema di Minc-Orevkov (english summary), *Boll. Un. Mat. Ital.* 1, 1–8.

Church, A.

(1932) A set of postulates for the foundation of logic, *Ann. of Math.* 33, 346–366.

(1936) An unsolvable problem of elementary number theory, *Ann. of Math.* 33, 346–366.

Cleaveland, W.R. see *Constable et al. (1986)*.

Constable, R.L., S.F. Allen, H.M. Bromley, W.R. Cleaveland, J.F. Cremer, R.W. Harper, D.J. Howe, T.B. Knoblock, N.P. Mendler, P. Panangaden, J.T. Sasaki and S.F. Smith

(1986) *Implementing Mathematics with the Nuprl Proof Development System* (Prentice-Hall, Englewood Cliffs, New Jersey).

Cremer, J.F. see *Constable et al. (1986).*

Curry, H.B.

(1930) Grundlagen der kombinatorischen Logik, *Amer. J. Math.* 52, 509–536, 789–834.

(1941) A formalisation of recursive arithmetic, *Amer. J. Math.* 63, 263–282.

Dalen, D. van

(1977) The use of Kripke's schema as an reduction principle, *JSL* 42, 238–240.

(1978) An interpretation of intuitionistic analysis, *AML* 13, 1–43.

(1982) Singleton reals, in: *van Dalen et al. (1982) 83–94.*

(1984) How to glue analysis models, *JSL* 49, 1339–1349.

(1986) Intuitionistic logic, in: *Gabbay and Guenthner (1986) 225–339.*

Dalen, D. van, D. Lascar and T.J. Smiley

(1982) eds., *Logic Colloquium '80* (North-Holland, Amsterdam).

Dantzig, D. van

(1956) Is $10^{10^{10}}$ a finite number?, *Dialectica* 9, 273–277.

Davis, M.

(1965) ed., *The Undecidable* (Raven Press, New York).

Demuth, O.

(1969) The differentiability of constructive functions (Russian), *Commentat. Math. Univ. Carolinae* 10, 357–390.

Diaconescu, R.

(1975) Axiom of choice and complementation, *Proc. Amer. Math. Soc.* 51, 176–178.

Dijkman, J.G.

(1952) Convergentie en divergentie in de intuïtionistische wiskunde (Dutch), PhD thesis, University of Amsterdam.

Dragalin, A.G.

(1969) Transfinite completions of constructive arithmetical calculus (Russian), *Dokl. Akad. Nauk SSSR* 189, 10–12; translated in *Sov. Math. Dokl.* 10, 1417–1420.

(1980) New forms of realizability and Markov's rule (Russian), *Dokl. Akad. Nauk SSSR* 251, 534–537; translated in *Sov. Math. Dokl.* 21, 461–464.

Du Bois Reymond, D.

(1882) *Die Allgemeine Funktionentheorie I* (Verlag der H. Laupp'schen Buchhandlung, Tübingen).

Dummett, M.A.E.

(1977) *Elements of Intuitionism* (Clarendon Press, Oxford).

Engelking, R.

(1968) *Outline of General Topology* (North-Holland, Amsterdam; PWN, Warszawa; John Wiley and Sons, New York).

Esenin-Vol'pin, A.S.

(1961) Le programme ultra-intuitionniste des fondements des mathématiques, in: *Infinitistic methods. Proceedings of the Symposium on Foundations of Mathematics, September 1959, Warsaw* (PWN Warszawa) pp. 201–223.

(1970) The ultra-intuitionistic criticism and the anti-traditional program for foundations of mathematics. in: *Myhill et al. (1970) 3–45.*

(1981) About infinity, finiteness and finitization, in: *Richman (1981) 274–313.*

Feferman, S.

(1964) Systems of predicative analysis, *JSL* 29, 1–30.

(1975) A language and axioms for explicit mathematics, in: *Crossley (1975) 87–139.*

(1978) A more perspicuous system for predicativity, in: K. Lorenz, ed., *Spezielle Wissenschaftstheorie, Vol. 1: Konstruktionen versus Positionen. Beiträge zur Diskussion um die konstruktive Wissenschaftstheorie*, 58–93.

Feferman, S. see also *Buchholz. et al. (1981)*.

Fenstad, J.E.

(1971) ed., *Proceedings of the Second Scandinavian logic symposium* (North-Holland, Amsterdam).

(1980) *General Recursion Theory* (Springer, Berlin).

Fitting, M.C.

(1969) *Intuitionistic Logic, Model Theory and Forcing* (North-Holland, Amsterdam).

Flagg, R.C.

(1986) Integrating classical and intuitionistic type theory, *APAL* 32, 27–51.

Flagg, R.C. and H.M. Friedman

(1986) Epistemic and intuitionistic formal systems, *APAL* 53–60.

Fourman, M.P., C.J. Mulvey and Scott, D.S.

(1979) eds., *Applications of Sheaves* (Springer, Berlin).

Freudenthal, H.

(1937) Zur intuitionistischen Deutung logischer Formeln, *Compos. Math.* 4, 112–116.

Friedman, H.M.

(1973) Some applications of Kleene's methods for intuitionistic systems, in: *Mathias, Rogers (1973), 113–170.*

(1975) The disjunction property implies the numerical existence property, *Proc. Nat. Acad. Sci. USA* 72, 2877–2878.

(1975A) Intuitionistic completeness of Heyting's predicate calculus, *Notices Amer. Math. Soc.* 22, A–648.

(1977) On the derivability of instantiation properties, *JSL* 42, 506–514.

(1978) Classically and intuitionistically provably recursive functions, in: *Müller and Scott (1978) 21–27.*

(1980) A strong conservative extension of Peano arithmetic, in: *Barwise et al. (1980) 113–122.*

Friedman, H.M., see also *Flagg and Friedman (1986)*.

Friedman, H.M., and A. Scedrov

(1983) Set existence property for intuitionistic theories with dependent choice, *APAL* 25, 129–140 (corrections ibidem 26, 101).

Friedman, H.M., S.G. Simpson and R.L. Smith

(1983) Countable algebra and set existence axioms, *APAL* 25, 141–181 (errata ibidem 28, 319–320).

Gabbay, D.

(1981) *Semantical Investigations in Heyting's Intuitionistic Logic* (Reidel, Dordrecht).

Gabbay, D. and F. Guenthner

(1986) eds., *Handbook of Philosophical Logic III* (Reidel, Dordrecht).

Gandy, R.O.

(1962) Effective operations and recursive functionals (abstract), *JSL* 39, 81–87.

Geiser, J.

(1974) A formalization of Essenin-Volpin's proof theoretical studies by means of non-standard analysis, *JSL* 39, 81–87.

Gelfond, A.O.

(1960) *Transcendental and algebraic numbers* (Dover Publications, New York). Translation from the first Russian edition by Leo F. Boron.

Gentzen, G.
(1933) Über das Verhältnis zwischen intuitionistischer und klassischer Logik. Originally to appear in the Mathematische Annalen, reached the stage of galley proofs but was withdrawn. It was finally published in *Arch. ML* 16 (1974), 119–132; a translation is in *Gentzen (1969) 53–67.*
(1935) Untersuchungen über das logische Schliessen, I, II, *Math. Z.* 39, 176–210, 405–431.
(1969) *Collected Papers*, ed. M.E. Szabo (North-Holland, Amsterdam).

Giaretta, P. see *Martino and Giaretta (1981)*.

Girard, J.-Y.
(1971) Une extension de l'interprétation de Gödel à l'analyse, et son application à l'élimination des coupures dans l'analyse et la théorie des types, in: *Fenstad (1971) 63–92.*
(1973) Quelques résultats sur les interpretations fonctionelles, in: *Mathias and Rogers (1983) 232–252.*

Glivenko, V.I.
(1928) Sur la logique de M. Brouwer, *Acad. Roy. Belg. Bull. Cl. Sci.* (5) 14, 225–228.
(1929) Sur quelques points de la logique de M. Brouwer, *Bull. Soc. Math. Belg.* 15, 183–188.

Gödel, K.
(1931) Über formal unentscheidbare Sätze der Principia Mathematica und verwandte Systeme I, *Monatshefte für Mathematik und Physik* 38, 173–198.
(1932) Zum intuitionistischen Aussagenkalkül, *Anzeiger der Akademie der Wissenschaften in Wien* 69, 65–66.
(1933) Zur intuitionistischen Arithmetik und Zahlentheorie, *Ergebnisse eines mathematischen Kolloquiums* 4, 34–38.
(1933A) Eine interpretation des intuitionistischen Aussagenkalküls, *Ergebnisse eines mathematischen Kolloquiums 4*, 39–40.
(1934) On undecidable propositions of formal mathematical systems. Mimeographed lecture notes taken by S.C. Kleene and J. Barkley Rosser. Reprinted with revisions in *Davis (1965) 39–74*; also in *Gödel (1986) 346–371.*
(1958) Über eine bisher noch nicht benützte Erweiterung des finiten Standpunktes, *Dialectica* 12, 280–287.
(1986) *Collected works I*, eds. S. Feferman, J.W. Dawson, jr, S.C. Kleene, G.H. Moore, R.M. Solovay and J. van Heijenoort (Oxford University Press, Oxford).
(1987) *Collected works II*, eds. S. Feferman, J.W. Dawson, jr, S.C. Kleene, G.H. Moore, R.M. Solovay and J. van Heijenoort (Oxford University Press, Oxford). (Added in proof: publication delayed)

Goodman, N.D.
(1970) A theory of constructions equivalent to arithmetic, in: *Myhill et al. (1970) 101–120.*

Goodman, N.D. and J. Myhill
(1978) Choice implies excluded middle, *ZMLGM* 24, 461.

Goodstein, R.L.
(1945) Function theory in an axiom-free equation calculus, *Proc. London Math. Soc.* 52, 81–106.
(1957) *Recursive Number Theory* (North-Holland, Amsterdam).
(1961) *Recursive Analysis* (North-Holland, Amsterdam).

Gordeev, L.

(1982) Constructive models for set theory with extensionality, in: *Troelstra and van Dalen (1982) 123–147.*

Grayson, R.J.

(1981) Concepts of general topology in constructive mathematics and in sheaves, *AML* 20, 1–41 (corrections ibidem 23, 99).

Griss, G.F.C.

(1946) Negationless intuitionistic mathematics I, *Indagationes Math.* 8, 675–681.

(1955) La mathématique intuitionniste sans negation, *Nieuw Archief voor Wiskunde (3)*, 3, 134–142.

Harper, R.W. see *Constable et al. (1986)*.

Harrop, R.

(1956) On disjunctions and existential statements in intuitionistic statements of logic, *Math. Ann.* 132, 347–361.

(1960) Concerning formulas of the types $A \to B \vee C$, $A \to (Ex)B(x)$ in intuitionistic formal systems, *JSL* 25, 27–32.

Heijenoort, J. van

(1967) ed., *From Frege to Gödel. A Source Book in Mathematical Logic 1879–1931* (Harvard University Press, Cambridge Mass.) Reprinted 1970.

Henkin L.

(1960) On mathematical induction, *American Mathematical Monthly* 67, 323–338.

Herbrand, J.

(1931) Unsigned note on Herbrand's thesis written by Herbrand himself. Published in J. Herbrand, *Écrits Logiques* (Presses Universitaires de France, Paris 1968) pp. 209–214, and J. Herbrand, *Logical Writings* ed., W.D. Goldfarb (Harvard University Press, Cambridge Mass. 1971) pp. 271–276.

Heyting, A.

(1925) *Intuïtionistische axiomatiek der projectieve meetkunde* (Dutch), PhD thesis, University of Amsterdam (P. Noordhoff, Groningen).

(1930) Die formalen Regeln der intuitionistischen Logik, *Sitzungsberichte der Preussischen Akademie von Wissenschaften. Physikalisch-mathematische Klasse*, 42–56.

(1930A) Die formalen Regeln der intuitionistischen Mathematik II, *Sitzungsberichte der Preussischen Akademie von Wissenschaften. Physikalisch-mathematische Klasse*, 57–71.

(1930B) Die formalen Regeln der intuitionistischen Mathematik III, *Sitzungsberichte der Preussischen Akademie von Wissenschaften. Physikalisch-mathematische Klasse*, 158–169.

(1930C) Sur la logique intuitionniste, *Acad. Roy. Belg. Bull. Cl. Sci.* (5) 16, 957–963.

(1931) Die intuitionistische Grundlegung der Mathematik, *Erkenntnis* 2, 106–115.

(1934) *Mathematische Grundlagenforschung. Intuitionismus. Beweistheorie* (Springer, Berlin). Reprinted 1974. Considerably expanded and updated French translation *Heyting (1955)*.

(1937) Bemerkungen zu dem Aufsatz von Herrn Freudenthal "Zur intuitionistischen Deutung logischer Formeln", *Compos. Math.* 4, 117–118.

(1955) *Les fondements des mathématiques. Intuitionnisme. Théorie de la démonstration* (Gauthier-Villars, Paris and E. Nauwelaerts, Louvain). Expanded translation of *Heyting (1934)*.

(1956) *Intuitionism, An Introduction* (North-Holland, Amsterdam) 2nd rev. ed. (1966), 3rd rev. ed. (1971). There exist copies with on the cover "third edition", but the text is that of the second edition; in these copies the foreword to the third edition is missing.
(1958) Intuitionism in mathematics, in: R. Klibansky, ed., *Philosophy in the Mid-Century. A survey* (La Nuova Italia Editrice, Firenze) 101–115.
(1959) ed., *Constructivity in Mathematics* (North-Holland, Amsterdam).

Hilbert, D.
(1905) Über die Grundlagen der Mathematik, *Verhandlungen des Dritten Internationalen Mathematiker-Kongresses in Heidelberg vom 8. bis 13. August 1904* (Teubner Verlag, Leipzig) pp. 174–185. Translated in *van Heijenoort (1967) 130–185.*
(1922) Neubegründung der Mathematik (Erste Mitteilung), *Abhandlungen aus dem mathematischen Seminar der Hamburgischen Universität* 1, 157–177.

Hilbert, D. and P. Bernays
(1934) *Grundlagen der Mathematik I* (Springer, Berlin) 2nd ed. (1968).

Hölder, O.
(1924) *Die mathematische Methode* (Springer, Berlin).

Howard, W.A. and G. Kreisel
(1966) Transfinite induction and bar induction of types zero and one, and the role of continuity in intuitionistic analysis, *JSL* 31, 325–358.

Howe, D.J. see *Constable et al. (1986).*

Jaskowski, S.
(1936) Recherches sur le système de la logique intuitionniste, in: *Actes du Congrès International de Philosophie Scientifique, Septembre 1935, Paris* (Hermann, Paris) Vol. VI, 58–61. Translation: *Studia Logica* 34, 117–120.

Jervell, H.R.
(1971) A normal form in first-order arithmetic, in: *Fenstad (1971) 93–108.*

Johansson, I.
(1937) Der Minimalkalkül, ein reduzierter intuitionistischer Formalismus, *Compos. Math.* 4, 119–136.

Jongh, D.H.J. de and C.A. Smorynski
(1976) Kripke models and the intuitionistic theory of species, *AML* 9, 157–186.

Johnstone, P.T.
(1977) *Topos Theory* (Academic Press, London).
(1982) *Stone Spaces* (Cambridge University Press, Cambridge).

Joyal, A. see *Boileau and Joyal (1981).*

Julian, W. and F. Richman
(1984) A uniformly continuous function on [0, 1] that is everywhere different from its infimum, *Pac. J. Math.* 111, 333–340.

Kleene, S.C.
(1936) λ-definability and recursiveness, *Duke Math. J.* 2, 340–353 (errata *JSL* 2, 39).
(1945) On the interpretation of intuitionistic number theory, *JSL* 10, 109–124.
(1952) *Introduction to Metamathematics* (North-Holland, Amsterdam).
(1952A) Recursive functions and intuitionistic mathematics, in: L.M. Graves, E. Hille, P.A. Smith, O. Zariski, eds., *Proceedings of the International Congress of Mathematicians, August 1950. Cambridge, Mass.* (Amer. Math. Soc., Providence, RI) pp. 679–685.

(1957) Realizability, in: *Summaries of talks presented at the Summer Institute for Symbolic Logic, July 1957, Ithaca N.Y.*, (Institute for Defense Analyses, Communications Research Division) pp. 100–104; reprinted in *Heyting (1959) 285–289.*

(1958) Extension of an effectively generated class of functions by enumeration, *Colloquium Mathematicum* 6, 67–78.

(1960) Realizability and Shanin's algorithm for the constructive deciphering of mathematical sentences, *Logique et Analyse* 3, 154–165.

(1962) Disjunction and existence under implication in elementary intuitionistic formalisms, *JSL* 27, 11–18; Addendum ibidem 28, 154–156.

(1969) Formalized recursive functionals and formalized realizability, *Mem. Amer. Math. Soc.* 89.

(1973) Realizability: a retrospective survey, in: *Mathias and Rogers (1973) 95–112.*

Kleene, S.C. and R.E. Vesley

(1965) *The Foundations of Intuitionistic Mathematics, Especially in Relation to Recursive Functions* (North-Holland, Amsterdam).

Knoblock, T.B. see *Constable et al. (1986).*

Kolmogorov, A.N.

(1925) On the principle of the excluded middle (Russian), *Mat. Sb.* 32, 646–667; translated in *van Heijenoort (1967) 414–437.*

(1932) Zur Deutung der intuitionistischen Logik, *Math. Z.* 35, 58–65.

Kreisel, G.

(1958) Mathematical significance of consistency proofs, *JSL* 23, 155–182.

(1960) La prédicativité, *Bull. Soc. Math. Fr.* 88, 371–379.

(1962) Foundations of intuitionistic logic, in: E. Nagel, P. Suppes and A. Tarski, eds., *LMPS 1* (Stanford University Press, Stanford) pp. 198–210.

(1963) Reports of the seminar on the foundations of analysis. Stanford University. Mimeographed report in two parts.

(1965) Mathematical logic, in: T.L. Saaty, ed., *Lectures on Modern Mathematics III* (Wiley and Sons, New York) pp. 95–195.

(1967) Informal rigour and completeness proofs, in I. Lakatos, ed., *Problems in the Philosophy of Mathematics* (North-Holland, Amsterdam) pp. 138–186.

(1968) Lawless sequences of natural numbers, *Compos. Math.* 20, 222–248.

(1968A) Functions, ordinals, species, in: *van Rootselaar and Staal (1968) 145–159.*

(1970) Church's thesis: a kind of reducibility axiom for constructive mathematics, in: *Myhill et al. (1970) 121–150.*

(1971) A survey of proof theory II, in: *Fenstad (1971) 109–170.*

(1972) Which number-theoretic problems can be solved in recursive progressions on Π^1_1-paths through $\mathcal{O}$?, *JSL* 37, 311–334.

Kreisel, G. and D. Lacombe,

(1957) Ensembles récursivements mésurables et ensembles récursivements ouverts ou fermés, *C.R. Acad. Sci. Paris, Ser. A-B* 245, 1106–1109.

Kreisel, G., D. Lacombe and J. Shoenfield

(1957) Fonctionelles récursivement définissables et fonctionelles récursives, *C.R. Acad. Sci. Paris, Ser. A-B* 245, 399–402.

(1959) Partial recursive functionals and effective operations, in: *Heyting (1959) 290–297.*

Kreisel, G. and A.S. Troelstra

(1970) Formal systems for some branches of intuitionistic analysis, *AML* 1, 229–387.

Kreisel, G. see also *Howard and Kreisel 1966.*

Kripke, S.A.
(1963) Semantical considerations on modal and intuitionistic logic, *Acta Philosophica Fennica* 16, 83–94.
(1965) Semantical analysis of intuitionistic logic, in: J. Crossley and M.A.E. Dummett, eds., *Formal Systems and Recursive Functions* (North-Holland, Amsterdam) pp. 92–130.

Krol', M.D.
(1978) A topological model for intuitionistic analysis with Kripke's schema, *ZMLGM* 24, 427–436.
(1978A) Distinct variants of Kripke's schema in intuitionistic analysis, *Dokl. Akad. Nauk SSSR* 271, 33–36; translation *Sov. Math. Dokl.* 28, 27–30.

Kronecker, L.
(1887) Über den Zahlbegriff, *J. Reine Angew. Math.* 101, 337–355.
(1901) *Vorlesungen über Zahlentheorie I*, ed., K. Hensel (Teubner, Leipzig).

Kuroda, S.
(1951) Intuitionistische Untersuchungen der formalistischen Logik, *Nagoya Math. J.* 2, 35–47.

Kushner, B.A.
(1973) *Lectures on constructive mathematical analysis* (Russian) (Nauka, Moskva). Translation by E. Mendelson (American Mathematical Society, Providence R.I., 1984).
(1981) Behaviour of the general term of a Specker series (Russian), in: M. Gladkij, ed., *Mathematical Logic and Mathematical Linguistics* (Kalinin Gos. Univ., Kalinin) pp. 112–116; *Math. Reviews* 84i: 03111 (n.v.).
(1983) A class of Specker sequences (Russian), in: *Mathematical Logic, Mathematical Linguistics and Theory of Algorithms* (Kalinin Gos. Univ., Kalinin) pp. 62–65; *Math. Reviews* 85e: 03152 (n.v.)

Lacombe, D., see *Kreisel and Lacombe (1957)*, *Kreisel et al. (1957, 1959)*.

Lambek, J. and P.J. Scott
(1986) *Introduction to Higher-order Categorical Logic* (Cambridge University Press, Cambridge).

Leivant, D.M.E.
(1971) A note on translations from C into I, Report ZW5/71 (Mathematisch Centrum, Amsterdam).
(1985) Syntactic translations and provably recursive functions, *JSL* 50, 682–688.

Lifschitz, V.A.
(1979) CT_0 is stronger than $CT_0!$, *Proc. Amer. Math. Soc.* 73, 101–106.

Lorenzen, P.
(1955) *Einführung in die operative Logik und Mathematik* (Springer, Berlin) 2nd edition in 1969.
(1965) *Differential und Integral* (Akademische Verlagsgesellschaft, Frankfurt a. M.). Translated as: *Differential and integral*, by J. Bacon (University of Texas Press, Austin, Texas, 1970).

Luckhardt, H.
(1970) Ein Henkin-Vollständigkeitsbeweis für die intuitionistische Prädikatenlogik bezüglich der Kripke-Semantik, *Arch. ML* 13, 55–59.
(1973) *Extensional Gödel Functional Interpretation. A Consistency Proof of Classical Analysis* (Springer, Berlin).

(1977) Über das Markov-Prinzip, *Arch. ML* 18, 73–80.
(1977A) Über das Markov-Prinzip II, *Arch. ML* 18, 147–157.

Lusin, N.
(1930) *Leçons sur les ensembles projectives et leur applications* (Gauthier-Villars, Paris).

Mandelkern, M.
(1976) Connectivity of an interval, *Proc. Amer. Math. Soc.* 54, 170–172.
(1982) Continuity of monotone functions, *Pac. J. Math.* 99, 413–418.

Mannoury, G.
(1909) *Methodologisches und Philosophisches zur Elementar-Mathematik* (P. Visser, Haarlem).

Marcus, R.B., G.J.W. Dorn and P. Weingartner
(1986) eds., *LMPS VII* (North-Holland, Amsterdam).

Markov, A.A.
(1954) On the continuity of constructive functions (Russian), *Usp. Math. Nauk* 9/3 61, 226–230.
(1954A) *Theory of Algorithms* (Russian), Trudy Matematiceskog Istituta imeni V.A. Steklova, vol. 42 (Izdatel'stvo Akademii Nauk SSSR, Moskva). English translation by J.J. Schoor-Kan and PST staff, Israel Program for Scientific Translations, Jerusalem 1961.
(1962) On constructive mathematics (Russian), *Tr. Mat. Inst. Steklov.* 67, 8–14; translated in *Amer. Math. Soc., Transl., II Ser.* 98, 1–9.

Martino, E. and P. Giaretta
(1981) Brouwer, Dummett and the bar theorem, in: S. Bernini, ed., *Atti del congresso nazionale di logica, Montecatini Terme, 1–5 ottobre 1979* (Bibliopolis, Napoli) pp. 541–558.

Mathias, A.R.D. and H. Rogers
(1973) eds., *Cambridge Summer School in Mathematical Logic* (Springer, Berlin).

McCarty, D.C.
(1986) Subcountability under realizability, *Notre Dame Journal of Formal Logic* 27, 210–220.

Mendelson, E.
(1964) *Introduction to Mathematical Logic* (Van Nostrand, New York) 2nd edition (1979).

Mendler, N.P. see *Constable et al. (1986).*

Mints, G.E.
(1976) What can be done with PRA (Russian), *Zap. Nauchn. Semin. Leningr. Otd. Mat. Inst. Steklov.*, 60, 93–102; translation *J. Sov. Math.* 14, 1487–1492.

Mints, G.E. and V.P. Orevkov
(1967) On imbedding operators (Russian), *Zap. Nauchn. Semin. Leningr. Otd. Mat. Inst. Steklov.* 4, 160–167; translation *Sem. Math. V.A. Steklov* 4, 64–66.

Molk, J.
(1885) Sur une notion qui comprend celle de la divisibilité et sur la théorie de l'élimination. *Acta Mathematica* 6, 1–166.

Mooij, J.J.A.
(1966) *La philosophie des mathématiques de Henri Poincaré* (Gauthier-Villars, Paris, and E. Nauwelaerts, Louvain).

Moschovakis, J.R.
(1981) A disjunctive decomposition theorem for classical theories, in: *Richman (1981)*, 250–259.

Moschovakis, Y.N.
(1964) Recursive metric spaces, *Fundam. Math.* 55, 215–238.
Müller, G.H.
(1987) ed., *Ω-bibliography of Mathematical Logic*, edited by G.H. Müller in collaboration with W. Lenski, *Volume 6: Proof Theory and Constructive Mathematics*, eds. D. van Dalen, J.E. Kister and A.S. Troelstra (Springer, Berlin).
Müller, G.H. and D.S. Scott
(1978) eds., *Higher Set Theory* (Springer, Berlin).
Myhill, J.
(1963) The invalidity of Markov's schema, *ZMLGM* 9, 359–360.
(1967) Notes towards an axiomatization of intuitionistic analysis, *Logique et Analyse* 9, 280–297.
(1971) A recursive function, defined on a compact interval and having a continuous derivative that is not recursive, *Michigan Math. J.* 18, 97–98.
(1974) "Embedding classical type theory in 'intuitionistic' type theory", a correction, in T. Jech, ed., *Axiomatic Set Theory, Part II* (Amer. Math. Soc., Providence, RI) pp. 185–188.
Myhill, J. see also *Goodman and Myhill (1978)*.
Myhill, J., A. Kino and R.E. Vesley
(1970) eds., *Intuitionism and Proof Theory* (North-Holland, Amsterdam).
Nelson, D.
(1947) Recursive functions and intuitionistic number theory, *Trans. Amer. Math. Soc.* 61, 307–368, 556.
Nishimura, I.
(1960) On formulas of one variable in intuitionistic propositional calculus, *JSL* 25, 327–331.
Orevkov, V.P. see *Mints and Orevkov (1976)*.
Panangaden, P. see *Constable et al. (1986)*.
Parikh, R.J.
(1971) Existence and feasibility in arithmetic, *JSL* 36, 494–508.
Peter, R.
(1940) Review of Skolem 1939, *JSL* 5, 34–35.
Pohlers, W. see *Buchholz et al. (1981)*.
Poincaré, H.
(1902) *Science et Hypothèse* (Flammarion, Paris).
(1905) *La Valeur de la Science* (Flammarion, Paris).
(1908) *Science et Méthode* (Flammarion, Paris).
(1913) *Dernières Pensées* (Flammarion, Paris).
Prawitz, D.
(1965) *Natural Deduction. A Proof-theoretical Study* (Almquist and Wiksell, Stockholm).
(1971) Ideas and results in proof theory, in: *Fenstad (1971) 237–309*.
Rasiowa, H.
(1954) Constructive theories, *Bull. Acad. Polon. Sci. Ser. Sci. Math. Astron. Phys.* 1, 229–231.
(1955) Algebraic models of axiomatic theories, *Fundam. Math.* 41, 291–310.
Renardel de Lavalette, G.R.
(1984) Descriptions in mathematical logic, *Studia Logica* 43, 281–294.
(1984A) Theories with type-free application and extended bar induction, PhD thesis, University of Amsterdam.

Rice, H.G.
(1954) Recursive real numbers, *Proc. Amer. Math. Soc.* 5, 784–791.

Richman, F.
(1981) ed., *Constructive Mathematics. Proceedings of the New Mexico State Conference* (Springer, Berlin).
(1983) Church's thesis without tears, *JSL* 48, 797–803.

Richman, F. see also *Julian and Richman (1984)*.

Rieger, L.
(1949) On the lattice theory of Brouwerian propositional logic, *Acta Univ. Carol., Math. Phys.* 189.

Rootselaar, B. van
(1952) Un problème de M. Dijkman, *Indagationes Math.* 14, 405–407.

Rootselaar, B. van and J.F. Staal
(1968) eds., *LMPS III* (North-Holland, Amsterdam).

Sasaki, J.T. see *Constable et al. (1986)*.

Scedrov, A.
(1986) Embedding sheaf models for set theory into boolean-valued permutation models with an interior operator, *APAL* 32, 103–109.

Schmidt, H.A., K. Schütte and H.-J. Thiele
(1968) eds., *Contributions to Mathematical Logic* (North-Holland, Amsterdam).

Schütte, K.
(1968) *Vollständige Systeme modaler und intuitionistischer Logik* (Springer, Berlin).

Scott, D.S.
(1979) Identity and existence in intuitionistic logic, in: *Fourman et al. (1979) 660–696.*

Scott, P.J. see *Lambek and Scott (1986)*.

Segerberg, K.
(1974) Proof of a conjecture of McKay, *Fundam. Math.* 81, 267–270.

Shanin, N.A.
(1958) On the constructive interpretation of mathematical judgments (Russian), *Tr. Mat. Inst. Steklova* 52, 226–311. Translated in: *Amer. Math. Soc., Transl., II Ser.* 23, 109–189.
(1958A) Über einen Algorithmus zur konstruktiver Dechiffrierung mathematischer Urteile (Russian), *ZMLGM* 4, 293–303.
(1962) Constructive real numbers and constructive function spaces (Russian), *Tr. Mat. Inst. Steklov.*, 67, 15–294; translated (Amer. Math. Soc., Providence RI, 1968).

Shapiro, S.
(1981) Understanding Church's thesis, *J. Philos. Logic* 10, 353–365.
(1985) ed., *Intensional Mathematics* (North-Holland, Amsterdam).

Shoenfield, J.R., see *Kreisel et al. (1957, 1959)*.

Sieg, W.
(1985) Fragments of arithmetic, *APAL* 28, 33–71.

Sieg, W., see also *Buchholz et al. (1981)*.

Simpson, S.G., see *Friedman et al. (1983)*.

Skolem, T.
(1923) Begründung der elementaren Arithmetik durch die rekurrierende Denkweise ohne Anwendung scheinbarer Veränderlichen mit unendlichen Ausdehnungsbereich, *Videnskaps Selskapet i Kristiana. Skrifler Utgit (1)* 6, 1–38. Translated in *van Heijenoort (1967) 303–333.*

(1939) Eine Bemerkung über die Induktionsschemata in der rekursiven Zahlentheorie, *Monatshefte für Mathematik und Physik* 48, 268–276.

Smith, R.L. see *Friedman et al. (1983)*.

Smith, S.F. see *Constable et al. (1986)*.

Smorynski, C.A.

(1973) Applications of Kripke models, in: *Troelstra (1973) 324–391*,

(1973A) Investigations of intuitionistic formal systems by means of Kripke models, PhD thesis, University of Chicago.

(1982) Nonstandard models and constructivity, in: *Troelstra and van Dalen (1982) 459–464*.

Smorynski, C.A. see also *de Jongh and Smorynski (1976)*.

Specker, E.

(1949) Nicht konstruktiv beweisbare Sätze der Analysis, *JSL* 14, 145–158.

Spector, C.

(1962) Provably recursive functionals of analysis: a consistency proof of analysis by an extension of principles formulated in current intuitionistic mathematics, in: J.C.E. Dekker, ed., *Recursive Function Theory*, Proceedings of Symposia in Pure Mathematics V (American Mathematical Society, Providence, RI) pp. 1–27.

Staples, J.

(1971) On constructive fields, *Proc. London Math. Soc.* 23, 753–768.

Sundholm, G.

(1983) Constructions, proofs and the meaning of logical constants, *J. Philos. Logic* 12, 151–172.

Takeuti, G.

(1978) *Two Applications of Logic to Mathematics* (Princeton University Press, Princeton).

Thomason, R.H.

(1968) On the strong semantical completeness of the intuitionistic predicate calculus, *JSL* 33, 1–7.

Troelstra, A.S.

(1966) Intuitionistic general topology, PhD thesis, University of Amsterdam.

(1967) Intuitionistic continuity, *Nieuw Archief voor Wiskunde (3)* 15, 2–6.

(1968) The use of "Brouwer's principle" in intuitionistic topology, in: *H.A. Schmidt et al. (1968) 289–298*.

(1969) *Principles of Intuitionism* (Springer, Berlin).

(1971) Notions of realizability for intuitionistic arithmetic and intuitionistic arithmetic in all finite types, in: *Fenstad (1971) 369–405*.

(1973) *Metamathematical Investigation of Intuitionistic Arithmetic and Analysis* (Springer Verlag, Berlin). A list of errata appeared as report 74-16 (Department of Mathematics, University of Amsterdam, 1974).

(1973A) Notes on intuitionistic second-order arithmetic, in: *Mathias and Rogers (1973) 171–205*.

(1977) *Choice Sequences, a Chapter of Intuitionistic Mathematics* (Clarendon Press, Oxford).

(1977A) Axioms for intuitionistic mathematics incompatible with classical logic, in: R.E. Butts and J. Hintikka, eds., *Logic, Foundations of Mathematics, and Computability Theory. Part one of LMPS V* (Reidel, Dordrecht) pp. 59–84.

(1978) A. Heyting on the formalization of intuitionistic mathematics, in: E.M.J. Bertin, H.J.M. Bos and A.W. Grootendorst, eds., *Two Decades of Mathematics in the*

Netherlands 1920–1940. A Retrospection on the Occasion of the Bicentennial of the Wiskundig Genootschap (Mathematisch Centrum, Amsterdam) pp. 153–175.

(1978A) Some remarks on the complexity of Henkin-Kripke models, *Indagationes Math*. 40, 296–302.

(1980) Intuitionistic extensions of the reals, *Nieuw Archief voor Wiskunde (3)*, 28, 63–113.

(1981) The interplay between logic and mathematics: intuitionism, in: E. Agazzi, ed., *Modern Logic – A Survey* (Reidel, Dordrecht) pp. 197–221.

(1981A) Arend Heyting and his contribution to intuitionism, *Nieuw Archief voor Wiskunde (3)* 29, 1–23.

(1982) Intuitionistic extensions of the reals II, in: *van Dalen et al. (1982) 279–310.*

(1982A) On the origin and development of Brouwer's concept of choice sequence, in: *Troelstra and van Dalen (1982) 465–486.*

(1983) Analyzing choice sequences, *J. Philos. Logic* 12, 197–260.

(1983A) Logic in the writings of Brouwer and Heyting, in: V. M. Abrusci, E. Casari and M. Mugnai, eds., *Atti del Convegno Internazionale di Storia della Logica. San Gimignano, 4–8 dicembre 1982* (Cooperativa Libraria Universitaria Editrice Bologna, Bologna) pp. 193–210.

Troelstra, A.S. see also *Kreisel and Troelstra (1970)*.

Troelstra, A.S. and D. van Dalen

(1982) eds., *The L.E.J. Brouwer Centenary Symposium* (North-Holland, Amsterdam).

Tsejtin, G.S.

(1959) Algorithmic operators in constructive complete separable metric spaces (Russian), *Dokl. Akad. Nauk SSSR* 128, 49–52.

(1962) Algorithmic operators in constructive metric spaces (Russian), *Tr. Mat. Inst. Steklov*. 67, 295–361. Translated in *Amer. Math. Soc., Transl., II Ser*. 64, 1–80.

Tsejtin, G.S. and I.D. Zaslavskij

(1962) On singular coverings and properties of constructive functions connected with them, *Tr. Mat. Inst. Steklov*. 67, 458–502. Translated in *Amer. Math. Soc., Transl., II Ser*. 98, 41–89.

Turing, A.M.

(1937) On computable numbers, with an application to the Entscheidungsproblem, *Proc. London Math. Soc*. 42, 230–265. Corrections ibidem 43, 544–546. Reprinted in *Davis (1965) 116–154.*

Unterhalt, M.

(1986) Kripke-Semantik mit partieller Existenz. Dissertation, Münster i. Wf., W-Germany.

Veldman, W.

(1981) Investigations in intuitionistic hierarchy theory, PhD thesis, Katholieke Universiteit Nijmegen, 235 pp plus a loose leaflet "Stellingen".

Vesley, R.E. see *Kleene and Vesley (1965)*.

Visser, A.

(1981) Aspects of diagonalization and provability, PhD thesis, Rijksuniversiteit Utrecht.

Webb, J.C.

(1980) *Mechanism, Mentalism and Metamathematics* (Reidel, Dordrecht).

Weinstein, S.

(1983) The intended interpretation of intuitionistic logic, *J. Philos. Logic* 12, 261–270.

Weyl, H.
(1918) *Das Kontinuum. Kritische Untersuchungen über die Grundlagen der Analysis* (Veit, Leipzig); reprinted in: *Das Kontinuum und andere Monographien* (Chelsea, New York, 1966).
(1921) Über die neue Grundlagenekrise der Mathematik, *Math. Z.* 10, 39–70.
Zaslavskij, I.D.
(1962) Some properties of constructive real numbers and constructive functions (Russian), *Tr. Mat. Inst. Steklov.* 67, 385–457; translated in *Amer. Math. Soc., Transl., II Ser.* 57, 1–84.
Zaslavskij, I.D., see also *Tsejtin and Zaslavskij (1962)*.

INDEX

In this index we have included notions and terminology of more than strictly local significance. The page numbers following an entry refer to the places where a definition, modification or extension of an earlier definition, or a notational convention concerning the entry is to be found.

LIST OF SYMBOLS

Part I lists the abbreviations designating formal systems, part II the abbreviations for axioms or axiom schemas. The listing is alphabetical, disregarding mathematical symbols which are not roman letters. Part III lists other notations in order of appearance. Notations which are used only locally ('ad hoc') are not listed as a rule.

I. Formal systems

II. Axioms and rules

III. Other notations

n, m are used for natural numbers, α for a function from natural numbers to natural numbers, A, A', A'' for formulas, X, Y for sets, t for a term, $\mathfrak{a}$ for arbitrary expressions.

Printed and bound by CPI Group (UK) Ltd, Croydon, CR0 4YY

Printed and bound by CPI Group (UK) Ltd, Croydon, CR0 4YY
03/10/2024
01040428-0010